Optimal Control of Hydrosystems

Optimal Control of Hydrosystems

Larry W. Mays
Arizona State University
Tempe, Arizona

CRC Press
Taylor & Francis Group
Boca Raton London New York

CRC Press is an imprint of the
Taylor & Francis Group, an **informa** business

st published 1997 by CRC Press
ylor & Francis Group
)0 Broken Sound Parkway NW, Suite 300
ca Raton, FL 33487-2742

ssued 2018 by CRC Press

rary of Congress Cataloging-in-Publication Data

ys, Larry W.
Optimal control of hydrosystems / Larry W. Mays.
p. cm.
Includes bibliographical references and index.
ISBN 0-8247-9830-9 (hardcover : alk. paper)
1. Hydraulics—Mathematical models. 2. Mathematical optimization.
. Control theory. I. Title.
TC163.M38 1997
27'.042—dc21 96-29981

ibrary of Congress record exists under LC control number: 96029981

lisher's Note
publisher has gone to great lengths to ensure the quality of this reprint but points out that some imperfections he original copies may be apparent.

claimer
publisher has made every effort to trace copyright holders and welcomes correspondence from those they have n unable to contact.

N 13: 978-1-315-89610-6 (hbk)
N 13: 978-1-351-07520-6 (ebk)

Preface

Optimal Control of Hydrosystems addresses the mathematical modeling of the optimal operation of hydrosystems. The term *hydrosystems* was coined by V. T. Chow to describe collectively the technical areas of hydrology, hydraulics, and water resources. *Hydrosystems* has also been used in reference to types of water projects, including systems for groundwater management, surface water storage, water distribution, flood control, drainage, etc. As used in this book, *hydrosystems* applies to both definitions. Specifically, the types of hydrosystems in this book include river-reservoir systems, groundwater systems, bay and estuary systems, soil aquifer treatment systems, and water distribution systems. Operations for reservoir systems include both long-term operation for water supply, sediment control, and freshwater inflow to bays and estuaries, and short-term operation for flood control and sediment control.

Optimization problems are characterized or mathematically formulated to include an objective function that is optimized (maximized or minimized) subject to a set of constraints, which are algebraic equations and/or inequalities. *Optimal control problems* are optimization problems in which part or all of the constraints are differential equations. In particular, these systems are modeled on the framework of a certain type of optimization problem referred to as *discrete-time optimal control problems*. These types of optimization problems are unique in that the physics (laws of motion) of the problem are described through differential equations that simulate the physical behavior of the problem. The main theoretical approaches to solve optimal control problems are calculus of variations; the maximum or minimum principle, which may be regarded as a special application of calculus of variation; dynamic programming; and mathematical programming (nonlinear programming). The methods to solve these types of problems include the combined use of (1) the hydraulic simulation of the physical process and (2) operation research techniques such as nonlinear programming and differential dynamic programming. The methods to solve these types of hydrosystems problems have never before been presented in one book.

The book is divided into three major parts: Basic Concepts for Optimal Control, Applications Based on Mathematical Programming, and Applications Based on Optimal Control Procedures. The first part introduces hydrosystems control problems as discrete-time optimal control problems, discusses system and optimal control concepts, and explores nonlinear programming concepts. Part II has four chapters that apply the optimal control concepts using mathematical programming to develop models and solution algorithms for groundwater systems operation, real-time operation of river-reservoir systems for flood control, water distribution systems operation, and reservoir operation for optimizing freshwater inflows to bays and estuaries. Part III has chapters on optimal control using differential dynamic programming, estuarine management, reservoir operation for water supply, groundwater management, reservoir operation for sediment control in rivers and reservoirs, and optimal operation of soil aquifer treatment systems. The final two chapters are on real-time stochastic optimal control using linear quadratic feedback for lumped systems and stochastic optimal control for distributed parameter systems. The methodologies in the final two chapters are applied to estuarine management.

This book is written at an advanced level for those with some background in operations research, hydraulics, and water resources engineering. Both graduate students and practicing engineers will find this to be a valuable reference book and text. It can also be used in graduate-level water resources engineering courses. It is not a review of the literature, but instead is an introduction to the theory and application of optimal control. A major focus of the book is the way in which hydraulic simulators can be interfaced with optimizers in an optimal control framework to solve realistic, large-scale hydrosystems operation problems that are formulated as optimal control problems. I hope that the book will further the goal of better engineering and management practices in the hydrosystems field.

Much of the work presented here is based on the research work of my former doctoral students, in particular, Chapter 4, Dr. Nisai Wanakule; Chapter 5, Dr. Olcay Unver; Chapter 6, Dr. Lehar Brion; Chapter 7, Dr. Yixing Bao; Chapter 10, Dr. Guihua Li; Chapter 11, Dr. Carlos Carriaga; Chapter 12, Drs. Guihua Li and Zongwu Tang; and Chapters 13 and 14,

Dr. Bing Zhao. Dr. Leon Lasdon, a friend and former colleague at The University of Texas at Austin, has been rather influential over the years in teaching me and my former graduate students at The University of Texas many of the concepts presented in this book, particularly those related to the optimal control concepts using mathematical programming. I also need to thank Dr. Lasdon for his willingness to provide us with versions of his GRG2 code and for his friendly advice on its use over the years. Other former graduate students of mine also have helped in the development of many of the concepts used to solve the optimal control problems for hydrosystems, including, at The University of Texas, Dr. M. John Cullinane, Dr. Ning Duan, Dr. Joong-Hoon Kim, Dr. Kevin Lansey, Dr. Jungi Matsumoto, Dr. Gerardo Ocauas, Dr. Chang-Kang Taur, Dr. Yeou-Koung Tung; and at Arizona State University, Dr. Richard Skaggs, Dr. Kuan Tuncok, and Dr. Liang Xu.

Larry W. Mays

Contents

Preface *iii*

PART I. BASIC CONCEPTS FOR OPTIMAL CONTROL

Chapter 1 Systems Theory and Optimal Control **1**

1.1 Classifications of Systems 1
1.2 Concepts of Systems 2
1.3 Definition of Optimal Control Problems 8
References 10

Chapter 2 Introduction to Hydrosystems Problems **13**

2.1 Optimization of Hydrosystems 13
2.2 Groundwater Management Systems 14
2.3 Real-Time Operation of River-Reservoir Systems for Flood Control 19
2.4 Reservoir System Operation for Water Supply 23
2.5 Water Distribution System Operation 24
2.6 Freshwater Inflows to Bays and Estuaries 25
2.7 Sediment Control in Rivers and Reservoirs 25
2.8 Soil Aquifer Treatment Systems 27
2.9 General Problem Formulation 30
References 37

Chapter 3 Nonlinear Optimization Methods 39

3.1 Matrix Algebra for Nonlinear Programming 39
3.2 Unconstrained Nonlinear Programming 42
3.3 Constrained Optimization: Optimality Conditions 47
3.4 Constrained Nonlinear Optimization: Generalized Reduced Gradient (GRG) Method 50
3.5 Constrained Nonlinear Optimization: Penalty Function Methods 54
3.6 The Augmented Lagrangian Method 56
3.7 Solution of Discrete-Time Optimal Control Problems 62
Appendix: Computer Software 68
References 69

PART II. APPLICATIONS BASED ON MATHEMATICAL PROGRAMMING

Chapter 4 Groundwater Management Systems 71

4.1 Problem Identification 71
4.2 Problem Formulation 73
4.3 Problem Solution 77
4.4 Application 86
Appendix: Computation of Basis Matrix Elements 88
References 94

Chapter 5 Real-Time Operation of River-Reservoir Systems 97

5.1 Problem Identification 97
5.2 Problem Formulation 101
5.3 Problem Solution 110
5.4 Application 116
Appendix A: Solution of Simulation Model 121
Appendix B: Computation of Basis Matrix Elements 124
References 128

Chapter 6 Water Distribution System Operation 131

6.1 Problem Identification 131
6.2 Problem Formulation 132
6.3 Problem Solution 135

6.4 Application 145
Appendix A: Simulation Model 146
Appendix B: Computation of Basis Elements 150
References 159

Chapter 7 Freshwater Inflows to Estuaries 163

7.1 Problem Identification 163
7.2 Problem Formulation 165
7.3 Problem Solution 175
7.4 Application 185
Appendix A: HYD-SAL Simulation Models 187
Appendix B: Derivation of Deterministic Equivalent of Chance Constraints Based on Regression Equations 189
References 190

PART III. APPLICATIONS BASED ON OPTIMAL CONTROL PROCEDURES

Chapter 8 Optimal Control by Feedback Control Methods 193

8.1 Dynamic Programming 193
8.2 Feedback Method of Optimal Control for Linear Systems 196
8.3 Groundwater Management Problems 200
8.4 Feedback Method of Optimal Control for Nonlinear Systems 201
References 205

Chapter 9 Differential Dynamic Programming 207

9.1 Differential Dynamic Programming Algorithm 207
9.2 Convergence of Unconstrained DDP 213
9.3 Multireservoir Operation 215
9.4 Optimal Control of Groundwater Hydraulics 218
9.5 Groundwater Reclamation Models 221
Appendix: Derivation of Coefficients for Quadratic 225
References 230

Chapter 10 Estuarine Management Model Using SALQR 233

10.1 Problem Formulation 233
10.2 Problem Solution 236
10.3 Example Application 245
References 252

Chapter 11 Sediment Control in Rivers and Reservoirs **255**

11.1 Problem Identification 255
11.2 Problem Formulation 257
11.3 Problem Solution 263
11.4 Example Application to Agua Fria River 265
Appendix: Hyperbolic Penalty Function for the Violation of Storage Constraint 273
References 275

Chapter 12 Optimal Operation of Soil Aquifer Treatment Systems Using SALQR **277**

12.1 Problem Identification 277
12.2 Problem Formulation 278
12.3 Problem Solution 281
12.4 Application Examples 286
Appendix: Derivative Calculations 298
References 299

Chapter 13 Real-Time Optimal Control: Linear Quadratic Feedback for Lumped Systems **301**

13.1 Introduction 301
13.2 Discrete-Time, Linear, Quadratic Regulator (LQR) 305
13.3 Discrete-Time, Linear, Quadratic Tracker (LQT) 307
13.4 General, Discrete-Time, Stochastic, Linear, Quadratic, Optimal Control for Estuarine Management 308
13.5 Analytical Feedback Control Law for General, Discrete-Time, Stochastic, Linear, Quadratic, Optimal Control for Estuarine Management 309
13.6 Application of Stochastic, Linear, Quadratic Feedback Optimal Control to Water Resources Engineering 315
13.7 Parameter Estimation for Lumped-Parameter System 317
13.8 Applications to Estuary Management 322
References 328

Chapter 14 Stochastic Optimal Control for Estuarine Management **331**

14.1 Introduction 331
14.2 Stochastic Optimal Control for a Distributed-Parameter System 336
14.3 Parameter Estimation with Uncertainty Analysis 340
14.4 Application to Estuary Management 349
References 360

Index *365*

Chapter 1
Systems Theory and Optimal Control

1.1 CLASSIFICATIONS OF SYSTEMS

A system may be defined as "a collection of objects arranged in an ordered form, which is, in some sense, purpose or goal directed" (Sinha, 1991). A system is characterized by "(1) a system boundary which is a rule that determines whether an element is to be considered as a part of the system or of the environment; (2) statement of input and output interactions with the environment; and (3) statements of interrelationships between the system elements, inputs and outputs, called feedback" (Mays and Tung, 1992). Simply speaking, a system consists of inputs, an operator, outputs, and/or feedback. A system without feedback can be expressed as

$$\mathbf{y} = \phi\,\mathbf{u} \tag{1.1.1}$$

in which $\mathbf{u}$, $\mathbf{y}$, and ϕ are the input vector, output vector, and transfer function, respectively. The transfer function ϕ is also called an operator or kernel of the system.

Systems can be classified in many different ways. The classifications of systems were well summarized by Sinha (1991) as "(1) static and dynamic systems, (2) linear and nonlinear systems, (3) time-varying and time-invariant systems, (4) deterministic and stochastic systems, (5) continuous-time and discrete-time systems, and (6) lumped-parameter and distributed-parameter systems."

The outputs from a static system only depend on the current values of the inputs, whereas outputs from a dynamic system depend on the current and the previous values of the inputs. The inputs and outputs for a linear system satisfy the superposition theorem, whereas a system is nonlinear if it does not satisfy the superposition theorem. The superposition theorem includes two parts, additivity and homogeneity. Additivity is that the output due to the sum of inputs is equal to the sum of the outputs duc to cach of the inputs or $\phi(\Sigma_{n=1}^{N} \mathbf{u}_n) = (\Sigma_{n=1}^{N} \phi\, \mathbf{u}_n)$. Homogeneity is $\phi(c\mathbf{u}_n) = c(\phi\, \mathbf{u}_n)$ in which c is a constant.

A system is time-invariant if the kernel does not change with time, whereas a system is time-variant if the kernel changes with the time. A system is deterministic if the kernel and inputs are known exactly; a system is stochastic if either the parameters in the kernel or the inputs are not exactly known and are described by statistical concepts. A system is a continuous-time system if the time varies continuously and can take any value in the continuous set of real numbers. A system is a discrete-time system if the inputs, outputs, and the parameters in the kernel take values at discrete times. A system is a lumped-parameter system if the inputs, outputs, and the parameters in the kernel are only functions of time and there is no spatial variable involved in the system. A system is a distributed-parameter system if the inputs, outputs, and the parameters in the kernel are functions of time and space. A continuous-time (discrete-time) lumped-parameter system is described by ordinary differential (difference) equations, whereas a distributed-parameter system is described by partial differential (difference) equations.

Water resource systems are usually distributed and properly described by partial differential equations with respect to time and space. Because the nature of hydrosystems are inherently distributed, they are divided typically into several subsystems such that each individual subsystem is treated as a lumped system. It is possible, in many cases, to obtain a good approximation to distributed system behavior by using linked lumped systems.

1.2 CONCEPTS OF SYSTEMS

1.2.1 Concept of the State

In the so-called modern system theory, the system structure is given explicit representation as a vector $\mathbf{x}$, where $\mathbf{x} = (x_1, \ldots, x_n)$ and the state variables $x_1, \ldots, x_n$, are a function of time. The state of the system at any given time t_1, is given by the value of state variables $x_1(t_1), x_2(t_1), \ldots, x_n(t_1)$ which constitutes the state vector, $\mathbf{x}(t_1)$. This is the fundamental concept of state variable modeling. In hydrosystems, the state variables are usually expressed in volumetric or mass units and can represent, for example, the volume of water or the amount of prescribed pollutants contained in various parts of the system. The input and output variables commonly correspond to volume or mass flow rates, which may be expressed as rainfall intensity or the rate of discharge of the pollutant. The state of the system is a measure of the level of activity in each of its components and can be thought of as the interface between the past and the future of the system's time history.

Formally, the state vector may be defined as the minimum number of variables needed so that if the state at time t_1, $\mathbf{x}(t_1)$, is known and the input from time t_1 to some later time t_2, $\mathbf{u}(t)$, $t_1 < t < t_2$, is also known, then the state, $\mathbf{x}(t_2)$, is completely determined from this information. Sometimes the state variable methodology is also called "state space" analysis.

Comparison between the "classical" and "modern" approaches to dynamic systems modeling may be visualized in vector space mappings. If an "input space" for the input vector, $\mathbf{u}$, and an "output space" for the output vector, $\mathbf{y}$, are defined in the same way as the state space has been defined, the transfer function is seen as a mapping from the input space directly to the output space. In the "modern" approach the input space is first related to the state space through the so-called state equation, which is a differential or difference equation. Then the state space and, in some cases, the input space are related to the output space through the so-called output equation, which is algebraic.

Descriptions of continuous-time and discrete-time deterministic state variable models are given in the following two subsections. Where vector–matrix operations are used, the notation employed is that a lowercase letter, a, is a scalar; a bold lowercase letter, $\mathbf{a}$, is a vector, and an bold uppercase letter, $\mathbf{A}$, is a matrix.

1.2.2 Continuous-Time Deterministic State Variable Model

A deterministic model is one in which a given input always produces the same output. The continuous-time deterministic state variable model is mathematically formulated by means of two equations: the state equation and the output equation. The state equation is a set of ordinary, first-order differential equations, one for each state variable, which is written in vector–matrix form to simplify the notation. The state equation describes the change in the state of the system over time in response to the inputs. The output equation is a set of algebraic equations, one for each output variable relating the output to the state of the system and, in some cases, to the inputs. The output equation is also commonly written in vector-matrix form. For the most general case, the state and output equations may be expressed as in Eqs. (1.2.1a) and (1.2.1b), respectively:

$$\dot{\mathbf{x}}(t) = \mathbf{g}[\mathbf{x}(t), \mathbf{u}(t), t] \qquad (1.2.1a)$$

$$\mathbf{y}(t) = \mathbf{h}[\mathbf{x}(t), \mathbf{u}(t), t] \qquad (1.2.1b)$$

$$\mathbf{x}(t) = \begin{bmatrix} x_1(t) \\ x_2(t) \\ \vdots \\ x_n(t) \end{bmatrix}, \quad \dot{\mathbf{x}}(t) = \begin{bmatrix} \dot{x}_1(t) \\ \dot{x}_2(t) \\ \vdots \\ \dot{x}_n(t) \end{bmatrix}, \quad \mathbf{y}(t) = \begin{bmatrix} y_1(t) \\ y_2(t) \\ \vdots \\ y_r(t) \end{bmatrix}, \quad \mathbf{u}(t) = \begin{bmatrix} u_1(t) \\ u_2(t) \\ \vdots \\ u_p(t) \end{bmatrix}$$

and

$$\dot{x}_1(t) = \frac{d[x_1(t)]}{dt}, \quad \dot{x}_2(t) = \frac{d[x_2(t)]}{dt}, \quad \dot{x}_n(t) = \frac{d[x_n(t)]}{dt}$$

The functions **g**[ ] in Eq. (1.2.1a) and **h**[ ] in Eq. (1.2.1b) are nonlinear and time-variant. They are nonlinear because products or powers of the variables may occur, and time-variant because the time, t, is included as an explicit variable.

For practical purposes, this model is usually simplified to the form shown in the following equations, which is the basic continuous-time, time-invariant, deterministic, state variable model:

$$\dot{\mathbf{x}}(t) = \mathbf{A}\mathbf{x}(t) + \mathbf{B}\mathbf{u}(t) \qquad (1.2.2)$$

$$\mathbf{y}(t) = \mathbf{C}\mathbf{x}(t) + \mathbf{D}\mathbf{u}(t) \qquad (1.2.3)$$

where **A**, **B**, **C**, and **D** are the matrices

$$\mathbf{A} = \begin{bmatrix} a_{11} & a_{12} & \cdots & a_{1n} \\ a_{21} & a_{22} & \cdots & a_{2n} \\ \vdots & \vdots & & \vdots \\ a_{n1} & a_{n2} & \cdots & a_{nn} \end{bmatrix}, \quad \mathbf{B} = \begin{bmatrix} b_{11} & b_{12} & \cdots & b_{1p} \\ b_{21} & b_{22} & \cdots & b_{2p} \\ \vdots & \vdots & & \vdots \\ b_{n1} & b_{n2} & \cdots & b_{np} \end{bmatrix},$$

$$\mathbf{C} = \begin{bmatrix} c_{11} & c_{12} & \cdots & c_{1n} \\ c_{21} & c_{22} & \cdots & c_{2n} \\ \vdots & \vdots & & \vdots \\ c_{r1} & c_{r2} & \cdots & c_{rn} \end{bmatrix}, \quad \mathbf{D} = \begin{bmatrix} d_{11} & d_{12} & \cdots & d_{1p} \\ d_{21} & d_{22} & \cdots & d_{2p} \\ \vdots & \vdots & & \vdots \\ d_{r1} & d_{r2} & \cdots & d_{rp} \end{bmatrix}$$

The time rate of change of the state of the system, $\dot{\mathbf{x}}(t)$, is formed as the sum of the modified inputs, $\mathbf{Bu}(t)$, and the modified current state, $\mathbf{Ax}(t)$. The matrix **A** is the most important of the four system matrices because it represents the proportion of the current system state, $\mathbf{x}(t)$, which contributes to changing that state. This state feedback has a major role in determining the future behavior of the system. The elements of matrix **B** are scalars and represent the proportion of the value of each of the input variables that affects each of the state variables.

The rate of change of the state, $\dot{\mathbf{x}}(t)$, is continuously integrated with the current state to produce the new state. The outputs, $\mathbf{y}(t)$, are formed by summing the new state which has been scaled by matrix **C** with a direct contribution from the modified input, $\mathbf{Du}(t)$. The elements of **C** and **D** are scalars which represent the proportions of each of the state and the input variables which produce the outputs, respectively.

Suppose the initial state vector $\mathbf{x}(0)$ is given. The solutions to Eqs. (1.2.2) and (1.2.3) can be derived as (Balakrishnan, 1983)

$$\mathbf{x}(t) = e^{\mathbf{A}t}\mathbf{x}(0) + \int_0^t e^{\mathbf{A}(t-s)}\mathbf{Bu}(s)\,ds \qquad (1.2.4)$$

$$\mathbf{y}(t) = \mathbf{C}e^{\mathbf{A}t}\mathbf{x}(0) + \int_0^t \mathbf{C}e^{\mathbf{A}(t-s)}\,\mathbf{Bu}(s)\,ds + \mathbf{Du}(t) \qquad (1.2.5)$$

in which

$$e^{\mathbf{A}t} = \sum_{n=0}^{\infty} \frac{t^n\mathbf{A}^n}{n!} \qquad (1.2.6)$$

The behavior of water resources systems often changes with time. For example, as the urbanization proceeds, the proportion of the urbanized watershed area, which is impervious, increases, causing the relationship between storm rainfall and runoff changes. This time-variant behavior can be incorporated into state variable models by making some of the elements in the matrices **A**, **B**, **C**, and **D** functions of time. Nonlinear response occurs when changes in the system's inputs do not produce linearly proportional changes in the system's outputs. These effects may be accounted for in state variable models by formulating some of the elements in the four system matrices **A**, **B**, **C**, and **D** as functions of the current system state.

1.2.3 Discrete-Time Deterministic State Variable Model

Although the nature of water resources systems operate continuously in time, the data are often collected and analyzed using discrete-time intervals, especially when a digital computer is involved in the data storage and analysis. For this situation, it is advantageous to formulate a discrete-time version of the deterministic state variable model. To do this, the time horizon is divided into K intervals or stages, $k = 1, 2, \ldots, K$, of length Δt. Time intervals, Δt, are not necessarily equal. The state $\mathbf{x}(t + \Delta t)$ may be related to the state $\mathbf{x}(t)$ at time t by using a Taylor's expansion:

$$\mathbf{x}(t + \Delta t) = \mathbf{x}(t) + (\Delta t)\dot{\mathbf{x}}(t) + \frac{\Delta t^2}{2}\ddot{\mathbf{x}}(t) + \cdots \qquad (1.2.7)$$

where $\ddot{x}(t) = d^2[\mathbf{x}(t)]/dt^2$. If the terms of order $(\Delta t)^2$ and higher are neglected, Eq. (1.2.7) may be written as

$$\mathbf{x}(t + \Delta t) = \mathbf{x}(t) + \dot{\mathbf{x}}(t) \cdot \Delta t \tag{1.2.8}$$

and the output equation is

$$\mathbf{y}(t) = \mathbf{Cx}(t) + \mathbf{Du}(t) \tag{1.2.9}$$

Equations (1.2.8) and (1.2.9) form the basic discrete-time, deterministic, state variable model. In the situation when the time intervals are equal and set to one unit of time, then $t = k\Delta t$, where k is the state index. The state equation (1.2.8) and output equation (1.2.9) can then be expressed as Eqs. (1.2.10) and (1.2.11), respectively, by substituting $\mathbf{Ax}(t) + \mathbf{Bu}(t)$ in Eq. (1.2.2) for $\dot{\mathbf{x}}(t)$.

$$\mathbf{x}(k + 1) = (\mathbf{A} + \mathbf{I})\mathbf{x}(k) + \mathbf{Bu}(k) \tag{1.2.10}$$

$$\mathbf{y}(k) = \mathbf{Cx}(k) + \mathbf{Du}(k) \tag{1.2.11}$$

where $\mathbf{I}$ is an identity matrix of rank n.

In the discrete-time model, the input and output variables correspond to the volume or mass of flow across the system boundaries in the unit time interval, instead of being the volume or mass flow rate, as they are in the continuous-time model.

At first, the input instruction to the system is divided into several stages with total number N. The time interval between two adjacent stages is not necessarily equal. Once the initial state of the system, $\mathbf{x}(t_0)$, and initial input to the system, $\mathbf{u}(t_0)$, are given, Eqs. (1.2.8) and (1.2.9) may be solved to obtain the state $\mathbf{x}(t_0 + \Delta t)$ at the next stage and the output $\mathbf{y}(t_0)$ at the current stage. Because the input at each stage is known, the process can be performed recursively until the last stage is reached.

If the matrix $\mathbf{A}$ in Eqs. (1.2.10) and (1.2.11) is redefined, a general discrete-time state variable model is expressed as

$$\mathbf{x}(k + 1) = \mathbf{Ax}(k) + \mathbf{Bu}(k) \tag{1.2.12}$$

$$\mathbf{y}(k) = \mathbf{Cx}(k) + \mathbf{Du}(k) \tag{1.2.13}$$

The solutions to Eqs. (1.2.12) and (1.2.13) can be derived as (Balakrishnan, 1984)

$$\mathbf{x}(k) = \mathbf{A}^k\mathbf{x}(0) + \sum_{i=0}^{k-1} \mathbf{A}^{k-i-1}\,\mathbf{Bu}(i) \tag{1.2.14}$$

$$\mathbf{y}(k) = \mathbf{CA}^k\mathbf{x}(0) + \sum_{i=0}^{k-1} \mathbf{CA}^{k-i-1}\,\mathbf{Bu}(i) + \mathbf{Du}(k) \qquad (1.2.15)$$

1.2.4 Controllability, Observability, and Stability of State Variable Model

A state variable model [Eqs. (1.2.2) and (1.2.3) or Eqs. (1.2.12) and (1.2.13)] is controllable if any state can be reached in some time, starting from the initial state, by applying a suitable input $\mathbf{u}(t)$. One of the useful theorems for controllability is that a state variable model with state vector $\mathbf{x}(t)$ of dimension $(n \times 1)$ and input vector $\mathbf{u}(t)$ of dimension $(m \times 1)$ is controllable if and only if the combined matrix $[\mathbf{B}, \mathbf{AB}, \mathbf{A}^2\mathbf{B}, \ldots, \mathbf{A}^{n-1}\mathbf{B}]$ has full rank (Balakrishnan, 1983). It should be pointed out that "$[\mathbf{B}, \mathbf{AB}, \mathbf{A}^2\mathbf{B}, \ldots, \mathbf{A}^{n-1}\mathbf{B}]$ has full rank" is equivalent to "$[\mathbf{B}, \mathbf{AB}, \mathbf{A}^2\mathbf{B}, \ldots, \mathbf{A}^{n-1}\mathbf{B}]$ has n linearly independent columns." It may be noted that the dimension of the combined matrix is $(n \times nm)$

A state variable model is observable if the initial state $\mathbf{x}(0)$ can be uniquely determined after pairs $[\mathbf{u}(t), \mathbf{y}(t)]$ for $t \geq 0$ are given. The state variable model is observable if and only if the combined matrix $[\mathbf{C}', \mathbf{A}'\mathbf{C}', (\mathbf{A}^2)'\mathbf{C}', \ldots, (\mathbf{A}^{n-1})'\mathbf{C}']$ has full rank (Balakrishnan, 1983). The state variable model is stable if the state of the system for zero input tends to zero asymptotically. The state variable model is stable if and only if the real parts of all eigenvalues of $\mathbf{A}$ is less than zero (Balakrishnan, 1983).

1.2.5 Applications of State Variable Modeling in Water Resources

State variable modeling has been applied to only a few water resource systems. Fan et al. (1973) developed a model to find control strategies for biological waste treatment using a state variable model of a continuously stirred tank reactor. Young and Beck (1974) formulated a state variable model for dissolved oxygen and biochemical oxygen demand in a river. This model was used to determine control schemes for sewage effluent discharges to rivers. Erscheler et al. (1974) developed a control strategy for the operation of penstock inlet gates in a hydroelectric power station based on a state variable model of the system. State variable approaches have also been used to model the storm rainfall and runoff processes. Muzik (1974) used a state variable approach to model overland flow. Duong et al. (1975) applied stochastic estimation theory to fit the parameters of a state variable rainfall-runoff model. Maidment (1976) and

Maidment and Chow (1976) developed a stochastic state variable dynamic programming model for reservoir operation. Tung and Mays (1978) developed a kinematic wave model for sewer network flow routing based on the state variable approach. Tung and Mays (1981) developed a rainfall-runoff model using the concepts of state variable modeling as described below.

Because of the characteristics of a watershed, the system is inherently nonlinear. The linear reservoir storage–discharge relation

$$S = kQ \tag{1.2.16}$$

can be modified to the form

$$S = K_1Q^N + K_2\frac{dQ}{dt} \tag{1.2.17}$$

where K_1, K_2, and N are assumed to be constants. Combining Eq. (1.2.16) with Eq. (1.2.17) for the conservation of mass, the following differential equation is obtained:

$$K_2\frac{d^2Q}{dt^2} + NK_1Q^{N-1}\frac{dQ}{dt} + Q = I \tag{1.2.18}$$

Equation (1.2.18) is rearranged to

$$\frac{d^2Q}{dt^2} = -\frac{K_1}{K_2}NQ^{N-1}\frac{dQ}{dt} - \frac{1}{K_2}Q + \frac{1}{K_2}I \tag{1.2.19}$$

The state variable formulation of the state equation in matrix form is written as follows:

$$\begin{bmatrix} x_1(t) \\ x_2(t) \end{bmatrix} = \begin{bmatrix} 0 & 1 \\ -e_0 & -e_1 \end{bmatrix}\begin{bmatrix} x_1(t) \\ x_2(t) \end{bmatrix} + \begin{bmatrix} 0 \\ h \end{bmatrix} I(t) \tag{1.2.20}$$

where $e_0 = 1/K_2$, $e_1 = K_1/K_2\,NQ^{N-1}$, and $h = 1/K_2$ and the output equation in matrix form is

$$Q(t) = [1 \quad 0]\begin{bmatrix} x_1(t) \\ x_2(t) \end{bmatrix} \tag{1.2.21}$$

1.3 DEFINITION OF OPTIMAL CONTROL PROBLEMS

The essential elements of the control problem are "(1) a mathematical model (system) to be controlled, (2) a desired output of the system, (3) a set of admissible inputs or controls, and (4) a performance index (or cost functional) which measures the effectiveness of a given control action" (Athans and Falb, 1966). The optimal control problem is to "steer" the system by controlling the inputs which generate the desired outputs while a chosen performance index is optimized.

More specifically, optimal control problems can be stated as follows:

Given:

1. The state equations
2. A set of boundary conditions on the state variables at the initial time and the terminal time
3. A set of constraints on the state variables and control variables

Determine the admissible control (values of the control variable) so that a performance index (an objective function) is optimized (minimized or maximized).

Mathematically, the optimal control model in continuous form is to optimize the objective function

$$\text{Optimize } F(\mathbf{u}) = \int_0^T f(\mathbf{x}(t), \mathbf{u}(t), t)\, dt \tag{1.3.1}$$

subject to the state equations

$$\dot{\mathbf{x}} = \mathbf{g}(\mathbf{x}(t), \mathbf{u}(t)) \tag{1.3.2}$$

the set of boundary conditions

$$\underline{\mathbf{x}} \le \mathbf{x}(t) \le \bar{\mathbf{x}} \tag{1.3.3}$$

$$\underline{\mathbf{u}} \le \mathbf{u}(t) \le \bar{\mathbf{u}} \tag{1.3.4}$$

and the set of constraints on $\mathbf{x}(t)$ and $\mathbf{u}(t)$

$$\mathbf{w}(\mathbf{x}(t), \mathbf{u}(t)) = \mathbf{0} \tag{1.3.5}$$

where $\mathbf{u}(t)$ is the control variable $\mathbf{u} = [u_1, \ldots, u_m]^T$ with lower and upper bounds of $\underline{\mathbf{u}}$ and $\bar{\mathbf{u}}$, respectively, and $\mathbf{x}(t)$ is the state variable $\mathbf{x} = (x_1, \ldots, x_n)$ with lower and upper bounds of $\underline{\mathbf{x}}$ of $\bar{\mathbf{x}}$, respectively.

The objective function f is a given continuous real-valued function and the integral in Eq. (1.3.1) is interpreted as taking a control $\mathbf{u}(t)$ such that $\underline{\mathbf{u}} \le \mathbf{u}(t) \le \bar{\mathbf{u}}$, solving the state equation to obtain the corresponding $\mathbf{x}(t)$, and then calculating f as a function of time. The control variables and state variables are related through the state equation which is expressed as a differential equation in Eq. (1.3.2)

$$\frac{d\mathbf{x}}{dt} = \mathbf{g}(\mathbf{x}(t), \mathbf{u}(t)) \tag{1.3.6}$$

for a continuous system and as

$$\mathbf{x}(t + 1) = \mathbf{g}(\mathbf{x}(t), \mathbf{u}(t)) \tag{1.3.7}$$

for a discrete system.

In many applications of control theory, the objective function (or performance index) has the form

$$F(\mathbf{u}) = \phi(\mathbf{x}(T)) + \int_0^T f(\mathbf{x}(t), \mathbf{u}(t))\, dt \tag{1.3.8}$$

where $T > 0$ is fixed and $\phi(\mathbf{x}(t))$ is a given continuously differentiable function that represents the terminal objective value at the final time T.

The optimal control problem in discrete form is to optimize

$$F(\mathbf{u}) = \sum_{t=1}^{T} f(\mathbf{x}(t), \mathbf{u}(t))$$

subject to the state equation in discrete form

$$\mathbf{x}(t + 1) = \mathbf{g}(\mathbf{x}(t), \mathbf{u}(t))$$

and the set of boundary conditions, Eqs. (1.3.3) and (1.3.4), and other constraints on $\mathbf{x}(t)$ and $\mathbf{u}(t)$, Eq. (1.3.5).

REFERENCES

Athans, M. and Falb, P. L., *Optimal Control*, McGraw-Hill, New York, 1966.

Balakrishnan, A. V., *Elements of State Space Theory of Systems*, Optimization Software, Inc., New York, 1983.

Balakrishnan, A. V., *Kalman Filtering Theory*, Optimization Software, Inc., New York, 1984.

Duong, N., Wynn, C. B., and Johnson, G. R., Modern Control Concepts in Hydrology, *IEEE Transactions on Systems, Man and Cybernetics*, Vol. SMC-5, No. 1, 1975.

Erschler, J., Roubellat, F., and Vernhes, J. P., Automation of a Hydroelectric Power Station Using Variable Structure Control Systems, *Automatica*, Vol. 10, No, 1, 1974.

Fan, L. T., Shah, P. S., Periera, N. C., and Erickson, L. E., Dynamic Analysis and Optimal Feedback Control Synthesis Applied to Biological Waste Treatment, *Water Research*, Vol. 7, No. 11, 1973.

Maidment, D. R., Stochastic State Variable Dynamic Programming for Water Resources Systems Analysis, Ph.D. thesis, University of Illinois at Urbana–Champaign, 1976.

Maidment, D. R. and Chow, V. T., A New Approach to Urban Water Resources Systems Optimization. *Proceedings* of the World Environment and Resources Council (WERC) Conference on the Environment of Human Settlements, Brussels, March 1976.

Mays, L. W. and Tung, Y. K., *Hydrosystems Engineering and Management*, McGraw-Hill, Inc., New York, 1992.

Muzik, I., State Variable Model of Overland Flow, *Journal of Hydrology*, Vol. 22, No. 3/4, 1974.

Sinha, N. K., *Linear Systems*, John Wiley & Sons, New York, 1991.

Tung, Y. K. and L. W. Mays, State Variable Model for Sewer Network Flow Routing, *Journal of the Environmental Engineering Division*, Vol. 104, No. 1, pp. 15–30, 1978.

Tung, Y. K. and L. W. Mays, State Variable Model for Rainfall-Runoff Process, *Water Resources Bulletin*, Vol. 17, No. 2, pp. 181–189, 1981.

Young, P. and Beck, B., The Modeling and Control of Water Quality in a River System, *Automatica*, Vol. 10, No. 5, 1974.

Chapter 2
Introduction to Hydrosystems Problems

2.1 OPTIMIZATION OF HYDROSYSTEMS

Many problems for the operation of hydrosystems can be formulated in a general optimization framework in terms of state (or dependent) variables (**x**) and control (or independent) variables (**u**)

$$\text{Minimize } f(\mathbf{x}, \mathbf{u}) \tag{2.1.1}$$

subject to process simulation equations

$$\mathbf{G}(\mathbf{x}, \mathbf{u}) = \mathbf{0} \tag{2.1.2}$$

and additional constraints for operation on the dependent (**u**) and independent (**x**) variables

$$\underline{\mathbf{w}} \leq \mathbf{w}(\mathbf{x}, \mathbf{u}) \leq \overline{\mathbf{w}} \tag{2.1.3}$$

The process simulation equations for hydrosystems applications basically consist of the governing physical equations (2.1.2) that simulate a physical process such as conservation of mass, energy, and momentum. These equations are typically large in number, sparse, and nonlinear in terms of the state and control variables. In most hydrosystem applications, these governing equations are ordinary or partial differential equations. Conceptually, the simplest approach is to have the optimizer solve the above optimization problem directly by embedding finite differences or finite element equations for the governing process equations. Unfortunately, many of the real-world problems cannot be solved in this manner because of their size and nonlinearity. The existing nonlinear programming (NLP) codes cannot solve such large, sparse problems.

An alternative approach is to use the appropriate process simulator to solve the constraints process simulation equations (2.1.2) each time the constraints need to be evaluated for the optimizer. The major advantage of such an approach is the reduced size of problem solved by the nonlinear optimizer, so that only a small subset of the complete set of constraint equations is evaluated by the optimizer. The basic idea is that the optimizer only sees the following reduced problem:

$$\text{Minimize } F(\mathbf{u}) = f(\mathbf{x}(\mathbf{u}), \mathbf{u}) \tag{2.1.4}$$

$$\text{subject to } \underline{\mathbf{w}} \leq \mathbf{w}(\mathbf{x}(\mathbf{u})) \leq \overline{\mathbf{w}} \tag{2.1.5}$$

as opposed to the much larger problem defined by Eqs. (2.1.1)–(2.1.3).

The class of problems that are being considered in this book essentially have a large number of partial differential equations or other simulation equations as part of the constraint set (process simulation equations) making them more complex than the standard type of optimization problem. These optimization problems are referred to herein as optimal control problems. Examples of hydrosystem optimal control problems are presented in Secs. 2.2–2.8. Each of these are nonlinear programming problems that can be solved by interfacing the appropriate simulator (simulation model) with the optimizer to solve a reduced nonlinear programming problem. Applications are presented for systems such as groundwater systems (Fig. 2.1), river-reservoir systems (Fig. 2.2) for flood-control, reservoir systems (Fig. 2.3) for water supply, water distribution systems (Fig. 2.4) operation, estuary systems (Fig. 2.5) for salinity control, for sediment control in rivers and reservoirs, and for the operation of soil aquifer treatment (SAT) systems (Figs. 2.6, 2.7).

2.2 GROUNDWATER MANAGEMENT SYSTEMS

The general groundwater management problem (GGMP) can be expressed mathematically as follows :

Objective

$$\text{Optimize } Z = f(\mathbf{h}, \mathbf{q}) \tag{2.2.1}$$

where $\mathbf{h}$ and $\mathbf{q}$ are vectors of heads and pumpages (or recharge), respectively. The objective function may be either maximization (e.g., sum of heads) or minimization (e.g., minimize pumpage) and can be a linear or

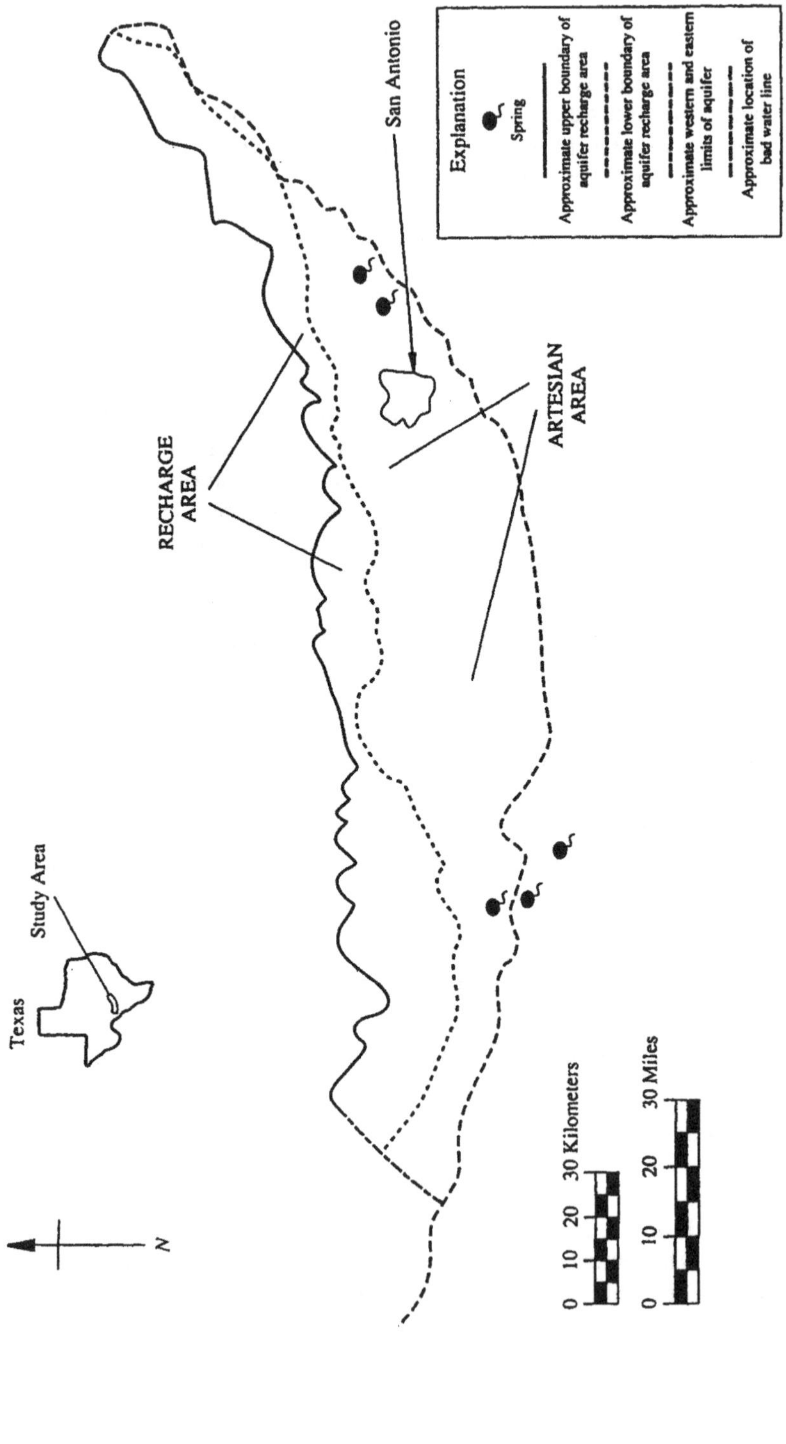

Figure 2.1 Edwards (Balcones fault zone) aquifer, San Antonio region.

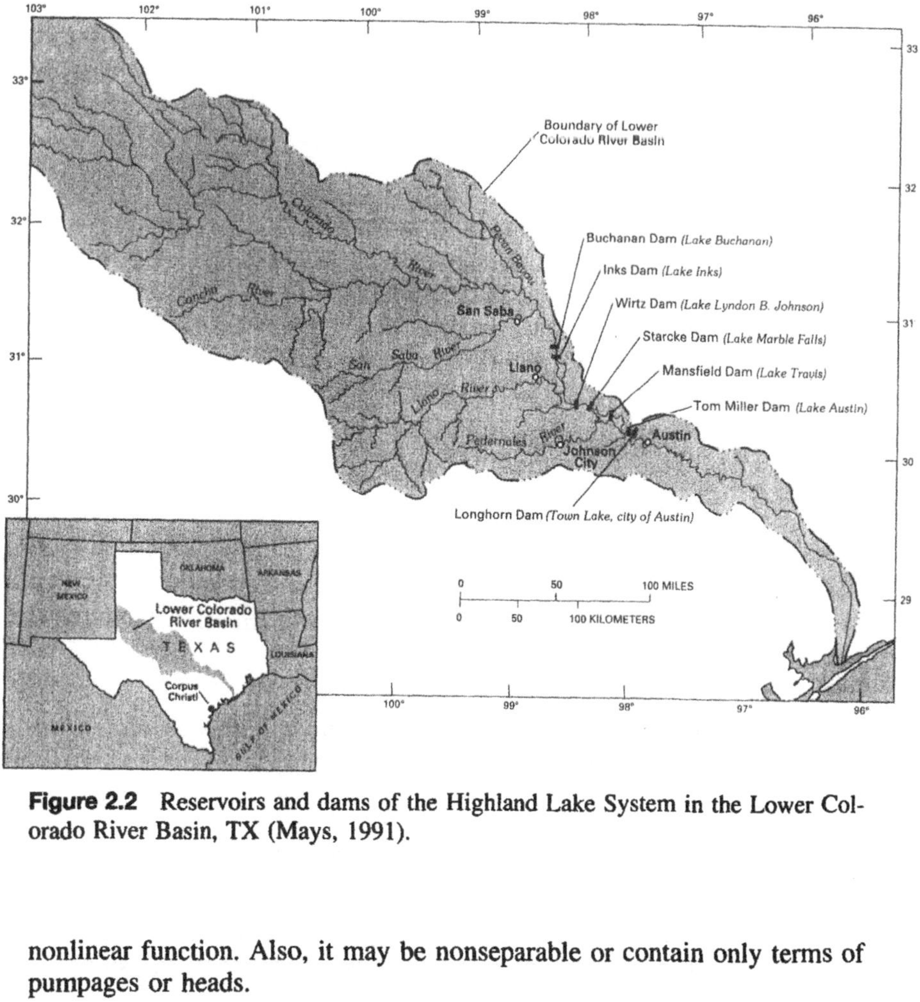

Figure 2.2 Reservoirs and dams of the Highland Lake System in the Lower Colorado River Basin, TX (Mays, 1991).

nonlinear function. Also, it may be nonseparable or contain only terms of pumpages or heads.

Constraints

(a) The general groundwater flow constraints represent a system of equations governing groundwater flow which are finite difference or simulator equations when $\mathbf{q}$ is unknown.

$$\mathbf{G}(\mathbf{h}, \mathbf{q}) = \mathbf{0} \tag{2.2.2}$$

(b) The upper ($\overline{\mathbf{q}}$) and lower ($\underline{\mathbf{q}}$) bounds on pumpages physically may or may not exist. Unlike pumpage, the lower bound on heads ($\underline{\mathbf{h}}$) can be viewed as the bottom elevation of the aquifer, whereas

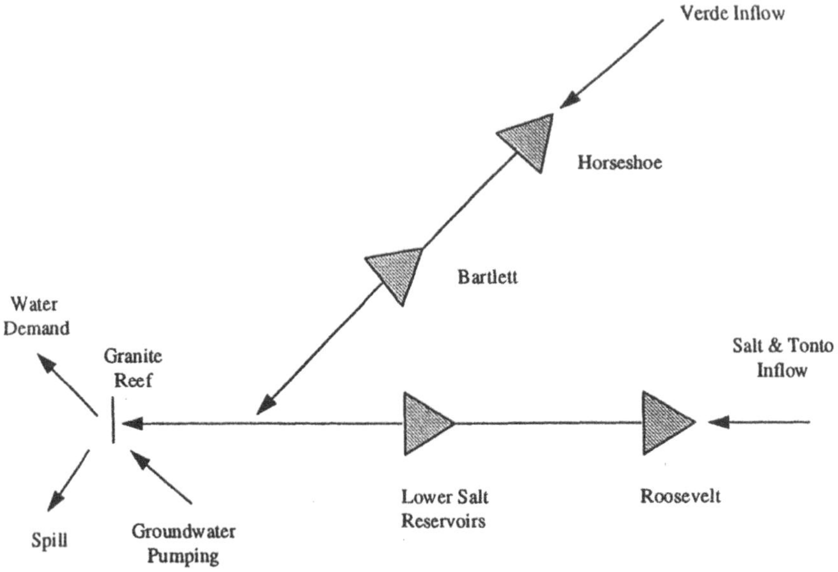

Figure 2.3 Salt River Project reservoir system in Arizona.

the upper head bound ($\bar{\mathbf{h}}$) can be viewed as ground surface elevations for the unconfined cells.

$$\underline{\mathbf{q}} \leq \mathbf{q} \leq \bar{\mathbf{q}} \tag{2.2.3}$$

$$\underline{\mathbf{h}} \leq \mathbf{h} \leq \bar{\mathbf{h}} \tag{2.2.4}$$

(c) In addition to constraint equations (2.2.2)–(2.2.4), other constraints may be included to impose restrictions such as water demands, operating rules, budgetary limitations, and so on.

$$\mathbf{w}(\mathbf{h}, \mathbf{u}) \leq \mathbf{0} \tag{2.2.5}$$

Both head, $\mathbf{h}$, and pumpage (or recharge), $\mathbf{q}$, are vectors of decision variables which have maximum dimensions equal to the product of the number of active nodes within the aquifer boundary and time steps. Fixed pumpages or recharges are considered to be constants. By convention, available pumpages have a positive value and the elements of $\mathbf{q}$ have a negative value where there is available recharge. Usually, the number of variable pumpages and/or recharges (hereafter, the terms pumpages that refer to $\mathbf{q}$ will imply both pumpages and/or recharges) is small and results in a much smaller dimension of $\mathbf{q}$ and $\mathbf{h}$.

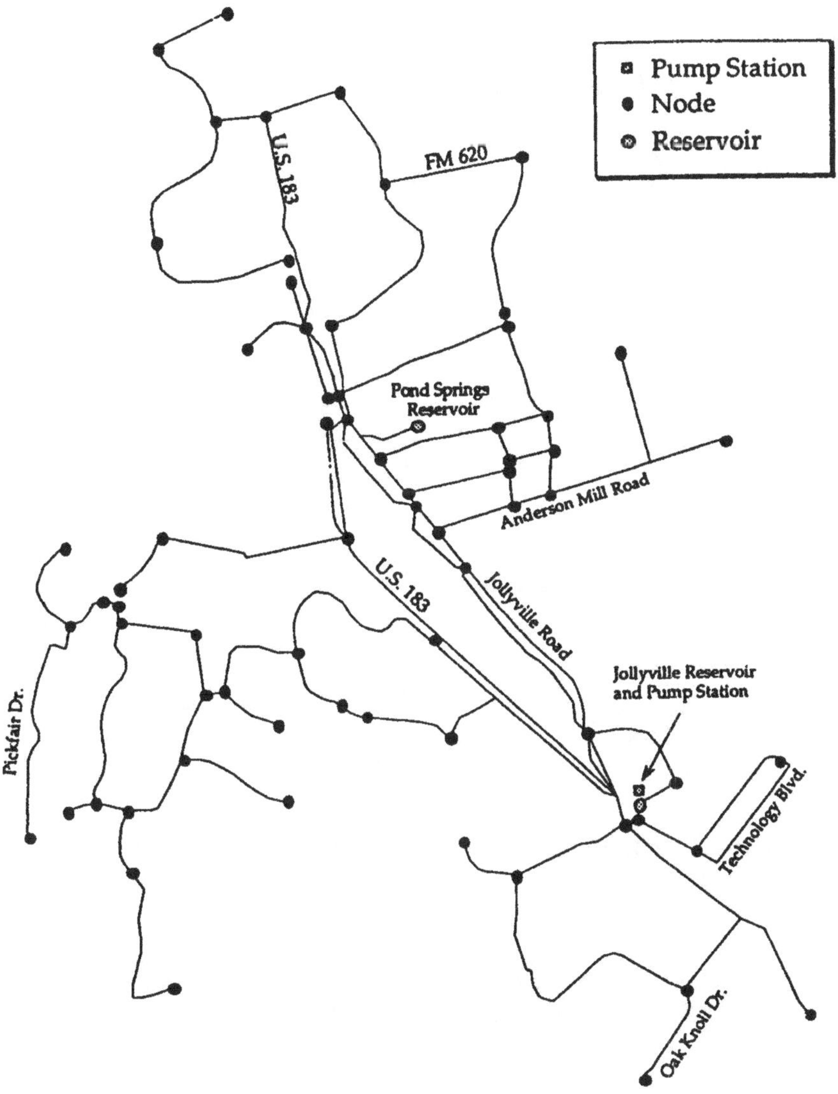

Figure 2.4 Water distribution system for City of Austin, Northwest B Pressure Zone.

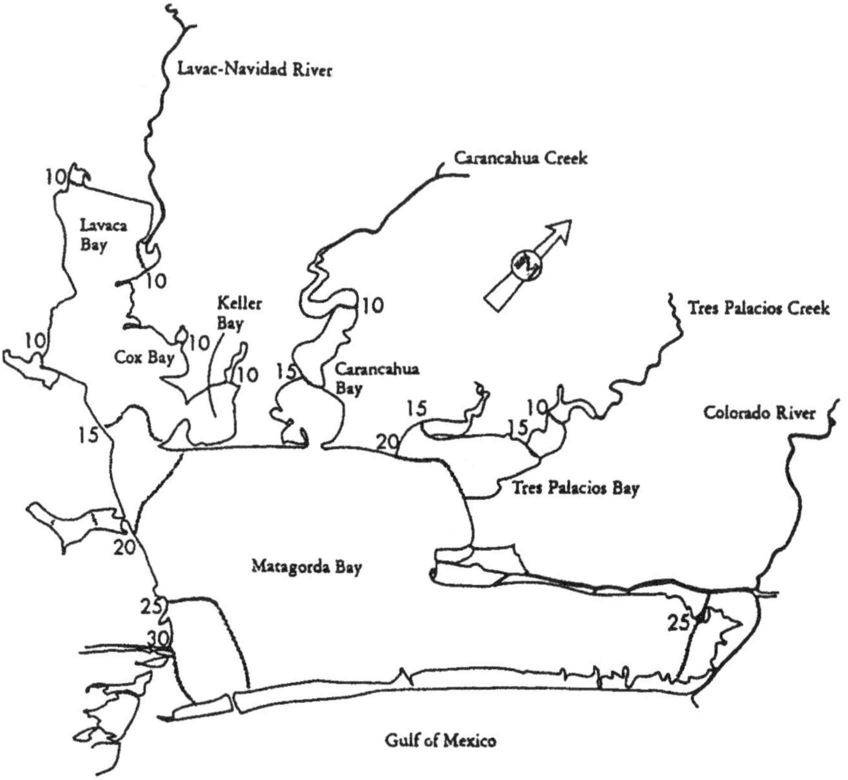

Figure 2.5 Bay System map and example of simulated salinity contour (ppt) in Lavaca–Tres Palacios Estuary.

2.3 REAL-TIME OPERATION OF RIVER-RESERVOIR SYSTEMS FOR FLOOD CONTROL

The optimization problem for the real-time operation of multireservoir systems under flooding conditions can be stated as follows:

Objective

$$\text{Minimize } Z = f(\mathbf{h}, \mathbf{Q}) \tag{2.3.1}$$

where **h** and **Q** are the vectors of water surface elevations and discharges, respectively. The objective is defined by minimizing (a) the total flood damages, (b) deviations from target levels, (c) water surface elevations in

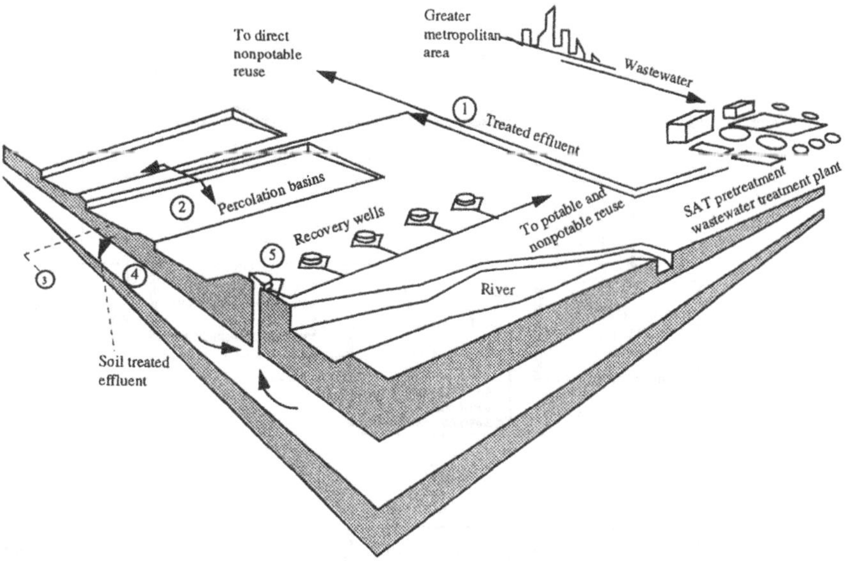

Figure 2.6 SAT system consists of five major components: (1) pipelines that carry the treated effluent from the wastewater treatment plant to the SAT system; (2) percolation (infiltration) basins where the treated effluent infiltrates into the ground; (3) the soil immediately below the infiltration basins (vadose zone); (4) the aquifer where water is stored for a long duration; and (5) the recovery wells where water is pumped from the aquifer for potable and nonpotable use.

the flood areas, or (d) spills from reservoirs or maximizing storage in reservoirs.

Constraints

(a) Hydraulic constraints are defined by the Saint-Venant equations for one-dimensional gradually varied unsteady flow and other relationships such as upstream, downstream, and internal boundary conditions and initial conditions that describe the flow in the different components of a river-reservoir system,

$$\mathbf{G}(\mathbf{h}, \mathbf{Q}, \mathbf{r}) = \mathbf{0} \tag{2.3.2}$$

where $\mathbf{h}$ is the vector of water surface elevations, $\mathbf{Q}$ is the vector of discharges, and $\mathbf{r}$ is the matrix of gate settings for spillway structures, all given in matrix form to consider the time and space dimensions of the problem.

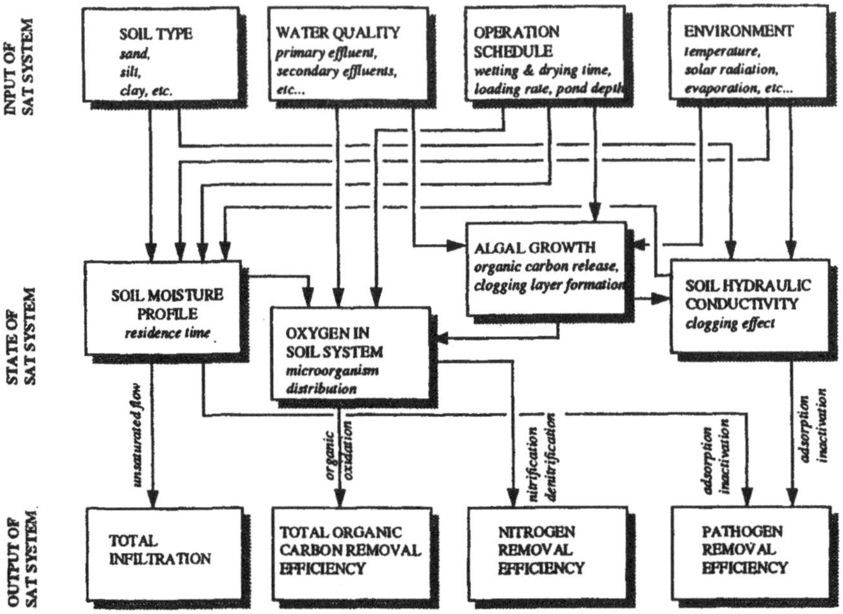

Figure 2.7 Soil aquifer treatment system dynamics.

(b) Bounds on discharges defined by minimum and maximum allowable reservoir releases and flow rates at specified locations,

$$\underline{\mathbf{Q}} \leq \mathbf{Q} \leq \overline{\mathbf{Q}} \tag{2.3.3}$$

Bars above and below a variable denote the upper and lower bounds, respectively, for that variable.

(c) Bounds on elevations defined by minimum and maximum allowable water surface elevations at specified locations (including reservoir levels).

$$\underline{\mathbf{h}} \leq \mathbf{h} \leq \overline{\mathbf{h}} \tag{2.3.4}$$

(d) Physical and operational bounds on spillway gate operations.

$$\mathbf{0} \leq \underline{\mathbf{r}} \leq \mathbf{r} \leq \overline{\mathbf{r}} \leq \mathbf{0} \tag{2.3.5}$$

where **r** represents the fraction of gate openings.

(e) Other constraints such as operating rules, target storages, storage capacities, and so forth.

$$\mathbf{W(r)} \leq \mathbf{0} \tag{2.3.6}$$

The constraints of the model can be divided into two groups: the hydraulic constraints [Eq. (2.3.2)] and the operational constraints [Eqs. (2.3.3)–(2.3.6)]. The hydraulic constraints are equality constraints consisting of the equations that describe the flow in the system. These are (a) the Saint-Venant equations for all computational reaches except internal boundary reaches, (b) relationships to describe the upstream and downstream conditions for the extremities, and (c) internal boundary conditions which describe the flow that cannot be described by the Saint-Venant equations, such as critical flow resulting from flow over a spillway or waterfall.

The operational constraints are basically "greater than" or "less than" types of constraints that define the variable bounds, operational targets, structural limitations, and capacities. Options for an operator to set the limits of certain variables are also classified under this category. Bound constraints are used to impose operational or optimization-related requirements. Non-negativity constraints on discharges are not used because discharges are allowed to take negative values in order to be able to realistically represent the reverse-flow phenomena (backwater effects) due to a rising lake or large tributaries into a lake or tidal condition. Non-negativity of water surface elevations is always satisfied because the system hydraulics are solved implicitly by the simulation model, DWOPER. The lower limits on elevations and discharges can be used to indirectly impose water quality considerations, minimum required reservoir releases, and other policy requirements. The upper bound on elevations and discharges can be used to set the maximum allowable levels (values beyond are either catastrophic or physically impossible) such as the overtopping elevations for major structures, spillway capacities, and so forth. Damaging elevations and/or discharges must be given to the model through the constraints, as the objective functions do not have any terms to control them.

The third model variable, $\mathbf{r}$, gate opening, can be allowed to vary between 0 and 1, which corresponds to 0% and 100% opening of the available total spillway gate areas, respectively. The bounds on gate settings are intended primarily to reflect the limitations on gate operations as well as to enable the operator to prescribe any portion(s) of the operation for any reservoir(s). Operational constraints other than bounds can be imposed for various purposes. The maximum allowable rates of change of gate openings, for instance, for a given reservoir can be specified through this formulation as a time-dependent constraint. This particular formulation may be very useful, especially for cases where sharp changes in gate operations (i.e., sudden openings and closures) are not desirable or physically impossible. It is handled by setting an upper bound to the change in the percentage of gate opening from one time step to the next. This constraint can also be used to model another aspect of gate operations for very short time intervals; that is, the gradual settings that have to be followed when opening or closing a gate. For this case, the gate cannot be opened (or closed) by more than a certain percentage during a given time interval.

2.4 RESERVOIR SYSTEM OPERATION FOR WATER SUPPLY

Reservoir system operation is for the purposes of meeting water supply demand and recreation demands, maintaining minimum flow levels for navigation and environmental concerns, and providing flood protection, power production, and flood control. The mathematical formulation of the reservoir system operation problem can be stated in general form as follows:

Objective

$$\text{Maximize Benefits} = \text{Max} \sum_{0}^{T} f(\mathbf{S}_t, \mathbf{U}_t, t) \tag{2.4.1}$$

Constraints

(a) The system equations which are the conservation of mass equations for the reservoirs and river reaches are

$$\mathbf{G}(\mathbf{S}_{t+1}, \mathbf{S}_t, \mathbf{U}_t, \mathbf{I}_t, \mathbf{L}_t) = \mathbf{0} \quad t = 0, \ldots, T-1 \tag{2.4.2}$$

where $\mathbf{S}_{t+1}$, and $\mathbf{S}_t$ are the vectors of reservoir storages at the beginning of time period $t + 1$ and t, respectively, $\mathbf{U}_t$ is the vector of the reservoir releases for M reservoirs during time t, $\mathbf{I}_t$ is the vector of hydrologic inputs (such as inflow to reservoirs), and $\mathbf{L}_t$ is the vector of reservoir losses.

(b) The bound constraints on reservoir releases, $\mathbf{U}_t$, are

$$\underline{\mathbf{U}}_t \le \mathbf{U}_t \le \overline{\mathbf{U}}_t, \quad t = 1, \ldots, T \tag{2.4.3}$$

where $\underline{\mathbf{U}}_t$ and $\overline{\mathbf{U}}_t$ are lower and upper bounds on the reservoir releases.

(c) The bound constraints on reservoir storages are deterministically defined as

$$\underline{\mathbf{S}}_t \le \mathbf{S}_t \le \overline{\mathbf{S}}_t, \quad t = 1, \ldots, T \tag{2.4.4}$$

where $\underline{\mathbf{S}}_t$ and $\overline{\mathbf{S}}_t$ are the lower and upper bounds, respectively, on storage.

(d) The bound constraints on reservoir storage could be defined in probabilistic form as storage reliability constraints as

$$P[\mathbf{S}_t \geq \underline{\mathbf{S}}_t] \leq \alpha_t^{\min}, \quad t = 1, \ldots, T \tag{2.4.5}$$

and

$$P[\mathbf{S}_t \leq \overline{\mathbf{S}}_t] \leq \alpha_t^{\max}, \quad t = 1, \ldots, T \tag{2.4.6}$$

where $P[\ ]$ denotes the probability and $\alpha_t^{\min}$ and $\alpha_t^{\max}$ represent the minimum and maximum reliability or tolerance levels, respectively, on storage.

(e) Other reservoir operational constraints are expressed as

$$\mathbf{w}(\mathbf{S}_t, \mathbf{U}_t) = \mathbf{0} \tag{2.4.7}$$

2.5 WATER DISTRIBUTION SYSTEM OPERATION

The optimization problem for water distribution system operation can be stated in terms of the nodal pressure heads, **H**, pipe flows, **Q**, tank water surface elevations, **E**, and pump operating times, **D**. The objective is to minimize energy costs:

Objective

$$\text{Minimize energy costs} = f(\mathbf{H}, \mathbf{Q}, \mathbf{D}) \tag{2.5.1}$$

Constraints

(a) Conservation of flow and energy constraints

$$\mathbf{G}(\mathbf{H}, \mathbf{Q}, \mathbf{D}, \mathbf{E}) = \mathbf{0} \tag{2.5.2}$$

(b) Pump operation constraints

$$\mathbf{w}(\mathbf{E}) = \mathbf{0} \tag{2.5.3}$$

(c) Nodal pressure head bounds

$$\underline{\mathbf{H}} \leq \mathbf{H} \leq \overline{\mathbf{H}} \tag{2.5.4}$$

(d) Bounds on pump operating times

$$\underline{\mathbf{D}} \leq \mathbf{D} \leq \overline{\mathbf{D}} \tag{2.5.5}$$

(e) Bounds on tank water surface elevation

$$\underline{\mathbf{E}} \leq \mathbf{E} \leq \overline{\mathbf{E}} \tag{2.5.6}$$

2.6 FRESHWATER INFLOWS TO BAYS AND ESTUARIES

The overall optimization model can be stated in the following general nonlinear programming format using an objective to minimize freshwater inflows or to maximize harvest.

Objective

$$\text{Optimize } z = f(\mathbf{Q}, \mathbf{s}, \mathbf{H}) \tag{2.6.1}$$

The general mathematical model can consider the following objective functions:

(a) Minimize the sum of freshwater inflows into the bay and estuary over an operational time frame, such as a year
(b) Maximize the harvest over an operational time frame, such as a year;
(c) Multiobjective to minimize freshwater inflows and maximize the harvest over an operational time frame, such as a year

Constraints

(a) Hydrodynamic transport equations that relate salinity, $\mathbf{s}$, at a given point in an estuary to inflow, $\mathbf{Q}$,

$$\mathbf{G}(\mathbf{Q}, \mathbf{s}) = \mathbf{0} \tag{2.6.2}$$

where $\mathbf{Q}$ is the independent variable (control variable) as a function of time and $\mathbf{s}$ is the dependent variable (state variable) as a function of time and location
(b) Regression equations that relate inflow to fish harvest

$$\mathbf{h}(\mathbf{Q}, \mathbf{H}) = \mathbf{0} \tag{2.6.3}$$

(c) Constraints that define limitations on freshwater inflows due to upstream demands and water uses, and historical ranges

$$\underline{\mathbf{Q}} \le \mathbf{Q} \le \overline{\mathbf{Q}} \tag{2.6.4}$$

(d) Constraints that define limitations on salinity

$$\underline{\mathbf{s}} \le \mathbf{s} \le \overline{\mathbf{s}} \tag{2.6.5}$$

2.7 SEDIMENT CONTROL IN RIVERS AND RESERVOIRS

The optimization problem for controlling sediment effects in rivers and reservoirs can be stated as follows.

Objective

$$\text{Minimize } z = \sum_{i=1}^{T} \mathbf{a}_t + \mathbf{b}_t \tag{2.7.1}$$

where z is the objective function (sum of aggradation and degradation depths in downstream river reaches) or the overall cost associated with the control policy (reservoir releases), t is the simulation time step, and T is the total number of simulation time steps. The state vector consists of the reservoir storage levels and downstream river bed elevations at the starting time t (having dimension I); $\mathbf{a}_t$ is the aggradation depth at time t (having dimension $I - 1$), and $\mathbf{b}_t$ is the degradation depth at time t (having dimension $I - 1$).

Constraints

Reservoir Operation Constraints

(a) The reservoir mass balance equation is

$$\mathbf{S}_{t+1} = \mathbf{S}_t + \mathbf{Q}_t + \mathbf{R}_t - \mathbf{L}_t + \mathbf{P}_t - \mathbf{D}_t \tag{2.7.2}$$

where $\mathbf{S}_t$ and $\mathbf{S}_{t+1}$ are the beginning and ending storage states, respectively, of the reservoir during time (t), $\mathbf{Q}_t$ is the inflow, $\mathbf{R}_t$ is the release, $\mathbf{L}_t$ is the seepage flow, $\mathbf{P}_t$ is the excess rainfall/precipitation, and $\mathbf{D}_t$ is the evaporation loss.

(b) Bound constraints on reservoir releases ($\mathbf{R}$) are

$$\underline{\mathbf{R}}_t \le \mathbf{R}_t \le \overline{\mathbf{R}}_t \tag{2.7.3}$$

(c) Bound constraints on reservoir storage levels are

$$\underline{\mathbf{S}}_t \le \mathbf{S}_t \le \overline{\mathbf{S}}_t \tag{2.7.4}$$

River Hydraulic and Sedimentation Constraints

(d) The steady-state continuity equation and energy equations for the river flow are expressed in general form as

$$\mathbf{G}_1(\mathbf{R}_t, \mathbf{E}_t) = \mathbf{0} \tag{2.7.5}$$

where $\mathbf{E}_t$ is the bed elevation at the stations at time (t) measured from a designated reference datum. These equations also include the depth of flow at station i during time t, the energy distribution coefficient associated with station i and at time t, the energy loss from station $i + 1$ to station i during time t, and the flow velocity at station i during time t.

(e) The sediment continuity equation is expressed in general form as

$$\mathbf{G}_2(\mathbf{G}_{s_t}) = \mathbf{0} \tag{2.7.6}$$

where $\mathbf{G}_{s_t}$ is the sediment discharge at time t. These equations also include the lateral or local sediment input during time t from bank or tributaries per unit length, the specific weight of bed material, the porosity of the bed sediment material, and the average width of the movable bed.

(f) The sediment transport equations can be expressed in general form as

$$\mathbf{G}_{s_t} = \mathbf{f}_t(\mathbf{R}) \tag{2.7.7}$$

The parameters that also define sediment discharge are the concentration of wash load, the hydraulic radius of the channel, the energy gradient, the Darcy–Weisbach friction factor, the fluid kinematic viscosity, the fluid density, the particle density, the geometric mean size of the bed material sediments, the geometrical standard deviation of the bed material, the mean settling velocity of the sediment particle, the plan-form geometry, the apparent dynamic viscosity, the shape factor of the sediment particles, the shape factor of the channel reach, the seepage force in the channel bed, the concentration of bed-material discharge, and the fine material concentration.

State variables for this problem are the upstream reservoir storages and bed elevations, and the decision variable is the reservoir release.

2.8 SOIL AQUIFER TREATMENT SYSTEMS

Soil aquifer treatment (SAT) systems consist of five components as shown in Figure 2.6. Figure 2.7 illustrates the SAT system dynamics including inputs to the system, the state of the system, and outputs from the system. Inputs to the system include soil type, water quality, operation of the SAT system, and environment, where each input affects the state of the system. The state of an SAT system includes the soil moisture profile within the vadose zone, level of oxygen, algal growth, and soil hydraulic conductivity, where the state of the system controls the residence time of water in the vadose zone and the level of microbial activity. Of particular importance is the level of oxygen, which is related to the microorganism distribution and oxygen demanding substrates. Algal growth is related to the clogging layer formation on the soil surface and effective soil hydraulic conductivity.

The soil moisture is directly affected by inputs (soil type and the environment) and indirectly by the water quality which affects algal growth and the soil hydraulic conductivity. The oxygen in the vadose zone is affected by the soil moisture profile, the water quality of treated effluent, the operation schedule, and the algal growth. Algal growth is affected by the water quality of the treated effluent, the operation schedule, and the environment. Soil hydraulic conductivity is affected by the algal growth, the soil type, and the environment.

The critical outputs of the system are the total infiltration, the total organic carbon removal efficiency, the nitrogen removal efficiency, and the pathogen removal efficiency. The total organic carbon removal efficiency is primarily affected by the amount of oxygen in the system through the organic oxidation process and the presence of acclimated microorganisms. Nitrogen removal efficiency is dependent on oxygen levels and the availability of biodegradable organic carbon through the nitrification–denitrification process. Pathogen removal efficiency is dependent on the soil moisture profile and the soil hydraulic conductivity through the adsorption–inactivation process.

The major purification processes occurring in the soil aquifer system are filtration, chemical precipitation/dissolution, organic biodegradation, nitrification, denitrification, disinfection, ion exchange, and adsorption/desorption. Soil aquifer treatment is a key component of overall water reuse strategies that provide for (a) mechanical filtration of suspended particles, (b) biologically mediated transformation of organics and nitrogen, and (c) physical–chemical retention of inorganic and organic dissolved constituents (e.g., phosphorus, potassium, trace elements) from the biologically treated wastewater.

An operation model is used to determine optimal values of the control variables (independent variable), the application time (**X**) and the drying time (**Z**) for the pond, in order to maximize the infiltration. The simulator determines the state of the SAT system for these decisions. The state variable (dependent variable) is the average water content (**ω**). The draining time (**Y**) and infiltration volume (**F**) are functions of the control variables (**X** and **Z**) and the state variable (**ω**). An operation period **CT** (cycle time) is divided into three periods: the application time **X**, the draining time **Y**, and the drying time **Z** for each pond.

A mathematical formulation of the overall optimization model (without water quality constraints) is stated as follows:

$$\text{Maximize V} = \sum_{t=1}^{N} F_t(\boldsymbol{\omega}_t, \mathbf{X}_t, \mathbf{Z}_t, t) \tag{2.8.1}$$

subject to the following constraints:

(1) Simulator equations to describe the water content distribution, the infiltration process, and the draining process are respectively

$$\boldsymbol{\omega}_{t+1} = \mathbf{T}_1(\boldsymbol{\omega}_t, \mathbf{X}_t, \mathbf{Z}_t, t) \tag{2.8.2}$$

$$\mathbf{F}_t = \mathbf{T}_2(\boldsymbol{\omega}_t, \mathbf{X}_t, \mathbf{Z}_t, t) \tag{2.8.3}$$

$$\mathbf{Y}_t = \mathbf{T}_3(\boldsymbol{\omega}_t, \mathbf{X}_t, t) \tag{2.8.4}$$

(2) The bound constraint on water content during drying

$$\boldsymbol{\omega}_{t+1} \leq \boldsymbol{\omega}_{\max} \tag{2.8.5}$$

(3) The bound constraint on cycle time

$$\mathbf{X}_t + \mathbf{Y}_t + \mathbf{Z}_t \leq \mathbf{CT}_{\max} \tag{2.8.6}$$

where V is the total infiltration volume during N cycles, $\mathbf{F}_t(\ )$ is the infiltration volume over given cycles, $\boldsymbol{\omega}_t$ is the average water content at the beginning of cycle t and which equals the average water content at the end of cycle $t - 1$, $\mathbf{X}_t$ is the application time, defined as the duration of time when water is being discharged to the pond, $\mathbf{Z}_t$ is the drying time, which is the time after infiltration ends to the beginning of next application period, $\boldsymbol{\omega}_{t+1}$ is the average water content, which is computed from the water content distribution obtained from the simulator at the end of cycle t, $\mathbf{Y}_t$ is the draining time, which is the time period after inflow ends until no water remains on the surface of the pond, and $\boldsymbol{\omega}_{\max}$ and $\mathbf{CT}_{\max}$ is the upper bounds for $\boldsymbol{\omega}_{t+1}$ and cycle time.

The objective function, Eq. (2.8.1), maximizes the infiltration volume (unit area) over N cycles. Equation (2.8.2) is the transition equation which relates the average water content $\boldsymbol{\omega}_{t+1}$ to the average water content $\boldsymbol{\omega}_t$, given the application time $\mathbf{X}_t$ and the drying time $\mathbf{Z}_t$. Equations (2.8.3) and (2.8.4) represent the relations of the infiltration volume and the draining time with respect to the water content, the application time, and the drying time. Equations (2.8.2)–(2.8.4) are solved by the simulator HYDRUS. Equations (2.8.5) and (2.8.6) are the bound constraints on the average water content and cycle time. Equations (2.8.1)–(2.8.6) constitute an optimization model for the SAT system operation with the structure of a discrete-time optimal control problem which is solved by the optimal control algorithm SALQR. Solution of this model will yield the optimal application time $\mathbf{X}$ and drying time $\mathbf{Z}$, which maximize the total infiltration volume (unit area) over N cycles. The maximum cycle time $\mathbf{CT}_{max}$ and the maximum water content $\boldsymbol{\omega}_{max}$ are known.

2.9 GENERAL PROBLEM FORMULATION

Each of the above optimization problems in Secs. 2.2–2.8 can be written in the following general form:

Objective

$$\text{Optimize } z = f(\mathbf{x}, \mathbf{u}) \tag{2.9.1}$$

Constraints

$$\mathbf{x}_{t+1} = \mathbf{g}(\mathbf{x}_t, \mathbf{u}_t, t) \quad t = 0, \ldots, T-1 \tag{2.9.2}$$

$$\underline{\mathbf{x}}_t \leq \mathbf{x}_t \leq \bar{\mathbf{x}}_t, \quad t = 0, \ldots, T \tag{2.9.3}$$

$$\underline{\mathbf{u}}_t \leq \mathbf{u}_t \leq \bar{\mathbf{u}}_t, \quad t = 0, \ldots, T-1 \tag{2.9.4}$$

where $\mathbf{x}_t$ is a column vector of dependent (state) variables at time t, $\mathbf{u}_t$ is the column vector of independent (control) variables at time t, $\underline{\mathbf{x}}_t$ and $\underline{\mathbf{u}}_t$ are column vectors of lower bounds, and $\bar{\mathbf{x}}_t$ and $\bar{\mathbf{u}}_t$ are column vectors of upper bounds. The objective function is assumed to be continuously differentiable in $(\mathbf{x}_t, \mathbf{u}_t)$. Time t can take only a finite number of discrete values.

The above optimization problem defined by Eqs. (2.9.1)–(2.9.4) is a discrete-time optimal control problem. Note that in each of the different hydrosystems problems in Secs. 2.2–2.8, there is a set of hydraulic processes (simulation) equations $\mathbf{G}(\ ) = \mathbf{0}$. These process simulation equations define the physics of the problem; that is, the governing physical equations that simulate the physical processes. These are the conservation of mass, conservation of energy, and/or conservation of momentum.

2.9.1 Groundwater Management

In the case of the groundwater management model, Eq. (2.2.2), $\mathbf{G}(\mathbf{h}, \mathbf{q}) = \mathbf{0}$ is the set of general groundwater flow equations. For non-steady-state groundwater flow, the governing physical equations for two-dimensional flow are

$$\frac{\partial}{\partial x_i}\left(T_{i,j}\frac{\partial h}{\partial x_j}\right) = S\frac{\partial h}{\partial t} + W, \quad i, j = 1, 2 \tag{2.9.5}$$

where $T_{i,j}$ is an element of the transmissivity vector, h is the hydraulic head, W is the volume flux per unit area, S is the storage coefficient, x_i and x_j are Cartesian coordinates, and t is time. The above partial differential equations can be written in a finite difference form:

$$\mathbf{G}(\mathbf{h}_{t+1}, \mathbf{h}_t, \mathbf{q}_{t+1}) = \mathbf{0} \quad t = 0, \ldots, T-1 \tag{2.9.6}$$

letting the volume flux to be replaced by the pumpage or recharge $\mathbf{q}$. Alternatively, Eq. (2.9.6) can be written as

$$\mathbf{h}_{t+1} = \mathbf{g}(\mathbf{h}_t, \mathbf{q}_{t+1}, t) \quad t = 0, \ldots, T-1 \tag{2.9.7}$$

which is in the form of Eq. (2.9.2). In this case, the state variable is the hydraulic head, $\mathbf{h}_t$, and the control variable is the pumpage or recharge $\mathbf{q}_t$. The finite difference cell map for an aquifer is shown in Figure 2.8.

2.9.2 Real-Time Operation of River-Reservoir System for Flood Control

In the case for the real-time operation of river-reservoir systems for flood control, the set of governing physical equations, $\mathbf{G}(\mathbf{h}, \mathbf{Q}, \mathbf{r}) = \mathbf{0}$ are the Saint-Venant equations for one-dimensional unsteady flow:

Continuity:

$$\frac{\partial Q}{\partial t} + \frac{\partial (A + A_0)}{\partial t} - q = 0 \tag{2.9.8}$$

Momentum:

$$\frac{\partial Q}{\partial t} + \frac{\partial (\beta Q^2/A)}{\partial x} + gA\left(\frac{\partial h}{\partial x} + S_f + S_e\right) - \beta q v_x + W_f B = 0 \tag{2.9.9}$$

where

x = longitudinal distance along the channel or river
t = time
A = cross-sectional area of flow
A_o = cross-sectional area of off-channel dead storage (contributes to continuity but not to momentum)
q = lateral inflow per unit length along the channel
h = water surface elevation
v_x = velocity of lateral flow in the direction of channel flow
S_f = friction slope
S_e = eddy loss slope
B = width of the channel at the water surface
W_f = wind shear force

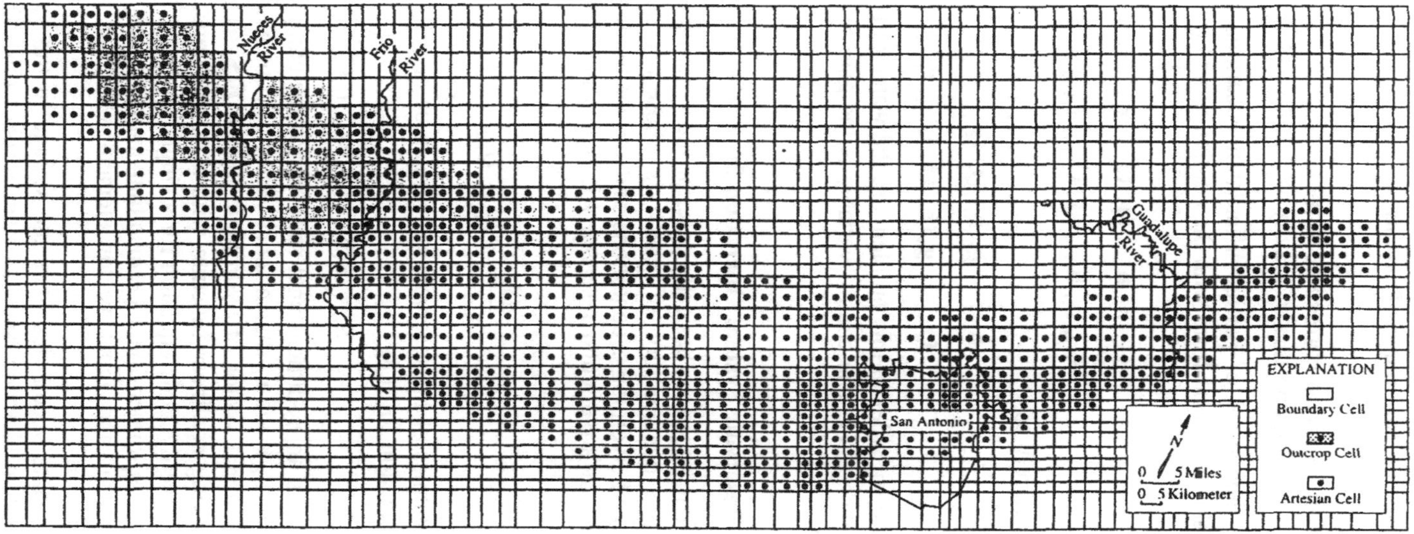

Figure 2.8 Cell map used for the digital computer model of the Edwards (Balcones fault zone) aquifer.

β = momentum correction factor
g = acceleration due to gravity

The above set of partial differential equations can be expressed respectively in general form with the continuity and momentum respectively as

$$\mathbf{G}_C(\mathbf{h}_t, \mathbf{h}_{t+1}, \mathbf{Q}_t, \mathbf{Q}_{t+1}, \mathbf{r}_{t+1}) = \mathbf{0} \tag{2.9.10}$$

$$\mathbf{G}_M(\mathbf{h}_t, \mathbf{h}_{t+1}, \mathbf{Q}_t, \mathbf{Q}_{t+1}, \mathbf{r}_{t+1}) = \mathbf{0} \tag{2.9.11}$$

or respectively as

$$\mathbf{h}_{t+1} = \mathbf{g}_C(\mathbf{h}_t, \mathbf{Q}_t, \mathbf{Q}_{t+1}, \mathbf{r}_{t+1}) \tag{2.9.12}$$

$$\mathbf{Q}_{t+1} = \mathbf{g}_M(\mathbf{h}_t, \mathbf{Q}_t, \mathbf{Q}_{t+1}, \mathbf{r}_{t+1}) \tag{2.9.13}$$

The state variables in this problem are the water surface elevations, **h**, and the discharge, **Q**. The control variable is the gate setting (spillway operation), **r**.

2.9.3 Reservoir System Operation for Water Supply

The system equations basically describe the dynamics of a reservoir system which is a configuration of reservoirs whose coordinated operation is a function of hydrologic conditions and/or institutional requirements. The dynamics of a particular reservoir j is represented by the conservation of mass

$$\frac{dS_{j,t}}{dt} = I_{j,t} - U_{j,t} - L_{j,t} \tag{2.8.14}$$

The dynamics of the reservoir system can be described in a vector differential equation form as

$$\mathbf{S}_t = \mathbf{F}(\mathbf{S}_t, t) + \mathbf{B}\mathbf{U}_t + \mathbf{C}\mathbf{I}_t \tag{2.8.15}$$

where $\mathbf{S}_t$ is an n_s-dimensional state vector including all reservoir storage variables, $\mathbf{U}_t$ is the n_u-dimensional vector of controllable releases, $\mathbf{I}_t$ is the n_I-dimensional vector of hydrologic inputs, $\mathbf{F}(\mathbf{S}_t, t)$ is an n_s-dimensional time-varying nonlinear function with a storage-dependent term as shown in the above dynamic equation for a single reservoir, **B** and **C** are $n_s \times n_u$- and $n_s \times n_u$-dimensional permutation matrices, respectively, associating each control and input vector element with the pertinent differential equation.

The state vector describes the storage in the various reservoirs and other system elements such as river reaches throughout the system as a function of time. At a particular time t_k when the state vector is known and for a known or specified set of inputs $\mathbf{I}_t$ and release $\mathbf{U}_t$ over the time interval $t \leq [t_k, T]$, then the state trajectory $\{\mathbf{S}_t, t \leq [t, T]\}$ can be computed by integrating the above equation. A state vector summarizes the knowledge or information from the system history prior to time t_k. This information is necessary to compute (predict) the reservoir system's future resource to input and output sequences. The purpose of a reservoir operation control model is to identify control schedules (reservoir releases) which generate optimum (desirable) state trajectories (storage).

2.9.4 Freshwater Inflows to Bays and Estuaries

In the case of the optimization of freshwater inflow to estuaries, the governing physical equations (2.6.2), $\mathbf{G}(\mathbf{Q}, \mathbf{s}) = \mathbf{0}$ for a two-dimensional (plan) formulation (see Fig. 2.9) are the vertical-mean equations of momentum, continuity, and salinity mass budget: the momentum equation in the x-direction,

$$\frac{\partial q_x}{\partial t} - \Omega q_y = -gd\frac{\partial h}{\partial y} - fqq_x + X_w \tag{2.9.16}$$

the momentum equation in the y-direction,

$$\frac{\partial q_y}{\partial t} - \Omega q_x = -gd\frac{\partial h}{\partial y} - fqq_y + Y_w \tag{2.9.17}$$

the continuity equation,

$$\frac{\partial q_x}{\partial x} + \frac{\partial q_y}{\partial y} + \frac{\partial h}{\partial t} = r - e \tag{2.9.18}$$

and the conservation (transport) equation,

$$\frac{\partial s}{\partial t} + \frac{\partial (Us)}{\partial x} + \frac{\partial (Vs)}{\partial y} = \frac{\partial}{\partial x} E_x \frac{\partial s}{\partial x} - \frac{\partial}{\partial y} E_y \frac{\partial s}{\partial y} \tag{2.9.19}$$

where

t = time
x and y = horizontal Cartesian coordinates
q_x, q_y = depth-averaged flow components in the x- and y-directions, respectively, per unit width
Ω = Coriolis parameter equal to $2\omega \sin \varphi$
ω = angular rotation of the earth
φ = latitude

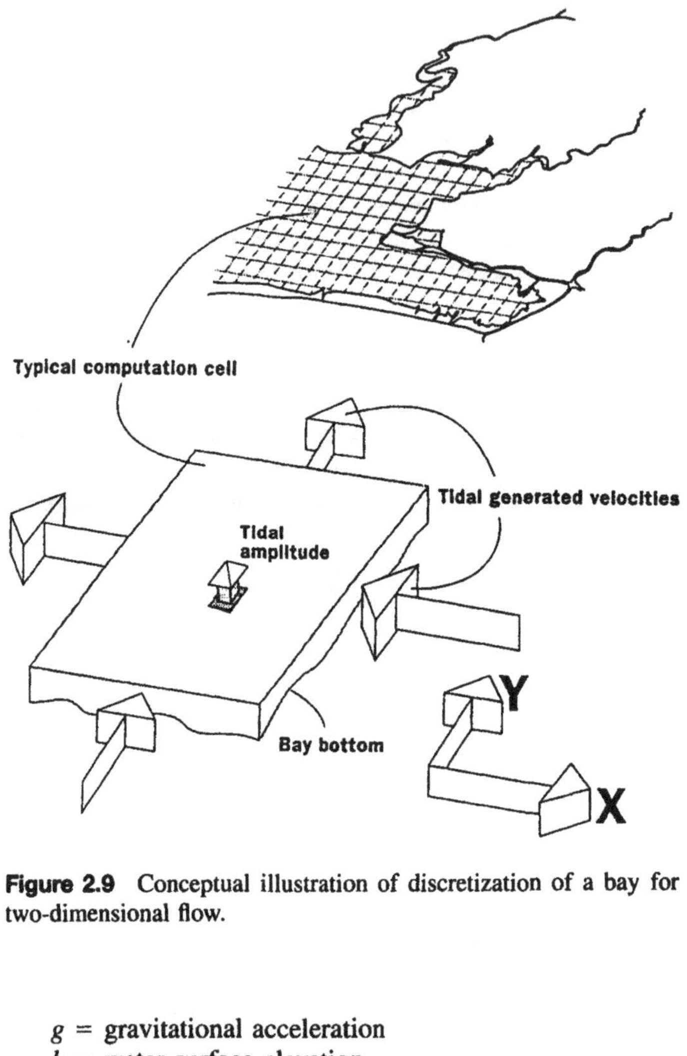

Figure 2.9 Conceptual illustration of discretization of a bay for depth-averaged two-dimensional flow.

g = gravitational acceleration
h = water surface elevation
d = water depth equal to $h - z$
z = bottom elevation
f = bottom friction term from the Manning equation
q = flow per unit width equal to $\sqrt{q_x^2 + q_y^2}$
X_w = wind stress per unit density of water in the x-direction equal to $KV_w^2 \cos \theta$
Y_w = wind stress per unit density of water in the y-direction equal to $KV_w^2 \sin \theta$

K = a wind stress coefficient
V_w = wind velocity at 10 m above the water surface
q = wind direction with respect to the x-axis
r = rainfall intensity
e = evaporation rate
U, V = net velocities over a tidal cycle
s = vertical-averaged salinity
E_x, E_y = horizontal dispersion coefficients in the x- and y-directions, respectively

The finite difference grid for the Lavaca–Tres Palacios Estuary along the Gulf of Mexico in Texas is shown in Figure 2.10.

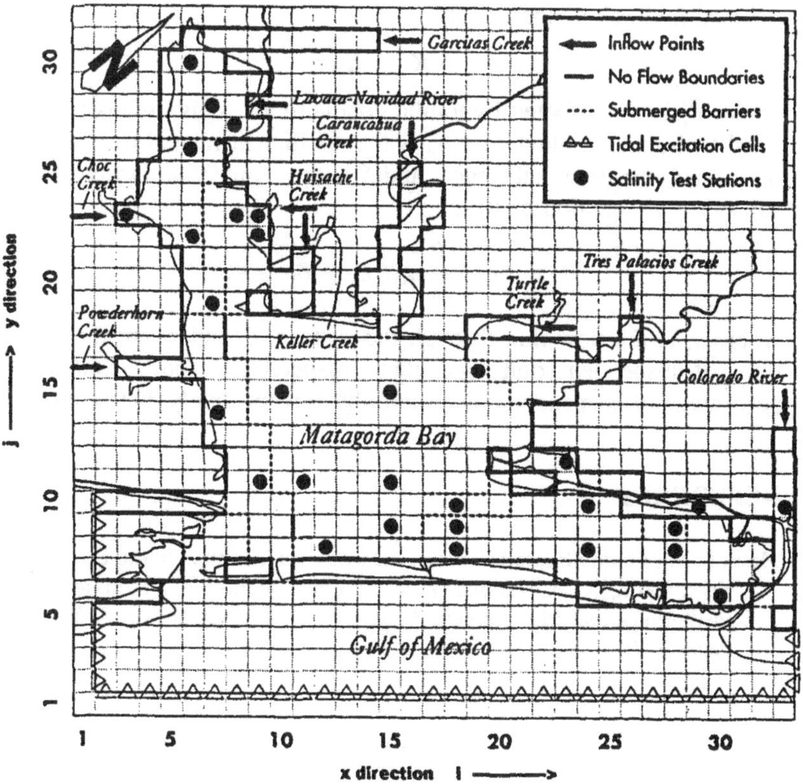

Figure 2.10 Finite difference grid of Lavaca–Tres Palacios Estuary with salinity test station sites.

2.9.5 Soil Aquifer Treatment Systems

Unsaturated flow is commonly modeled using flow equations based on the conservation of mass and application of Darcy's law using parameters dependent on the water content of the soil. Generally, the one-dimensional Richard's equation is used to simulate unsteady flow in soil aquifer treatment systems, which has the form

$$c\frac{\partial h}{\partial t} = \frac{\partial}{\partial z}\left(K\frac{\partial h}{\partial z} - K\right) \tag{2.9.20}$$

where h is the pressure head, $c = d\omega/dh$ is the soil water capacity in which ω is the volumetric water content, K is the hydraulic conductivity, z is the soil depth assumed to increase in the downward direction, and t is time.

The general expression for the clogging effect is modeled through the affected hydraulic conductivity

$$K_c = K_s e^{\alpha_c S} \tag{2.9.21}$$

where K_c is the affected hydraulic conductivity, K_s is the unaffected saturated hydraulic conductivity, α_c is the coefficient related to characteristics of soil and clogging substance, and S is the cumulative influx of clogging substance as function of time

$$S = \int_0^t q_s(\delta_s c_s + \delta_a c_a)\, d\tau \tag{2.9.22}$$

where q_s is the flow flux through the surface, c_s and c_a are the concentrations of suspended solids and algae, respectively, and d_s and d_a are the percentage of suspended solids and algae, respectively, being intercepted.

REFERENCE

Mays, L. W., Flood Simulation for a Large Reservoir System in the Lower Colorado River Basin, Texas, National Water Summary 1988–89—Floods and Droughts: Institutional and Management Aspects, U.S. Geological Survey Water-Supply Paper 2375, 1991.

Chapter 3
Nonlinear Optimization Methods

Earlier hydrosystems applications of operations research techniques relied mainly on the use of linear and dynamic programming techniques. The use of these techniques applied to solving hydrosystem problems has been rather widespread in the literature. Linear programming codes are widely available, whereas dynamic programming requires a specific code for each application. The use of nonlinear programming in solving hydrosystems problems has not been as widespread, even though most of the problems requiring solutions are nonlinear problems. The recent development of new nonlinear programming techniques and the availability of nonlinear programming codes have attracted new applications of nonlinear programming in hydrosystems. Unconstrained and constrained nonlinear optimization procedures are described, followed by a description of the augmented Lagrangian penalty function method and its application to discrete-time optimal control problems. (Sections 3.1–3.5 of this chapter were adapted from *Hydrosystems Engineering and Management*, by L. M. Mays and Y. K. Tung, by permission of The McGraw-Hill Companies © 1992.)

3.1 MATRIX ALGEBRA FOR NONLINEAR PROGRAMMING

To explain the concepts of nonlinear programming, various techniques of matrix algebra and numerical linear algebra are used. A brief introduction to some of the concepts is provided in this section.

A function of many variables $f(\mathbf{x})$ at point $\mathbf{x}$ is also an important concept. For a function that is continuous and continuously differentiable, there is a vector of first partial derivatives called the gradient or gradient vector

$$\nabla f(\mathbf{x}) = \left[\frac{\partial f}{\partial \mathbf{x}}\right] = \left(\frac{\partial f}{\partial x_1}, \frac{\partial f}{\partial x_2}, \ldots, \frac{\partial f}{\partial x_n}\right)^T \tag{3.1.1}$$

where ∇ is the vector of the gradient operator $(\partial/\partial x_1, \ldots, \partial/\partial x_n)^T$. Geometrically, the gradient vector at a given point represents the direction along which the maximum rate of increase in function value would occur. For $f(\mathbf{x})$ twice continuously differentiable, there exists a matrix of second partial derivatives called the Hessian matrix or Hessian

$$\mathbf{H}(\mathbf{x}) = \nabla^2 f(\mathbf{x}) = \begin{bmatrix} \dfrac{\partial^2 f}{\partial x_1^2} & \dfrac{\partial^2 f}{\partial x_1\, \partial x_2} & \cdots & \dfrac{\partial^2 f}{\partial x_1\, \partial x_n} \\ \dfrac{\partial^2 f}{\partial x_2\, \partial x_1} & \dfrac{\partial^2 f}{\partial x_2^2} & \cdots & \dfrac{\partial^2 f}{\partial x_2\, \partial x_n} \\ \vdots & \cdots & \cdots & \vdots \\ \dfrac{\partial^2 f}{\partial x_n\, \partial x_1} & \cdots & \cdots & \dfrac{\partial^2 f}{\partial x_n^2} \end{bmatrix} \tag{3.1.2}$$

The Hessian is a square and symmetric matrix.

The concepts of convexity and concavity are used to establish whether a local optimum, local minimum, or local maximum is also the global optimum, which is the best among all solutions. In the univariate case, a function $f(\mathbf{x})$ is said to be convex over a region if for every x_a and x_b, $x_a \neq x_b$, the following holds:

$$f[\theta x_a + (1 - \theta)x_b] \leq \theta f(x_a) + (1 - \theta)\, f(x_b), \quad 0 \leq \theta \leq 1 \tag{3.1.3}$$

The function is strictly convex when the above relation holds with a less than (<) sign.

Conversely, a function is concave over a region if for every x_a and x_b, $x_a \neq x_b$, the following holds:

$$f[\theta x_a + (1 - \theta)\, x_b] \geq \theta f(x_a) + (1 - \theta)\, f(x_b), \quad 0 \leq \theta \leq 1 \tag{3.1.4}$$

The function is strictly concave when the above relation holds with a greater than (>) sign.

Equations (3.1.3) and (3.1.4) are not convenient to use in testing for convexity or concavity of a univariate function. Instead, it is easier to examine the sign of its second derivative, $d^2f(\mathbf{x})/dx^2$. From fundamental calculus, if $d^2f/dx^2 < 0$, then the function is concave, and if $d^2f/dx^2 > 0$, then the function is convex.

The convexity and concavity of multivariable functions $f(\mathbf{x})$ can also be determined using the Hessian matrix. First, the definitions of positive definite, negative definite and indefinite are used to identify the type of Hessian, i.e.,

Positive definite **H**:	$\mathbf{x}^T \mathbf{Hx} > 0$ for all $\mathbf{x} \neq \mathbf{0}$
Negative definite **H**:	$\mathbf{x}^T \mathbf{Hx} < \mathbf{0}$ for all $\mathbf{x} \neq \mathbf{0}$
Indefinite **H**:	$\mathbf{x}^T \mathbf{Hx} < 0$ for some $\mathbf{x}$
	> 0 for other $\mathbf{x}$
Positive semidefinite **H**:	$\mathbf{x}^T \mathbf{Hx} \geq 0$ for all $\mathbf{x}$
Negative semidefinite **H**:	$\mathbf{x}^T \mathbf{Hx} \leq 0$ for all $\mathbf{x}$

The basic rules for convexity and concavity of a multivariate function $f(\mathbf{x})$ with continuous second partial derivatives are as follows:

1. $f(\mathbf{x})$ is concave, $\mathbf{H}(\mathbf{x})$ is negative semidefinite;
2. $f(\mathbf{x})$ is strictly concave, $\mathbf{H}(\mathbf{x})$ is negative definite;
3. $f(\mathbf{x})$ is convex, $\mathbf{H}(\mathbf{x})$ is positive semidefinite;
4. $f(\mathbf{x})$ is strictly convex, $\mathbf{H}(\mathbf{x})$ is positive definite.

To test the status of $\mathbf{H}(\mathbf{x})$ for strict convexity, two tests are available (Edgar and Himmelblau, 1988). The first is that all diagonal elements of $\mathbf{H}(\mathbf{x})$ must be positive and the determinants of all leading principal minors, $\det\{M_i(\mathbf{H})\}$, and also of $\mathbf{H}(\mathbf{x})$, $\det(\mathbf{H})$ are positive (> 0) . Another test is that all eigenvalues of $\mathbf{H}(\mathbf{x})$ are positive (> 0). For strict concavity, all diagonal elements must be negative and $\det(\mathbf{H})$ and $\det\{M_i(\mathbf{H})\} > 0$ if i is even ($i = 2, 4, 6, \ldots$); $\det(\mathbf{H})$ and $\det\{M_i(\mathbf{H})\} < 0$ if i is odd ($i = 1, 3, 5, \ldots$). The strict inequalities $>$ or $<$ in these tests are replaced by $\geq$ or $\leq$, respectively, to test for convexity and concavity.

Convex regions or sets are used to classify constraints. A convex region exists if for any two points in the region, $x_a \neq x_b$, all points $\mathbf{x} = \theta x_a + (1 - \theta)x_b$, where $0 \leq \theta \leq 1$, on the line connecting x_a and x_b are in the set. Figure 3.1 illustrates convex and nonconvex regions.

The convexity of a feasible region and the objective function in nonlinear optimization has an extremely important implication with regard to the type of optimal solution to be obtained. For linear programming problems, the objective function and feasible region both are convex; therefore, the optimal solution is global. On the other hand, the convexity of both the objective function and feasible region in a nonlinear programming problem cannot be ensured; the optimal solution achieved, therefore, cannot be guaranteed to be global.

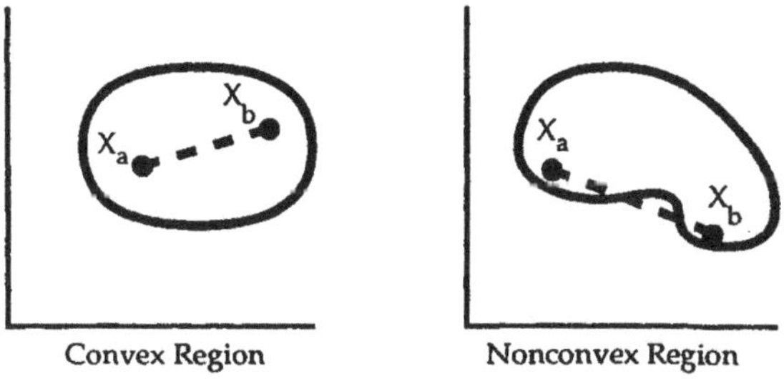

Figure 3.1 Illustration of convex and nonconvex regions (Mays and Tung, 1992).

3.2 UNCONSTRAINED NONLINEAR PROGRAMMING

This section describes the basic concepts of unconstrained nonlinear optimization, including the necessary and sufficient conditions of a local optimum. Further, unconstrained optimization techniques for univariate and multivariate problems are described. Understanding unconstrained optimization procedures is important because these techniques are the fundamental building blocks in many of the constrained nonlinear optimization algorithms.

3.2.1 Basic Concepts

The problem of unconstrained minimization can be stated as

$$\underset{x \in E^n}{\text{Minimize}}\ f(\mathbf{x}) \tag{3.2.1}$$

in which $\mathbf{x}$ is a vector of n decision variables, $\mathbf{x} = (x_1, x_2, \ldots, x_n)^T$, defined over the entire Euclidean space E^n. Because the feasible region is infinitely extended without bound, the optimization problem does not contain any constraints.

Assume that $f(\mathbf{x})$ is a nonlinear function and twice differentiable; it could be convex, concave, or a mixture of the two over E^n. In the one-dimensional case, the objective function $f(\mathbf{x})$ could behave as Figure 3.2a, consisting of peaks, valleys, and inflection points. The necessary conditions for a solution to Eq. (3.2.1) at $\mathbf{x}^*$ are (1) $\nabla f(\mathbf{x}^*) = \mathbf{0}$ and (2) $\nabla^2 f\ (\mathbf{x}^*) = \mathbf{H}(\mathbf{x}^*)$ is semipositive definite. The sufficient conditions for an unconstrained minimum are (1) $\nabla f(\mathbf{x}^*) = \mathbf{0}$ and (2) $\nabla^2 f\ (\mathbf{x}^*) = \mathbf{H}(\mathbf{x}^*)$ is strictly positive definite.

In theory, the solution to Eq. (3.2.1) can be obtained by solving the following system of n nonlinear equations with n unknowns:

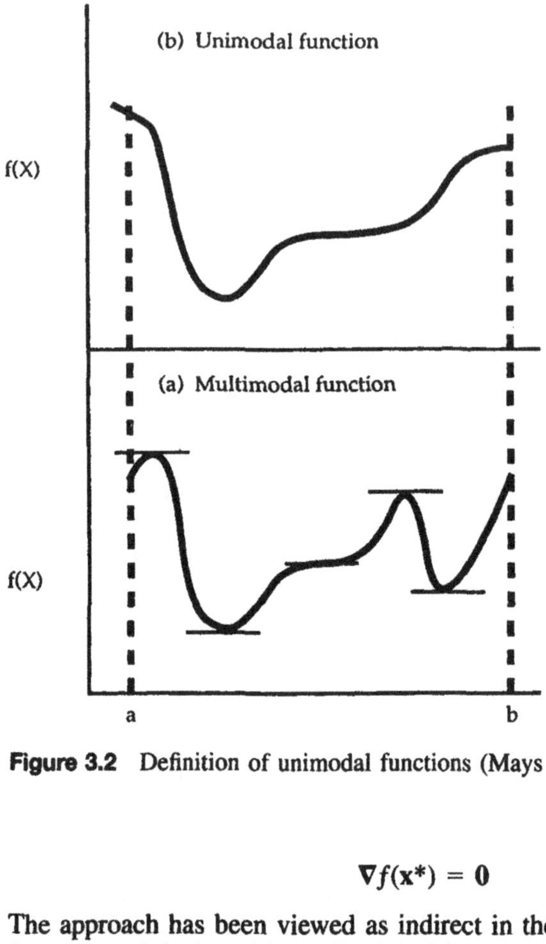

Figure 3.2 Definition of unimodal functions (Mays and Tung, 1992).

$$\nabla f(\mathbf{x}^*) = \mathbf{0} \tag{3.2.2}$$

The approach has been viewed as indirect in the sense that it backs away from the original problem of minimizing $f(\mathbf{x})$. Furthermore, an iterative numerical procedure is required to solve the system of nonlinear equations, which tends to be computationally inefficient.

By contrast, the preference is given to those solution procedures which directly attack the problem of minimizing $f(\mathbf{x})$. Direct solution methods, during the course of iteration, generate a sequence of solution points in E^n that terminate or converge to a solution to Eq. (3.2.1). Such methods can be characterized as search procedures.

In general, all search algorithms for unconstrained minimization consist of two basic steps. The first step is to determine the search direction along which the objective function value decreases. The second step is called a line search (or one-dimensional search) to obtained the optimum solution point along the search direction determined by the first step. Mathematically, minimization for the line search can be stated as

$$\underset{\beta}{\text{Min}}\ f(\mathbf{x}^0 + \beta\mathbf{d}) \tag{3.2.3}$$

in which $\mathbf{x}^0$ is the current solution point, $\mathbf{d}$ is the vector indicating the search direction, and β is a scalar, $-\infty < \beta < \infty$, representing the step size whose optimal value is to be determined. There are many search algorithms whose differences primarily lie in the way the search direction $\mathbf{d}$ is determined.

Due to the very nature of search algorithms, it is likely that different starting solutions might converge to different local minima. Hence, there is no guarantee of finding the global minimum by any search technique applied to solve Eq. (3.2.1) unless the objective function is a convex function over E^n.

In implementing search techniques, specification of convergence criteria or stopping rules is an important element that affects the performance of the algorithm and the accuracy of the solution. Several commonly used stopping rules in an optimum seeking algorithm are

$$\|\mathbf{x}^k - \mathbf{x}^{k+1}\| < \varepsilon_1 \tag{3.2.4a}$$

$$\frac{\|\mathbf{x}^k - \mathbf{x}^{k+1}\|}{\|\mathbf{x}^k\|} < \varepsilon_2 \tag{3.2.4b}$$

$$|f(\mathbf{x}^k) - f(\mathbf{x}^{k+1})| < \varepsilon_3 \tag{3.2.4c}$$

$$\left|\frac{f(\mathbf{x}^k) - f(\mathbf{x}^{k+1})}{f(\mathbf{x}^k)}\right| < \varepsilon_4 \tag{3.2.4d}$$

in which the superscript k is the index for iteration, ε represents the tolerance or accuracy requirement, $\|\mathbf{x}\|$ is the length of the vector $\mathbf{x}$, and $|x|$ is the absolute value. The specification of the tolerance depends on the nature of the problem and on the accuracy requirement. Too small a value of ε (corresponding to the high accuracy requirement) could result in excessive iterations, wasting computer time. On the other hand, too large a value of ε could make the algorithm terminate prematurely at a nonoptimal solution.

3.2.2 Unconstrained Optimization: One-Dimensional Search

The line search techniques for solving one-dimensional optimization problems form the backbone of nonlinear programming algorithms. Multidimensional problems are ultimately solved by executing a sequence of successive line searches. One-dimensional search techniques can be classified as curve fitting (approximation) techniques or as interval elimination techniques. Interval elimination techniques for a one-dimensional search essentially eliminate or delete a calculated portion of the range of the variable from consideration in each successive iteration of the search for the optimum of $f(\mathbf{x})$. After a number of iterations when the remaining interval is sufficiently small, the search procedure terminates. These methods determine the minimum value of a function over a closed interval $[a, b]$ assuming that a function is unimodal; that is, it has only one minimum value in the interval (Fig. 3.2). Two interval elimination techniques commonly used are the golden section method and the Fibonacci search method (Mays and Tung, 1992).

3.2.3 Unconstrained Optimization: Multivariable Methods

Unconstrained optimization problems can be stated in a general form as

$$\text{Minimize } z = f(\mathbf{x}) = f(x_1, x_2, \ldots, x_n) \tag{3.2.5}$$

For maximization, the problem is to minimize $-f(\mathbf{x})$. The solution of these types of problems can be stated in an algorithm involving the following basic steps or phases:

Step 0: Select an initial starting point $\mathbf{x}^{k=0} = (x_1^0, x_2^0, \ldots, x_n^0)$.
Step 1: Determine a search direction, $\mathbf{d}^k$.
Step 2: Find a new point $\mathbf{x}^{k+1} = \mathbf{x}^k + \beta^k \mathbf{d}^k$, where β^k is the step size, a scalar, which minimizes $f(\mathbf{x}^k + \beta^k \mathbf{d}^k)$.
Step 3: Check the convergence criteria such as Eqs. (3.2.4a)–(3.2.4d) for termination; if not satisfied, set $k = k + 1$ and return to Step 1.

The various unconstrained multivariate methods differ in the way the search directions are determined. The recursive line search for an unconstrained minimization problem is expressed in Step 2 as

$$\mathbf{x}^{k+1} = \mathbf{x}^k + \beta^k \mathbf{d}^k \tag{3.2.6}$$

Table 3.1 lists the equations for determining the search direction for four basic groups of methods: descent methods, conjugate direction methods, quasi-Newton methods, and Newton's method. The simplest are the steepest descent methods, whereas the Newton methods are the most computationally intensive.

In the steepest descent method, the search direction is $-\nabla f(\mathbf{x})$. $\nabla f(\mathbf{x})$ points in the direction of the maximum rate of increase in objective function value; therefore, a negative sign is associated with the gradient vector

Table 3.1 Computation of Search Directions

Search direction	Definition of terms
Steepest Descent $d^{k+1} = -\nabla f(x^{k+1})$	
Conjugate Gradient Methods (1) Fletcher–Reeves	
$\mathbf{d}^{k+1} = -\nabla f(\mathbf{x}^{k+1}) + a_1\,\mathbf{d}^k$	$a_1 = \dfrac{\nabla^T f(\mathbf{x}^{k+1})\nabla f(\mathbf{x}^{k+1})}{\nabla^T f(\mathbf{x}^k)\nabla f(\mathbf{x}^k)}$ $\mathbf{d}^0 = -\nabla f(\mathbf{x}^0)$
(2) Polak–Ribiere	
$\mathbf{d}^{k+1} = -\nabla f(\mathbf{x}^{k+1}) + a_2\,\mathbf{d}^k$	$a_2 = \dfrac{\nabla^T f(\mathbf{x}^{k+1})\mathbf{Y}^{k+1}}{\nabla^T f(\mathbf{x}^k)\nabla f(\mathbf{x}^k)}$ $\mathbf{Y}^{k+1} = \nabla f(\mathbf{x}^{k+1}) - \nabla f(\mathbf{x}^k)$ $\mathbf{d}^0 = -\nabla f(\mathbf{x}^0)$
(3) One-Step BFGS	
$\mathbf{d}^{k+1} = -\nabla f(\mathbf{x}^{k+1})$ $+ a_3(a_4\mathbf{S}^{k+1} + a\mathbf{Y}^k)$	$a_3 = \dfrac{1}{(\mathbf{S}^{k+1})^T\mathbf{Y}^{k+1}}$ $a_4 = -\left(1 + \dfrac{(\mathbf{Y}^{k+1})^T\mathbf{Y}^{k+1}}{(\mathbf{S}^{k+1})^T\mathbf{Y}^{k+1}}\right)(\mathbf{S}^{k+1})^T\nabla f(\mathbf{x}^{k+1})$ $+ (\mathbf{Y}^{k+1})^T\nabla f(\mathbf{x}^{k+1})$ $a_5 = (\mathbf{S}^{k+1})^T\nabla f(\mathbf{x}^k)$ $\mathbf{S}^{k+1} = \mathbf{x}^{k+1} - \mathbf{x}^k$ $\mathbf{d}^0 = -\nabla f(\mathbf{x}^0)$
Quasi-Newton Methods[a] (1) Davidon–Fletcher–Powell (DFP) Method (Variable Metric Method)	
$\mathbf{d}^{k+1} = \mathbf{G}^{k+1}\,\nabla f(\mathbf{x}^{k+1})$	$\mathbf{G}^{k+1} = \mathbf{G}^k + \dfrac{\mathbf{S}^k(\mathbf{S}^k)^T}{(\mathbf{S}^k)^T\mathbf{Y}^k} - \dfrac{\mathbf{G}^k\mathbf{Y}^k(\mathbf{G}^k\mathbf{Y}^k)^T}{(\mathbf{Y}^k)^T\mathbf{G}^k\mathbf{Y}^k}$
(2) Broyden–Fletcher–Goldfarb–Shanno (BFGS) Method	
$\mathbf{d}^{k+1} = \mathbf{G}^{k+1}\,\nabla f(\mathbf{x}^{k+1})$	$\mathbf{G}^{k+1} = \mathbf{G}^k + \left(\dfrac{1 + (\mathbf{Y}^k)^T\mathbf{G}^k\mathbf{Y}^k}{(\mathbf{Y}^k)^T\mathbf{S}^k}\right)\dfrac{\mathbf{S}^k(\mathbf{S}^k)^T}{(\mathbf{S}^k)^T\mathbf{Y}^k}$ $- \dfrac{\mathbf{Y}^k(\mathbf{S}^k)^T\mathbf{G}^k + \mathbf{G}^k\mathbf{S}^k(\mathbf{Y}^k)^T}{(\mathbf{Y}^k)^T\mathbf{S}^k}$
(3) Broyden Family $\mathbf{G}^\phi = (1 - \phi)\mathbf{G}^{DFP} + \phi\mathbf{G}^{BFGS}$	
Newton Method $\mathbf{d}^{k+1} = -\mathbf{H}^{-1}(\mathbf{x}^k)\nabla f(\mathbf{x}^k)$	

Note: Formulas for other search directions can be found in Luenberger (1984).

[a] $\mathbf{G}^0 = \mathbf{0}$.

Source: Mays and Tung (1992).

in Eq. (3.2.6) because the problem is a minimization type. The recursive line search equation for the steepest descent method is, then, reduced to

$$\mathbf{x}^{k+1} = \mathbf{x}^k - \beta^k \nabla f(\mathbf{x}^k) \tag{3.2.7}$$

Using Newton's method, the recursive equation for a line search is

$$\mathbf{x}^{k+1} = \mathbf{x}^k - \mathbf{H}^{-1}(\mathbf{x}^k) \nabla f(\mathbf{x}^k) \tag{3.2.8}$$

Although Newton's method converges faster than most other algorithms, the major disadvantage is that it requires inverting the Hessian matrix in each iteration, which is a computationally cumbersome task.

The conjugate direction methods and quasi-Newton's methods are intermediate between the steepest descent and Newton's method. The conjugate direction methods are motivated by the need to accelerate the typically slow convergence of the steepest descent methods. Conjugate direction methods, as can be seen in Table 3.1, define the search direction by utilizing the gradient vector of the objective function of the current iteration and the information on the gradient and search direction of the previous iteration. The motivation of quasi-Newton methods is to avoid inverting the Hessian matrix as required by Newton's method. These methods use approximations to the inverse Hessian with a different form of approximation for the different quasi-Newton methods. Detailed descriptions and theoretical development can be found in textbooks such as Luenberger (1984), Fletcher (1980), Dennis and Schnable (1983), Gill et al. (1981), and Edgar and Himmelblau (1988).

3.3 CONSTRAINED OPTIMIZATION: OPTIMALITY CONDITIONS

3.3.1 Lagrange Multiplier

Consider the general nonlinear programming problem with the nonlinear objective

$$\text{Minimize } f(\mathbf{x}) \tag{3.3.1a}$$

subject to

$$g_i(\mathbf{x}) = 0, \quad i = 1, \ldots, m \tag{3.3.1b}$$

and

$$\underline{x}_j \le x_j \le \bar{x}_j, \quad j = 1, 2, \ldots, n \tag{3.3.1c}$$

in which Eq. (3.3.1c) is a bound constraint for the jth decision variable x_j with $\underline{x}_j$ and $\bar{x}_j$ being the lower and upper bounds, respectively.

In a constrained optimization problem, the feasible space is not infinitely extended, unlike an unconstrained problem. As a result, the solution that satisfies the optimality condition of the unconstrained optimization problem does not guarantee to be feasible in constrained problems. In other words, a local optimum for a constrained problem might be located on the boundary or a corner of the feasible space at which the gradient vector is not equal to zero. Therefore, modifications to the optimality conditions for unconstrained problems must be made.

The most important theoretical results for nonlinear constrained optimization are the Kuhn–Tucker conditions. These conditions must be satisfied at any constrained optimum, local or global, of any linear and nonlinear programming problems. They form the basis for the development of many computational algorithms.

Without losing generality, consider a nonlinear constrained problem stated by Eqs. (3.3.1a–c) with no bounding constraints. Note that constraint equations (3.3.1b) are all equality constraints. Under this condition, the Lagrange multiplier method converts a constrained nonlinear programming problem into an unconstrained one by developing an augmented objective function, called the Lagrangian. For a minimization, the Lagrangian function $L(\mathbf{x}, \boldsymbol{\lambda})$ is defined as

$$L(\mathbf{x}, \boldsymbol{\lambda}) = f(\mathbf{x}) + \boldsymbol{\lambda}^T \mathbf{g}(\mathbf{x}) \tag{3.3.2}$$

in which $\boldsymbol{\lambda}$ is the vector of Lagrange multipliers and $\mathbf{g}(\mathbf{x})$ is a vector of constraint equations. Algebraically, Eq. (3.3.2) can be written

$$L(x_1, \ldots, x_n, \lambda_1, \ldots, \lambda_m) = f(x_1, \ldots, x_n) + \sum_{i=1}^{m} \lambda_i g_i(x_1, \ldots, x_n) \tag{3.3.3}$$

$L(\mathbf{x}, \boldsymbol{\lambda})$ is the objective function, with $m + n$ variables, that is to be minimized. The necessary and sufficient conditions for $\mathbf{x}^*$ to be the solution for minimization are:

$f(\mathbf{x}^*)$ is convex and $\mathbf{g}(\mathbf{x}^*)$ is convex in the vicinity of $\mathbf{x}^*$

$$\frac{\partial L(\mathbf{x}^*)}{\partial x_j} = \frac{\partial f}{\partial x_j} + \sum_{i=1}^{m} \lambda_i \frac{\partial g_i}{\partial x_j} = 0, \quad j = 1, \ldots, n \tag{3.3.4a}$$

$$\frac{\partial L}{\partial \lambda_i} = g_i(\mathbf{x}) = 0, \quad i = 1, \ldots, m \tag{3.3.4b}$$

$$\lambda_i \text{ is unrestricted in sign}, \quad i = 1, \ldots, m \tag{3.3.4c}$$

Solving Eqs. (3.3.4a) and (3.3.4b) simultaneously provides the optimal solution.

Lagrange multipliers have an important interpretation in optimization. For a given constraint, these multipliers indicate how much the optimal objective function value will change for a differential change in the right-hand side of the constraint; that is, $\partial f / \partial b_i|_{\mathbf{x}=\mathbf{x}^*} = \boldsymbol{\lambda}$ illustrating that the Lagrange multiplier λ_i is the rate of change of the optimal value of the original objective function with respect to a change in the value of the right-hand side of the ith constraint. The λ_i's are called dual variables or shadow prices.

3.3.2 Kuhn–Tucker Conditions

Equations (3.3.4a)–(3.3.4c) form the optimality conditions for an optimization problem involving only equality constraints. The Lagrange multipliers associated with the equality constraints are unrestricted in sign. Using the Lagrange multiplier method, the optimality conditions for the following generalized nonlinear programming problem can be derived.

$$\text{Minimize } f(\mathbf{x})$$

subject to

$$g_i(\mathbf{x}) = 0, \quad i = 1, \ldots, m$$

and

$$\underline{x}_j \leq x_j \leq \overline{x}_j, \quad j = 1, \ldots, n$$

In terms of the Lagrangian method, the above nonlinear minimization problem can be written as

$$\text{Min } L = f(\mathbf{x}) + \boldsymbol{\lambda}^T \mathbf{g}(\mathbf{x}) + \underline{\boldsymbol{\lambda}}^T(\underline{\mathbf{x}} - \mathbf{x}) + \overline{\boldsymbol{\lambda}}^T(\mathbf{x} - \overline{\mathbf{x}}) \qquad (3.3.5)$$

in which $\boldsymbol{\lambda}$, $\underline{\boldsymbol{\lambda}}$, and $\overline{\boldsymbol{\lambda}}$ are vectors of Lagrange multipliers corresponding to constraints $\mathbf{g}(\mathbf{x}) = \mathbf{0}$, $\underline{\mathbf{x}} - \mathbf{x} \leq \mathbf{0}$, respectively. The Kuhn–Tucker conditions for the optimality of the above problem are

$$\nabla_x L = \nabla_x f + \boldsymbol{\lambda}^T \nabla_x \mathbf{g} - \underline{\boldsymbol{\lambda}} + \overline{\boldsymbol{\lambda}} = \mathbf{0} \qquad (3.3.6a)$$

$$g_i(\mathbf{x}) = 0 \qquad (3.3.6b)$$

$$\underline{\lambda}_j(\underline{x}_j - x_j) = \overline{\lambda}_j(x_j - \overline{x}_j) = 0 \quad j = 1, 2, \ldots, n \qquad (3.3.6c)$$

$$\boldsymbol{\lambda} \text{ unrestricted in sign, } \underline{\boldsymbol{\lambda}} \geq 0, \overline{\boldsymbol{\lambda}} \geq 0 \qquad (3.3.6d)$$

3.4 CONSTRAINED NONLINEAR OPTIMIZATION: GENERALIZED REDUCED GRADIENT (GRG) METHOD

3.4.1 Basic Concepts

Similar to the linear programming simplex method, the fundamental idea of the generalized reduced gradient (GRG) method is to express m (number of constraint equations) of the variables, called basic variables, in terms of the remaining $n - m$ variables, called nonbasic variables. The decision variables can then be partitioned into the basic variables, $\mathbf{x}_B$, and the nonbasic variables, $\mathbf{x}_N$,

$$\mathbf{x} = (\mathbf{x}_B, \mathbf{x}_N)^T \tag{3.4.1}$$

Nonbasic variables not at their bounds are called superbasic variables (Murtaugh and Saunders, 1978).

The optimization problem can now be restated in terms of the basic and nonbasic variables.

$$\text{Minimize } f(\mathbf{x}_B, \mathbf{x}_N) \tag{3.4.2a}$$

subject to

$$\mathbf{g}(\mathbf{x}_B, \mathbf{x}_N) = \mathbf{0} \tag{3.4.2b}$$

and

$$\underline{\mathbf{x}}_B \le \mathbf{x}_B \le \bar{\mathbf{x}}_B \tag{3.4.2c}$$

$$\underline{\mathbf{x}}_N \le \mathbf{x}_N \le \bar{\mathbf{x}}_N \tag{3.4.2d}$$

The m basic variables in theory can be expressed in terms of the $n - m$ nonbasic variables as $\mathbf{x}_B(\mathbf{x}_N)$. Assume that constraints $\mathbf{g}(\mathbf{x}) = \mathbf{0}$ is differentiable and the $m \times m$ basis matrix $\mathbf{B}$ can be obtained as

$$\mathbf{B} = \left[\frac{\partial \mathbf{g}(\mathbf{x})}{\partial \mathbf{x}_B}\right]$$

which is nonsingular such that there exists a unique solution of $\mathbf{x}_B(\mathbf{x}_N)$. Nonsingular means that the determinant of $\mathbf{B} \neq \mathbf{0}$.

The objective called a reduced objective can be expressed in terms of the nonbasic variables as

$$F(\mathbf{x}_N) = f(\mathbf{x}_B(\mathbf{x}_N), \mathbf{x}_N) \tag{3.4.3}$$

The original nonlinear programming problem is transformed into the following reduced problem:

$$\text{Minimize } F(\mathbf{x}_N) \tag{3.4.4a}$$

subject to

$$\underline{\mathbf{x}}_N \leq \mathbf{x}_N \leq \bar{\mathbf{x}}_N \tag{3.4.4b}$$

which can be solved by an unconstrained minimization technique with slight modification to account for the bounds on nonbasic variables. Generalized reduced gradient algorithms, therefore, solve the original problem (3.3.1) by solving a sequence of reduced problems (3.4.4), using unconstrained minimization algorithms.

3.4.2 General Algorithm and Basis Changes

Consider solving the reduced problem (3.4.4) starting from an initial feasible point $\mathbf{x}^0$. To evaluate $F(\mathbf{x}_N)$ by Eq. (3.4.3), the values of the basic variables $\mathbf{x}_B$ must be known. Except for a very few cases, $\mathbf{x}_B(\mathbf{x}_N)$ cannot be determined in closed form; however, it can be computed for any $\mathbf{x}_N$ by an iterative method which solves a system of m nonlinear equations with the same number of unknowns as equations. A procedure for solving the reduced problem starting from the initial feasible solution $\mathbf{x}^{k=0}$ is as follows:

Step 0: Start with initial feasible solution $\mathbf{x}^{k=0}$ and set $\mathbf{x}_N^k = \mathbf{x}^{k=0}$.

Step 1: Substitute $\mathbf{x}_N^k$ into Eq. (3.4.2b) and determine the corresponding values of $\mathbf{x}_B$ by an iterative method for solving m nonlinear equations $\mathbf{g}(\mathbf{x}_B(\mathbf{x}_N^k), \mathbf{x}_N^k) = \mathbf{0}$.

Step 2: Determine the search direction $\mathbf{d}^k$ for the nonbasic variables by a line search scheme.

Step 3: Choose a step size for the line search scheme, β^k such that

$$\mathbf{x}_N^{k+1} = \mathbf{x}_N^k + \beta^k \mathbf{d}^k \tag{3.4.5}$$

This is done by solving the one-dimensional search problem: Minimize $F(\mathbf{x}_N^k + \beta \mathbf{d}^k)$ on $\mathbf{x}_n$. This one-dimensional search requires repeated applications of Step 1 to evaluate F for the different β values.

Step 4: Test the current point $\mathbf{x}^k = (\mathbf{x}_B^k, \mathbf{x}_N^k)$ for optimality; if not optimal, set $k = k + 1$ and return to Step 1.

3.4.2 The Reduced Gradient

Computation of the reduced gradient is required in the generalized reduced gradient method in order to define the search direction. Consider the simple problem

$$\text{Minimize } f(x_1, x_2)$$

subject to

$$g(x_1, x_2) = 0$$

The total derivative of the objective function is

$$df(\mathbf{x}) = \frac{\partial f(\mathbf{x})}{\partial x_1} dx_1 + \frac{\partial f(\mathbf{x})}{\partial x_2} dx_2 \tag{3.4.6}$$

and the total derivative of the constraint function is

$$dg(\mathbf{x}) = \frac{\partial g(\mathbf{x})}{\partial x_1} dx_1 + \frac{\partial g(\mathbf{x})}{\partial x_2} dx_2 = 0 \tag{3.4.7}$$

The reduced gradients are $\nabla f(\mathbf{x})$ and $\nabla g(\mathbf{x})$ defined by the coefficients in the total derivatives,

$$\nabla f(\mathbf{x}) = \left[\frac{\partial f}{\partial x_1}, \frac{\partial f}{\partial x_2}\right]^T \tag{3.4.8}$$

$$\nabla g(\mathbf{x}) = \left[\frac{\partial g}{\partial x_1}, \frac{\partial g}{\partial x_2}\right]^T \tag{3.4.9}$$

Consider the basic (dependent) variable to be x_1 and the nonbasic (independent) variable to be x_2. Equation (3.4.7) can be used to solve for dx_1,

$$dx_1 = \frac{\partial g(\mathbf{x})/\partial x_2}{\partial g(\mathbf{x})/\partial x_1} dx_2 \tag{3.4.10}$$

which is then substituted into Eq. (3.4.6) in order to eliminate dx_1. The resulting total derivative of the objective function $f(\mathbf{x})$ can be expressed as

$$df(\mathbf{x}) = \left\{-\left(\frac{\partial f(\mathbf{x})}{\partial x_1}\right)\left(\frac{\partial g(\mathbf{x})}{\partial x_1}\right)^{-1}\left(\frac{\partial g(\mathbf{x})}{\partial x_2}\right) + \left(\frac{\partial f(\mathbf{x})}{\partial x_2}\right)\right\} dx_2 \tag{3.4.11}$$

The reduced gradient is the expression in the outer brackets and can be reduced to

$$\frac{df(\mathbf{x})}{dx_2} = \frac{\partial f(\mathbf{x})}{\partial x_2} - \left(\frac{\partial f(\mathbf{x})}{\partial x_1}\right)\left(\frac{\partial x_1}{\partial x_2}\right) \tag{3.4.12}$$

which is scalar because there is only one nonbasic variable, x_2.

The reduced gradient can be written in vector form for the multiple variable case as

$$\nabla_N F = \left[\frac{\partial F}{\partial \mathbf{x}_N}\right] = \left[\frac{\partial f(\mathbf{x})}{\partial \mathbf{x}_N}\right] - \left[\frac{\partial f(\mathbf{x})}{\partial \mathbf{x}_B}\right]^T \left[\frac{\partial \mathbf{g}(\mathbf{x})}{\partial \mathbf{x}_B}\right]^{-1} \left[\frac{\partial \mathbf{g}(\mathbf{x})}{\partial \mathbf{x}_N}\right] \quad (3.4.13)$$

in which

$$\left[\frac{\partial \mathbf{x}_B}{\partial \mathbf{x}_N}\right] = \left[\frac{\partial \mathbf{g}(\mathbf{x})}{\partial \mathbf{x}_B}\right]^{-1} \left[\frac{\partial \mathbf{g}(\mathbf{x})}{\partial \mathbf{x}_N}\right] = \mathbf{B}^{-1} \left[\frac{\partial \mathbf{g}(\mathbf{x})}{\partial \mathbf{x}_N}\right] \quad (3.4.14)$$

The Kuhn–Tucker multiplier vector $\boldsymbol{\pi}$ is defined by

$$\left[\frac{\partial f(\mathbf{x})}{\partial \mathbf{x}_B}\right]^T \left[\frac{\partial \mathbf{g}(\mathbf{x})}{\partial \mathbf{x}_B}\right]^{-1} = \left[\frac{\partial f(\mathbf{x})}{\partial \mathbf{x}_B}\right]^T \mathbf{B}^{-1} = \boldsymbol{\pi}^T \quad (3.4.15)$$

Using these definitions the reduced gradient in Eq. (3.4.13) can be expressed as

$$\nabla_N F = \left[\frac{dF}{d\mathbf{x}_N}\right] = \left[\frac{\partial f(\mathbf{x})}{\partial \mathbf{x}_N}\right] - \boldsymbol{\pi}^T \left[\frac{\partial \mathbf{g}(\mathbf{x})}{\partial \mathbf{x}_N}\right] \quad (3.4.16)$$

3.4.4 Optimality Conditions for the GRG Method

Consider the nonlinear programming problem

$$\text{Minimize } f(\mathbf{x})$$

subject to

$$g_i(\mathbf{x}) = 0, \quad i = 1, \ldots, m$$

$$\underline{x}_j \le x_j \le \overline{x}_j, \quad j = 1, \ldots, n$$

In terms of basic and nonbasic variables, the Lagrangian function for the problem can be stated as

$$\begin{aligned} L &= f(\mathbf{x}) + \boldsymbol{\lambda}^T \mathbf{g}(\mathbf{x}) + \underline{\boldsymbol{\lambda}}^T(\underline{\mathbf{x}} - \mathbf{x}) + \overline{\boldsymbol{\lambda}}^T(\mathbf{x} - \overline{\mathbf{x}}) \\ &= f(\mathbf{x}_B, \mathbf{x}_N) + \boldsymbol{\lambda}^T \mathbf{g}(\mathbf{x}_B, \mathbf{x}_N) + \underline{\boldsymbol{\lambda}}_B^T(\underline{\mathbf{x}}_B - \mathbf{x}_B) + \underline{\boldsymbol{\lambda}}_N^T(\underline{\mathbf{x}}_N - \mathbf{x}_N) \\ &\quad + \overline{\boldsymbol{\lambda}}_B^T(\mathbf{x}_B - \overline{\mathbf{x}}_B) + \overline{\boldsymbol{\lambda}}_N^T(\mathbf{x}_N - \overline{\mathbf{x}}_N) \end{aligned} \quad (3.4.17)$$

in which $\boldsymbol{\lambda}_N$ and $\boldsymbol{\lambda}_B$ are vectors of Lagrange multipliers for nonbasic and basic variables, respectively.

Based on Eq. (3.3.6), the Kuhn–Tucker conditions for optimality in terms of the basic and nonbasic variables are

$$\nabla_B L = \nabla_B f + \boldsymbol{\lambda}^T \nabla_B \mathbf{g} - \underline{\boldsymbol{\lambda}}_B + \overline{\boldsymbol{\lambda}}_B = \mathbf{0} \quad (3.4.18a)$$

$$\nabla_N L = \nabla_N f + \boldsymbol{\lambda}^T \nabla_N \mathbf{g} - \underline{\boldsymbol{\lambda}}_N + \overline{\boldsymbol{\lambda}}_N = \mathbf{0} \quad (3.4.18b)$$

$$\underline{\boldsymbol{\lambda}}_B \geq \mathbf{0}, \qquad \underline{\boldsymbol{\lambda}}_N \geq \mathbf{0} \tag{3.4.18c}$$

$$\bar{\boldsymbol{\lambda}}_B \geq \mathbf{0}, \qquad \bar{\boldsymbol{\lambda}}_N \geq \mathbf{0} \tag{3.4.18d}$$

$$\underline{\boldsymbol{\lambda}}_B^T(\underline{\mathbf{x}}_B - \mathbf{x}_B) = \bar{\boldsymbol{\lambda}}_B^T(\mathbf{x}_B - \bar{\mathbf{x}}_B) = \mathbf{0} \tag{3.4.18e}$$

$$\underline{\boldsymbol{\lambda}}_N^T(\underline{\mathbf{x}}_N - \mathbf{x}_N) = \bar{\boldsymbol{\lambda}}_N^T(\mathbf{x}_N - \bar{\mathbf{x}}_N) = \mathbf{0} \tag{3.4.18f}$$

If $\mathbf{x}_B$ is strictly between its bounds, then $\underline{\boldsymbol{\lambda}}_B = \bar{\boldsymbol{\lambda}}_B = \mathbf{0}$ by Eq. (3.4.18e), so that from Eq. (3.4.18a),

$$\boldsymbol{\lambda}^T = \left[-\frac{\partial f}{\partial \mathbf{x}_B}\right]^T \left[\frac{\partial \mathbf{g}}{\partial \mathbf{x}_B}\right]^{-1} = \left[-\frac{\partial f}{\partial \mathbf{x}_B}\right]^T \mathbf{B}^{-1} = -\boldsymbol{\pi}^T \tag{3.4.19}$$

In other words, when $\underline{\mathbf{x}}_B < \mathbf{x}_B < \bar{\mathbf{x}}_B$, the Kuhn–Tucker multiplier vector $\boldsymbol{\pi}$ is the Lagrange multiplier vector for the equality constraints $\mathbf{g}(\mathbf{x})=\mathbf{0}$. Then from Eqs. (3.4.16) and (3.4.18b): If $\mathbf{x}_N$ is strictly between its bounds, i.e., $\underline{\mathbf{x}}_N < \mathbf{x}_N < \bar{\mathbf{x}}_N$, then $\underline{\geq}_N = \bar{\boldsymbol{\lambda}}_N = \mathbf{0}$ by Eq. (3.4.18f), so that

$$\left[\frac{\partial F}{\partial \mathbf{x}_N}\right] = 0 \tag{3.4.20}$$

If $\mathbf{x}_N$ is at its lower bound, $\mathbf{x}_N = \underline{\mathbf{x}}_N$, then $\bar{\boldsymbol{\lambda}}_N = \mathbf{0}$, so

$$\left[\frac{\partial F}{\partial \mathbf{x}_N}\right] = \underline{\boldsymbol{\lambda}}_N \geq 0 \tag{3.4.21}$$

If $\mathbf{x}_N$ is at its upper bound, $\mathbf{x}_N = \bar{\mathbf{x}}_N$, then $\underline{\boldsymbol{\lambda}}_N = \mathbf{0}$, so that

$$\left[\frac{\partial F}{\partial \mathbf{x}_N}\right] = \bar{\boldsymbol{\lambda}}_N \leq 0 \tag{3.4.22}$$

Equations (3.4.20)–(3.4.22) define the optimality conditions for the reduced problem (3.4.4). The Kuhn–Tucker conditions for the original problem may be viewed as the optimality conditions for the reduced problem.

3.5 CONSTRAINED NONLINEAR OPTIMIZATION: PENALTY FUNCTION METHODS

The essential idea of penalty function methods is to transform constrained nonlinear programming problems into a sequence of unconstrained optimization problems. The basic idea of these methods is to add one or more functions of the constraints to the objective function and to delete the constraints. Basic reasoning for such approaches is that the unconstrained problems are much easier to solve. Using a penalty function, a constrained nonlinear programming problem is transformed to an unconstrained problem.

$$\left.\begin{array}{l}\text{Minimize } f(\mathbf{x})\\ \text{subject to } \mathbf{g}(\mathbf{x})\end{array}\right\} \Rightarrow \text{Minimize } L[f(\mathbf{x}), \mathbf{g}(\mathbf{x})]$$

where $L[f(\mathbf{x}), \mathbf{g}(\mathbf{x})]$ is a penalty function. Various forms of penalty functions have been proposed which can be found elsewhere (McCormick, 1983; Gill et al., 1981). The penalty function is minimized by stages for a series of values of parameters associated with the penalty. In fact, the Lagrangian function is one form of penalty function. For many of the penalty functions, the Hessian of the penalty function becomes increasingly ill-conditioned (i.e., the function value is extremely sensitive to a small change in the parameter value) as the solution approaches the optimum. This section briefly describes a penalty function method called the augmented Lagrangian method.

The augmented Lagrangian method adds a quadratic penalty function loss term to the Lagrangian function [Eq. (3.3.2)], to obtain

$$L_A(\mathbf{x}, \boldsymbol{\lambda}, \psi) = f(\mathbf{x}) + \sum_{i=1}^{m} \lambda_i g_i(\mathbf{x}) + \frac{\psi}{2} \sum_{i=1}^{m} g_i^2(\mathbf{x})$$

$$= f(\mathbf{x}) + \boldsymbol{\lambda}^T \mathbf{g}(\mathbf{x}) + \frac{\psi}{2} \mathbf{g}(\mathbf{x})^T \mathbf{g}(\mathbf{x}) \tag{3.5.1}$$

where ψ is a positive penalty parameter. Some desirable properties of Eq. (3.5.1) are discussed by (Gill et al., 1981).

For ideal circumstances, $\mathbf{x}^*$ can be computed by a single unconstrained minimization of the differentiable function [Eq. (3.5.1)]. However, in general, $\boldsymbol{\lambda}^*$ is not available until the solution has been determined. An augmented Lagrangian method, therefore, must include a procedure for estimating the Lagrange multipliers. Gill et al. (1981) present the following algorithm:

Step 0: Select initial estimates of the Lagrange multipliers $\boldsymbol{\lambda}^{k=0}$, the penalty parameter ψ, and an initial point $\mathbf{x}^{k=0}$. Set $k = k + 1$ and set the maximum number of iterations as J.

Step 1: Check to see if $\mathbf{x}^k$ satisfies optimality conditions or if $k > J$. If so, terminate the algorithm.

Step 2: Minimize the augmented Lagrangian function, Minimize $L_A(\mathbf{x}, \boldsymbol{\lambda}, \psi)$, in Eq. (3.5.1). Procedures to consider unboundedness must be considered. The best solution is denoted as $\mathbf{x}^{k+1}$.

Step 3: Update the multiplier estimate by computing $\boldsymbol{\lambda}^{k+1}$.

Step 4: Increase the penalty parameter ψ if the constraint violations at $\mathbf{x}^{k+1}$ have not decreased sufficiently from those at $\mathbf{x}^k$.

Step 5: Set $k = k + 1$ and return to Step 1.

Augmented Lagrangian methods can be applied to inequality constraints. For the set of violated constraints, $\mathbf{g}(\mathbf{x})$ at $\mathbf{x}^k$, the augmented Lagrangian function has discontinuous derivatives at the solution if any of the constraints are active (Gill et al., 1981). Buys (1972) and Rockafellar (1973a, 1973b, 1974) presented the augmented Lagrangian function for inequality-constrained problems

$$L_A(\mathbf{x}, \boldsymbol{\lambda}, \psi) = f(\mathbf{x}) + \sum_{i=1}^{m} \begin{cases} \lambda_i g_i(\mathbf{x}) + \dfrac{\psi}{2} [g_i(\mathbf{x})]^2 & \text{if } g_i(\mathbf{x}) \le \dfrac{\lambda_i}{\psi} \\ -\dfrac{\psi}{2} \lambda_i^2 & \text{if } g_i(\mathbf{x}) > \dfrac{\lambda_i}{\psi} \end{cases} \tag{3.5.2}$$

3.6 THE AUGMENTED LAGRANGIAN METHOD

3.6.1 Introduction

The nonlinear programming problem with equality constraints only can be stated as

$$\text{Minimize } f(\mathbf{x}) \tag{3.6.1}$$

$$\text{subject to } c_i(\mathbf{x}) = 0, \quad i = 1, 2, \ldots, \mathbf{m} \tag{3.6.2}$$

where $\mathbf{x}$ is a vector of n components and usually $n \ge m$. An optimal solution, $\mathbf{x}^*$, can be obtained by solving the corresponding related set of $n + m$ nonlinear equations,

$$\nabla_\lambda L(\mathbf{x}^*, \boldsymbol{\lambda}^*) = c_i(\mathbf{x}^*) = 0 \tag{3.6.3}$$

$$\nabla_x L(\mathbf{x}^*, \boldsymbol{\lambda}^*) = \nabla f(\mathbf{x}^*) - \sum_i \lambda_i^* \nabla c_i (\mathbf{x}^*) = 0 \tag{3.6.4}$$

where $\boldsymbol{\lambda}^*$ is the m-dimensional Lagrange multiplier vector and is part of the entire solution vector, besides the n-dimensional $\mathbf{x}^*$ vector. Equations (3.6.3) and (3.6.4) are the first-derivative vectors with respect to $\boldsymbol{\lambda}$ and $\mathbf{x}$, respectively, evaluated at $\mathbf{x}^*$ of the Lagrangian function

$$L(\mathbf{x}, \boldsymbol{\lambda}) = f(\mathbf{x}) - \sum_i \lambda_i c_i(\mathbf{x}) \tag{3.6.5}$$

In the classical sense, Eqs. (3.6.3) and (3.6.4) comprise the first-order necessary conditions for $\mathbf{x}^*$ to be at least a local minimum of $f(\mathbf{x})$. If the Lagrangian function (3.6.5) does not contain a saddle point, however, the solution obtained may not yield a minimum point (Lasdon, 1970).

Rewriting Eqs. (3.6.1)–(3.6.2) in an exterior penalty type form,

$$P(\mathbf{x}, \boldsymbol{\theta}, \boldsymbol{\sigma}) = f(\mathbf{x}) + \frac{1}{2}\sum_i \sigma_i[c_i(\mathbf{x}) - \theta_i]^2, \quad i = 1, 2, \ldots, m$$

$$= f(\mathbf{x}) + \tfrac{1}{2}[c(\mathbf{x}) - \theta]^T\mathbf{S}[c(\mathbf{x}) - \theta] \tag{3.6.6}$$

where $\boldsymbol{\theta}$ is an m-dimensional parameter vector and $\mathbf{S}$ is an $m \times m$ diagonal matrix whose elements are the penalty weights $\sigma_i > 0$. The solution procedure to the unconstrained minimization problem (3.6.6) involves the variation of σ_i and θ_i in such a way that $\mathbf{x}(\boldsymbol{\sigma}, \boldsymbol{\theta}) \rightarrow \mathbf{x}^*$. When $\boldsymbol{\theta} = \mathbf{0}$, the second term in (3.6.6) is sometimes called the penalty term. At each iteration, when $\mathbf{x}$ becomes infeasible, this penalty term is added to $f(\mathbf{x})$ and $\boldsymbol{\sigma}$ is increased for the next iteration. Convergence is guaranteed by letting σ_i approach infinity. When these penalty weights are allowed to grow without bound, an ill-conditioned matrix may arise even before $\mathbf{x}$ gets close to $\mathbf{x}^*$. An attractive feature of the augmented Lagrangian method is that σ_i need not approach infinity and may, in fact, be held constant. Instead, $\boldsymbol{\theta}$ is varied, such that $\boldsymbol{\theta} \rightarrow \boldsymbol{\theta}^*$, an optimum parameter vector, while satisfying the condition

$$\theta_i^*\sigma_i = \lambda_i^*, \quad i = 1, 2, \ldots, m \tag{3.6.7}$$

If σ_i is sufficiently large, each iteration needs to update only θ_i. A further increase in σ_i is only required when the rate of convergence of $\mathbf{x}(\boldsymbol{\theta}, \boldsymbol{\sigma}) \rightarrow \mathbf{x}^*$ is small. A satisfactory value of σ_i is usually obtained near the early steps of calculation and can be held constant throughout the remaining iterations (Powell, 1978). The augmented Lagrangian function, formed from the Lagrangian function (3.6.5) augmented by the penalty term defined earlier, would then be

$$L_A(\mathbf{x}, \boldsymbol{\lambda}, \boldsymbol{\sigma}) = f(\mathbf{x}) - \sum_i \lambda_i c_i(\mathbf{x}) + \frac{1}{2}\sum_i \sigma_i[c_i(\mathbf{x})]^2$$

$$= f(\mathbf{x}) - \lambda^T c(\mathbf{x}) + \tfrac{1}{2}c(\mathbf{x})^T\mathbf{S}c(\mathbf{x}) \tag{3.6.8}$$

By letting

$$\theta_i = \frac{\lambda_i}{\sigma_i}, \quad i = 1, 2, \ldots, m \tag{3.6.9}$$

and expanding (3.6.6) results in

$$P(\mathbf{x}, \boldsymbol{\theta}, \boldsymbol{\sigma}) = f(\mathbf{x}) + \frac{1}{2}\left\{\sum_i \sigma_i \left[c_i^2(\mathbf{x}) - 2c_i(\mathbf{x})\frac{\lambda_i}{\sigma_i} + \frac{\lambda_i^2}{\sigma_i^2}\right]\right\}$$

$$= f(\mathbf{x}) + \frac{1}{2}\sum_i \sigma_i[c_i(\mathbf{x})]^2 - \sum_i \lambda_i[c_i(\mathbf{x})] + \frac{1}{2}\sum_i \frac{\lambda_i^2}{\sigma_i}$$

$$= L_A(\mathbf{x}, \lambda, \boldsymbol{\sigma}) + \frac{1}{2}\sum_i \frac{\lambda_i^2}{\sigma_i} \tag{3.6.10}$$

Because the second term of the right-hand side of (3.6.10) is not a function of x_i, then $\mathbf{x}(\boldsymbol{\theta}, \boldsymbol{\sigma}) = \mathbf{x}(\boldsymbol{\lambda}, \boldsymbol{\sigma})$ for any $\boldsymbol{\sigma}$ as long as (3.6.9) holds. For a well-scaled problem, a single scalar value, say r, can replace all σ_i's in $\mathbf{S}$ such that $\mathbf{S} = r\mathbf{I}$. As such, considerable reduction in the number of unknowns can be realized.

Now, consider the nonlinear programming problem with inequality constraints only

$$\text{Minimize } f(\mathbf{x}) \tag{3.6.11}$$

$$\text{subject to } h_i(\mathbf{x}) \geq 0, \quad i = 1, 2, \ldots, m' \tag{3.6.12}$$

The constraint set (3.6.12) can be modified in the form of inequality (3.6.2) by incorporating the slack variable z_i

$$h_i(\mathbf{x}) - z_i = 0, \quad z_i \geq 0, i = 1, 2, \ldots, m' \tag{3.6.13}$$

The set $(\mathbf{x}, \mathbf{z})$ forms the new feasible space. The new augmented Lagrangian function would be (3.6.11) plus

$$\sum_i t_i(\mathbf{x}, \mathbf{z}, \boldsymbol{\mu}, \boldsymbol{\sigma}) = \sum_i - \mu_i[h_i(\mathbf{x}) - z_i] + \frac{1}{2}\sum_i \sigma_i[h_i(\mathbf{x}) - z_i]^2 \tag{3.6.14}$$

where $i = 1, 2, \ldots, m'$. The slack variable z_i can be eliminated from the calculations by performing minimization on the function over $\mathbf{z}$ (Powell, 1978). As only Eq. (3.6.14) depends on $\mathbf{z}$, by the first-order necessary conditions are

$$z_i^* = h_i(\mathbf{x}) - \frac{\mu_i}{\sigma_i}, \quad z_i \geq 0$$

$$= \max\left[0, h_i(\mathbf{x}) - \frac{\mu_i}{\sigma_i}\right] \tag{3.6.15}$$

Equation (3.6.14) is transformed into

$$\sum_i t_i(\mathbf{x}, \boldsymbol{\mu}, \boldsymbol{\sigma}) = \sum_i \begin{cases} -\mu_i h_i(\mathbf{x}) + \frac{1}{2}\sigma_i[h_i(\mathbf{x})]^2 & \text{if } h_i(\mathbf{x}) < \frac{\mu_i}{\sigma_i} \\ -\frac{1}{2}\frac{\mu_i^2}{\sigma_i} & \text{if } h_i(\mathbf{x}) \geq \frac{\mu_i}{\sigma_i} \end{cases} \tag{3.6.16}$$

or

$$\sum_i t_i(\mathbf{x}, \boldsymbol{\mu}, \boldsymbol{\sigma}) = \frac{1}{2}\sum_i \sigma_i \left\{\min\left[0, h_i(\mathbf{x}) - \frac{\mu_i}{\sigma_i}\right]\right\}^2 - \frac{1}{2}\sum_i \frac{\mu_i^2}{\sigma_i} \tag{3.6.17}$$

When the equality constraints (3.6.2) and inequality constraints (3.6.12) occur concurrently in a nonlinear programming problem, the augmented Lagrangian function becomes

$$\begin{aligned} L_A(\mathbf{x}, \boldsymbol{\lambda}, \boldsymbol{\mu}, \boldsymbol{\sigma}) &= f(\mathbf{x}) - \sum_{i=1}^{m} \lambda_i c_i(\mathbf{x}) + \frac{1}{2}\sum_{i=1}^{m} \sigma_i[c_i(\mathbf{x})]^2 \\ &\quad + \frac{1}{2}\sum_{i=1}^{m} \sigma_i \left\{\min\left[0, h_i(\mathbf{x}) - \frac{\mu_i}{\sigma_i}\right]\right\}^2 - \frac{1}{2}\sum_{i=1}^{m'} \frac{\mu_i^2}{\sigma_i} \end{aligned} \tag{3.6.18}$$

3.6.2 Optimality Results of the Lagrange Multipliers

Some important duality results will be discussed in this subsection, showing the optimum choice of the $\boldsymbol{\lambda}$ and $\boldsymbol{\mu}$ (or $\boldsymbol{\sigma}$) parameters which are determined by the maximization problem in terms of these parameters.

The first-order necessary condition for $\mathbf{x}^*$ to be a local minimum of $L_A(\mathbf{x}, \boldsymbol{\lambda}, \sigma)$ is that ∇L_A vanishes at $\mathbf{x}^*$. Deriving ∇L_A from the function in Eq. (3.6.8) produces

$$\nabla L_A(\mathbf{x}, \boldsymbol{\lambda}, \boldsymbol{\sigma}) = \nabla f(\mathbf{x}) - \sum_i \lambda_i \nabla c_i(\mathbf{x}) + \sum_i \sigma_i c_i(\mathbf{x}) \nabla c_i(\mathbf{x}) \tag{3.6.19}$$

On the other hand, the first-order necessary condition for $\mathbf{x}^*$ to be a local minimum of the original problem (3.6.1)–(3.6.2) is that it had to satisfy Eqs. (3.6.3) and (3.6.4). It follows, then, that $\nabla L_A(\mathbf{x}^*, \boldsymbol{\lambda}^*, \boldsymbol{\sigma}^*) = 0$.

The next and final step would be to prove that $\nabla^2 L_A(\mathbf{x}^*, \boldsymbol{\lambda}^*, \boldsymbol{\sigma})$ is positive definite; that is, the second-order sufficiency condition should hold. By taking the derivative of Eq. (3.6.19) with respect to x, we obtain

$$\nabla^2 L_A(\mathbf{x}, \boldsymbol{\lambda}, \boldsymbol{\sigma}) = \nabla^2 L(\mathbf{x}, \boldsymbol{\lambda}, \boldsymbol{\sigma}) + \sum_i \sigma_i[c_i(\mathbf{x})\nabla^2 c_i(\mathbf{x}) + \nabla c_i(\mathbf{x})\nabla c_i(\mathbf{x})^T] \tag{3.6.20}$$

where ∇^2 is the second derivative. At the optimum point $\mathbf{x}^*$, the matrix (3.6.20) becomes

$$\nabla^2 L_A(\mathbf{x}^*, \boldsymbol{\lambda}^*, \boldsymbol{\sigma}) = \nabla^2 L(\mathbf{x}^*, \boldsymbol{\lambda}^*, \boldsymbol{\sigma}) + \sum_i \sigma_i \nabla c_i(\mathbf{x}^*)\nabla c_i(\mathbf{x}^*)^T \quad (3.6.21)$$

Let $\mathbf{y}$ be a unit vector orthogonal to $\nabla c(\mathbf{x}^*)$, then the matrix $\nabla^2 L_A^*$ or $\nabla^2 L_A(\mathbf{x}^*, \boldsymbol{\lambda}^*, \boldsymbol{\sigma})$ is positive definite because

$$\mathbf{y}^T\nabla^2 L_A^*\mathbf{y} = \mathbf{y}^T\, \nabla^2 L^*\mathbf{y} + \boldsymbol{\sigma}[\mathbf{y}^T\nabla c(\mathbf{x}^*)]^2 \quad (3.6.22)$$

If $\nabla^2 L^*$ is not a positive definite matrix and thus $\mathbf{y}\nabla c(\mathbf{x}^*) \neq 0$, then $\boldsymbol{\sigma}$ has to be sufficiently large, say $\boldsymbol{\sigma} > \boldsymbol{\sigma}' > \mathbf{0}$ such that the second term on the right-hand side of Eq. (3.6.21) dominates the negative first term. If this is pursued, $\nabla^2 L_A^*$ is positive and the second-order condition is satisfied.

The augmented Lagrangian function comprising $f(\mathbf{x})$ plus Eq. (3.6.16) or (3.6.17) is discontinuous in its derivatives. A remedy to this problem would be to partition the function into two parts such that

$$I_- = \left\{ i \middle| h_i(\mathbf{x}) < \frac{\mu_i}{\sigma_i} \right\} \quad (3.6.23)$$

$$I_+ = \left\{ i \middle| h_i(\mathbf{x}) > \frac{\mu_i}{\sigma_i} \right\} \quad (3.6.24)$$

where I is a general index set, $i = 1, 2, \ldots, m'$, and $I = I_- \cup I_+$. The augmented Lagrangian function considering inequality constraints only would be

$$L_A(\mathbf{x}, \boldsymbol{\mu}, \boldsymbol{\sigma}) = f(\mathbf{x}) + \sum_{i=1}^{m'} \begin{cases} -\mu_i h_i(\mathbf{x}) + \frac{1}{2}\sigma_i[h_i(\mathbf{x})]^2 & \text{if } i \in I_- \\ -\dfrac{1}{2}\dfrac{\mu_i^2}{\sigma_i} & \text{if } i \in I_+ \end{cases} \quad (3.6.25)$$

If the second-order conditions on problem (3.6.11)–(3.6.12) are satisfied and $\boldsymbol{\mu} = \boldsymbol{\mu}^*$, then there exists a $\boldsymbol{\sigma}' > \mathbf{0}$ such that for all $\boldsymbol{\sigma} \geq \boldsymbol{\sigma}'$, $\mathbf{x}^*$ is a local minimum of $L_A(\mathbf{x}, \boldsymbol{\mu}, \boldsymbol{\sigma})$. Consider the first-order derivatives of Eqs. (3.6.25) which are

$$\nabla L_A(\mathbf{x}, \boldsymbol{\mu}, \boldsymbol{\sigma}) = \begin{cases} \nabla L(\mathbf{x}, \boldsymbol{\mu}) + \sum\limits_{i=1}^{m'} \sigma_i h_i(\mathbf{x})\nabla h_i(\mathbf{x}) & \text{if } i \in I_- \\ \nabla f(\mathbf{x}) & \text{if } i \in I_+ \end{cases} \quad (3.6.26)$$

where

$$\nabla L(\mathbf{x}, \boldsymbol{\mu}) = \nabla f(\mathbf{x}) - \sum_{i=1}^{m'} \mu_i \nabla h_i(\mathbf{x}), \quad \mu_i \geq 0 \text{ for } i \in I_- \quad (3.6.27)$$

The first-order necessary conditions for $\mathbf{x}^*$ minimize $L_A(\mathbf{x}, \boldsymbol{\mu}^*, \boldsymbol{\sigma})$ can be proved to hold in both sides of the partition. For $i \in I_-$, it follows from Eqs. (3.6.16) that $z_i^* = 0$ or $h_i(\mathbf{x}^*) = 0$ and thus Eqs. (3.6.26) gives $\nabla L_A(\mathbf{x}^*, \boldsymbol{\mu}^*, \boldsymbol{\sigma}) = \nabla L(\mathbf{x}^*, \boldsymbol{\mu}^*) = 0$. For $i \in I_+$, the necessary conditions for unconstrained minimization of $f(\mathbf{x})$ implies that $\nabla f(\mathbf{x}^*) = 0$ and so does $L_A(\mathbf{x}^*, \boldsymbol{\mu}^*, \boldsymbol{\sigma})$.

Using the Kuhn–Tucker conditions, $\nabla^2 L_A(\mathbf{x}^*, \boldsymbol{\mu}^*, \boldsymbol{\sigma})$ is positive definite in either of the two cases discussed above. [Note that the derivatives of the function are undefined at $h_i(\mathbf{x}) = \mu_i/\sigma_i$.] Further details on dual theorems of the function can be found in Fletcher (1975) and Rockafellar (1973a,b).

3.6.3 Updating Formula and Convergence

The following discussion is based on the form of the augmented Lagrangian function for inequality constraints given in Eqs. (3.6.25). For the case with equality constraints, however, it can also be implied from Eqs. (3.6.25) when $i \in I_-$. The general form for the updating formula for multipliers is

$$\boldsymbol{\mu}^{(k+1)} = \boldsymbol{\mu}^{(k)} + \Delta\boldsymbol{\mu}^{(k)} \tag{3.6.28}$$

where the superscripts refer to the iteration number. The second term of the equation is continuously modified such that $\boldsymbol{\mu}^{(k+1)} \to \boldsymbol{\mu}^*$. The first-order necessary conditions at optimum for the original problem (3.6.11)–(3.6.12) gives

$$\nabla L(\mathbf{x}^*, \boldsymbol{\mu}^*) = \nabla f(\mathbf{x}^*) - \sum_i \mu_i^* \nabla h_i(\mathbf{x}^*) = 0 \tag{3.6.29}$$

$$\mu_i^* h_i(\mathbf{x}^*) = 0, \quad \mu_i^* \geq 0 \tag{3.6.30}$$

The first derivatives of Eqs. (3.6.25) are defined as

$$\nabla L_A(\mathbf{x}, \boldsymbol{\mu}, \boldsymbol{\sigma}) = \begin{cases} \nabla f(\mathbf{x}) - \sum_i [\mu_i - \sigma_i h_i(\mathbf{x})] \nabla h_i(\mathbf{x}) & \text{if } i \in I_- \\ \nabla f(\mathbf{x}) & \text{if } i \in I_+ \end{cases} \tag{3.6.31}$$

When $i \in I_+$, the complementary slackness conditions, $\mu_i\, h_i(\mathbf{x}^*) = 0$, give $\mu_i^* = 0$. Because $\boldsymbol{\mu}^{(k+1)} \to \boldsymbol{\mu}^*$, substituting $\mu_i^* = 0$ in Eq. (3.6.28) for $\mu^{(k+1)}$ implies

$$\Delta\mu^{(k)} = -\mu^{(k)} \quad \text{for } i \in I_+ \tag{3.6.32}$$

When $i \in I_-$, Eqs. (3.6.29) can be equated to (3.6.31); then cancel similar terms and obtain $\mu_i^* = \mu_i - \sigma_i h_i(\mathbf{x})$. By using Eq. (3.6.28) again, the implication becomes

$$\Delta\mu_i^{(k)} = -\sigma_i h_i(\mathbf{x}) \quad \text{for } i \in I_- \tag{3.6.33}$$

Equations (3.6.32) and (3.6.33) are the simplest updating formulas which do not require any derivatives. Either updating formula represents a steepest ascent toward the maximum of the dual function of Eqs. (3.6.26) with a linear rate of convergence.

3.7 SOLUTION OF DISCRETE-TIME OPTIMAL CONTROL PROBLEMS

3.7.1 Discrete Optimal Control Problem

Consider the following discrete optimal control problem:

$$\min z = f(\mathbf{x}, \mathbf{u}) \tag{3.7.1}$$

subject to

$$\mathbf{x}_{t+1} = \mathbf{g}(\mathbf{x}_t, \mathbf{u}_t, t), \quad t = 0, \ldots, T-1 \tag{3.7.2}$$

$$\underline{\mathbf{x}}_t \le \mathbf{x}_t \le \bar{\mathbf{x}}_t, \quad t = 0, \ldots, T \tag{3.7.3}$$

$$\underline{\mathbf{u}}_t \le \mathbf{u}_t \le \bar{\mathbf{u}}_t \quad t = 0, \ldots, T \tag{3.7.4}$$

where $\mathbf{x}_t$ is the column vector of the state variable at time t; $\mathbf{u}_t$ is the column vector of control variables at time t; $\bar{\mathbf{x}}_t$ and $\bar{\mathbf{u}}_t$ are column vectors of upper bounds; and $\underline{\mathbf{x}}_t$ and $\underline{\mathbf{u}}_t$ are column vectors of lower bounds. f and $\mathbf{g}$ are assumed continuously differentiable in $(\mathbf{x}_t, \mathbf{u}_t)$ for each t. The time, t, can only take on a finite number of discrete values, $t = 0, 1, \ldots, T$. Equation (3.7.2) represent the process or simulator equation and inequalities (3.7.3) and (3.7.4) represent the bound constraints on the state and control variables, respectively.

The structure of the Jacobian of Eq. (3.7.2) is shown in Figure 3.3. Nonzero elements are only in the unit submatrices and in the submatrices defined by

$$\mathbf{B}_t = \frac{\partial \mathbf{g}}{\partial \mathbf{x}_t} \tag{3.7.5}$$

$$\mathbf{K}_t = \frac{\partial \mathbf{g}}{\partial \mathbf{u}_t} \tag{3.7.6}$$

3.7.2 Reduced Objective Problem

Considering a given $\mathbf{u}$, the system of Eq. (3.7.2) may be solved for a unique $\mathbf{x}$, $\mathbf{x}(\mathbf{u})$. The function $\mathbf{x}(\mathbf{u})$ can then be used to eliminate $\mathbf{u}$ in the objective (3.7.1) to yield the reduced objective function

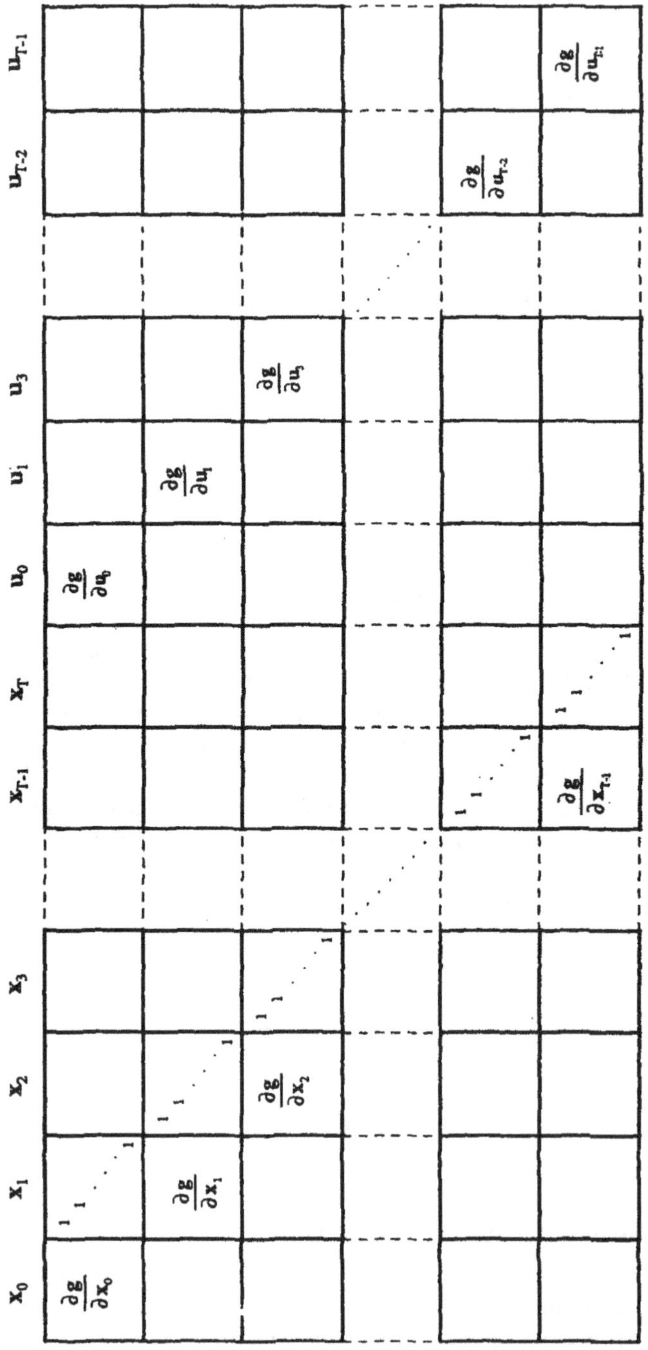

Figure 3.3 Structure of the Jacobian of the simulator equation.

$$F(\mathbf{u}) = f(\mathbf{x}(\mathbf{u}), \mathbf{u}) \tag{3.7.7}$$

By the implicit function theorem (Luenberger, 1984), $\mathbf{x}(\mathbf{u})$ is continuously differentiable, so that F is a differentiable function of $\mathbf{u}$.

Solving the process simulation equations (3.7.2) for a particular set of control variables, $\mathbf{u}$, each time these equations need to be evaluated, the reduced optimization problem takes the form

$$\text{Min } f(\mathbf{x}(\mathbf{u}), \mathbf{u}) = \text{Min } F(\mathbf{u}) \tag{3.7.8}$$

subject to

$$\underline{\mathbf{x}}_t \leq \mathbf{x}_t(\mathbf{u}_t) \leq \bar{\mathbf{x}}_t \tag{3.7.9}$$

$$\underline{\mathbf{u}}_t \leq \mathbf{u}_t \leq \bar{\mathbf{u}}_t \tag{3.7.10}$$

State (or dependent) variables and the control (or independent) variables are implicitly related through the simulator. In essence, the simulator equations are used to express the states in terms of the controls, yielding a much smaller optimization problem. The reduced gradient $\partial F/\partial \mathbf{u}$, where $F(\mathbf{u}) = f(\mathbf{x}(\mathbf{u}), \mathbf{u})$, is required to solve the reduced problem. In order to determine the reduced gradient, the following procedure can be used.

Step 1: Use the appropriate simulation model to solve the simulator (process) equations.

Step 2: Solve the following set of linear equations:

$$\boldsymbol{\pi} \left[\frac{\partial \mathbf{g}}{\partial \mathbf{x}} \right] = \frac{\partial f}{\partial \mathbf{x}} \tag{3.7.11}$$

or

$$\boldsymbol{\pi} \mathbf{B} = \frac{\partial f}{\partial \mathbf{x}}$$

for the row vector of Lagrange of multipliers $\boldsymbol{\pi}$.

Step 3: Evaluate the reduced gradient

$$\frac{\partial F}{\partial \mathbf{u}} = \frac{\partial f}{\partial \mathbf{u}} - \boldsymbol{\pi} \frac{\partial \mathbf{g}}{\partial \mathbf{u}} \tag{3.7.12}$$

In the above two equations all elements of $\partial \mathbf{g}/\partial \mathbf{x}$ and $\partial \mathbf{g}/\partial \mathbf{u}$ are evaluated at some model solution $\mathbf{u}$ for which the $\partial F/\partial \mathbf{u}$ is evaluated. Because the simulator equations (3.7.2) have a sequential form, $\partial \mathbf{g}/\partial \mathbf{u}$ is block lower triangular with square nonsingular blocks. The large linear system (3.7.11) decomposes into T smaller sequential systems, which are solved backward in time (see Fig. 3.4). The difference equations for the multipliers are

$$\boldsymbol{\pi}_T \mathbf{B}_T = \frac{\partial f}{\partial \mathbf{x}_T} \qquad (3.7.13a)$$

$$\boldsymbol{\pi}_t \mathbf{B}_t = \frac{\partial f}{\partial \mathbf{x}_t} - \boldsymbol{\pi}_{t+1} \frac{\partial \mathbf{g}_{t+1}}{\partial \mathbf{x}_t}, \quad t = T - 1, T - 2, \ldots, 1 \qquad (3.7.13b)$$

In these equations, the matrices $\mathbf{B}_t$ and $\partial \mathbf{g}_{t+1}/\partial \mathbf{x}_t$ are evaluated using the control and state variables obtained in step 1 when solving the simulator

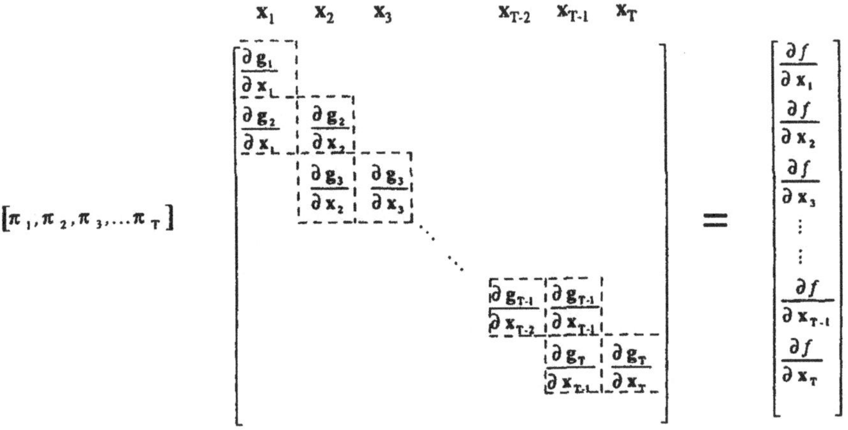

(a) Matrix $\pi \dfrac{\partial \mathbf{g}}{\partial \mathbf{x}} = \dfrac{\partial f}{\partial \mathbf{x}}$

$$\pi_1 \frac{\partial \mathbf{g}_1}{\partial \mathbf{x}_1} + \pi_2 \frac{\partial \mathbf{g}_2}{\partial \mathbf{x}_1} = \frac{\partial f}{\partial \mathbf{x}_1}$$

$$\pi_2 \frac{\partial \mathbf{g}_2}{\partial \mathbf{x}_2} + \pi_3 \frac{\partial \mathbf{g}_3}{\partial \mathbf{x}_2} = \frac{\partial f}{\partial \mathbf{x}_2}$$

$$\vdots$$

$$\pi_{T-1} \frac{\partial \mathbf{g}_{T-1}}{\partial \mathbf{x}_{T-1}} + \pi_T \frac{\partial \mathbf{g}_T}{\partial \mathbf{x}_{T-1}} = \frac{\partial f}{\partial \mathbf{x}_{T-1}}$$

$$\pi_T \frac{\partial \mathbf{g}_T}{\partial \mathbf{x}_T} = \frac{\partial f}{\partial \mathbf{x}_T}$$

(b) Equation of Submatrices

Figure 3.4 Computation of $\pi_T, \ldots, \pi_1$.

equations and all vectors in Eqs. (3.7.13a) and (3.7.13b) are row vectors. Then Eq. (3.7.13a) is solved for $\boldsymbol{\pi}_T$ and Eq. (3.7.13b) is solved sequentially for $\pi_{T-1}, \pi_{T-2}, \ldots, \pi_1$. Equations (3.7.13a) and (3.7.13b) are derived from the general reduced gradient equation $\boldsymbol{\pi}\mathbf{B} = \partial f/\partial \mathbf{x}$. The components of the $\partial F/\partial \mathbf{u}$ are evaluated by

$$\frac{\partial F}{\partial \mathbf{u}_t} = \frac{\partial f}{\partial \mathbf{u}_t} - \boldsymbol{\pi}_t \frac{\partial \mathbf{g}_t}{\partial \mathbf{u}_t} \tag{3.7.14}$$

The dynamic structure of the simulator equations could be of the form

$$\mathbf{g}(\mathbf{x}_t, \ldots, \mathbf{x}_{t-s}, \mathbf{u}_t, \ldots, \mathbf{u}_{t-c}) = \mathbf{0}, \quad t = 1, \ldots, T \tag{3.7.15}$$

where $\mathbf{g}(\ )$ is an m vector of function, assumed differentiable, and s and c are the maximum lags of $\mathbf{x}$ and $\mathbf{u}$, respectively. For many applications, $\mathbf{g}$ has the form $\mathbf{g} = -\mathbf{x}_t + \mathbf{h}(\mathbf{x}_t, \ldots, \mathbf{x}_{t-s}, \mathbf{u}_t, \ldots, \mathbf{u}_{t-c}) = \mathbf{0}$. The difference equations for the Lagrange multipliers are

$$\boldsymbol{\pi}_T \mathbf{B}_T = \frac{\partial F}{\partial \mathbf{x}_T} \tag{3.7.16}$$

$$\boldsymbol{\pi}_t \mathbf{B}_t = \frac{\partial F}{\partial \mathbf{x}_t} - \sum_{\tau=t+1}^{b} \boldsymbol{\pi}_\tau \mathbf{B}_{\tau,t}, \quad t = T-1, T-2, \ldots, 1 \tag{3.7.17}$$

where

$$b = \min(t + s, T) \tag{3.7.18}$$

$$\mathbf{B}_{\tau,t} = \frac{\partial \mathbf{g}_\tau}{\partial \mathbf{x}_t} \tag{3.7.19}$$

$$\mathbf{B}_t = \frac{\partial \mathbf{g}_t}{\partial \mathbf{x}_t} \tag{3.7.20}$$

$\partial \mathbf{g}/\partial x$ is nonsingular if and only if all matrices $\mathbf{B}_t$ are nonsingular. The components of $\partial F/\partial \mathbf{u}$ are evaluated using

$$\frac{\partial F}{\partial \mathbf{u}_t} = \frac{\partial f}{\partial \mathbf{u}_t} - \sum_{\tau=t}^{a} \boldsymbol{\pi}_\tau \frac{\partial \mathbf{g}_\tau}{\partial \mathbf{u}_t} \tag{3.7.21}$$

where

$$a = \min(t - c, T) \tag{3.7.22}$$

If the simulator equations are not simultaneous, each $\mathbf{B}_t$ is triangular so Eqs. (3.7.16) and (3.7.17) can be solved quickly.

3.7.3 GRG Algorithm to Solve Optimal Control Problem

Consider an optimal control problem of the form

$$\text{Min } z = \text{Min} \sum_{t=1}^{T} f_t\,(\mathbf{x}_t, \ldots, \mathbf{x}_{t-s'}, \mathbf{u}_t, \ldots, \mathbf{u}_{t-c'}) \tag{3.7.23}$$

subject to

$$\mathbf{g}_t\,(\mathbf{x}_t, \ldots, \mathbf{x}_{t-s}, \mathbf{u}_t, \ldots, \mathbf{u}_{t-c}) = \mathbf{0}, \quad t = 1, \ldots, T \tag{3.7.24}$$

$$\underline{\mathbf{u}}_t \le \mathbf{u}_t \le \overline{\mathbf{u}}_t, \quad t = 1, \ldots, T \tag{3.7.25}$$

For simplicity, bound constraints on the state variable have been suppressed. The state and control lags, s' and c', for the objective function may differ from s and c for the simulator equations. The vector of functions $\mathbf{g}$ and objective functions f_t may all be nonlinear and are assumed to be continuously differentiable. The recursive equations (3.7.24) are assumed to have a unique solution $\mathbf{x}_1, \ldots, \mathbf{x}_T$ for any set of control vectors $\mathbf{u}_1, \ldots, \mathbf{u}_T$ satisfying Eq. (3.7.24) and for any initial conditions.

Bounds on the state variables may be dealt with by penalty or augmented Lagrangian methods, which require no basis changes and, consequently, simplify the algorithm. Penalty or Lagrangian methods may not be as efficient as methods that deal with state bounds directly.

The algorithm presented by Mantell and Lasdon (1978) is stated as follows:

Step 0: Given are the initial control vector $\mathbf{u}^{(i)}$ and all initial values of lagged states and control variables, set $k = 0$.

Step 1: Simulate the system with $\mathbf{u} = \mathbf{u}^k$ to determine all state variables and the objective value $F(\mathbf{u}^k)$.

Step 2: Compute $\nabla F(\mathbf{u}^k)$ using Eq. (3.7.21).

Step 3: Check for convergence and stop if convergence criteria are satisfied, otherwise go to Step 4.

Step 4: Compute the search direction $\mathbf{d}^k$ using an unconstrained minimization algorithm.

Step 5: Perform a one-dimensional search along $\mathbf{d}^k$ to find β^k, the step size that minimizes $\mathbf{F}(\mathbf{u}^k + \beta\mathbf{d}^k)$ subject to $\beta > 0$ and $\underline{\mathbf{u}} \le \mathbf{u}^k + \beta\mathbf{d}^k \le \overline{\mathbf{u}}$. For each value of β in the search it is required to simulate the system by solving the simulator equation (3.7.24), compute the objective, and possibly compute the reduced gradient.

Step 6: Set $\mathbf{u}^{k+1} = \mathbf{u}^k + \beta^k\mathbf{d}^k$

Step 7: Replace k by $k + 1$ and return to Step 3 (to Step 2 if the reduced gradient is not computed in the one-dimensional search).

APPENDIX: COMPUTER SOFTWARE

This appendix briefly introduces four nonlinear programming computer codes that have been applied to solve NLP problems. They are (1) GRG2 (Generalized Reduced Gradient 2) developed by Lasdon and his colleagues (Lasdon et al., 1978; Lasdon and Waren, 1978); (2) GINO (Liebman et al., 1986); (3) MINOS (Modular In-core Nonlinear Optimiation System) developed by Murtagh and Saunders (1983; 1987); and (4) GAMS by Brooke et al. (1988).

GRG2 Computer Code

The GRG2 computer code utilizes the fundamental idea of the generalized reduced gradient algorithm. GRG2 requires that the user provide a subroutine, GCOMP, specifying the objective function and constraints of the nonlinear programming (NLP) problem. It is optimal for the user to provide the subroutine that contains derivatives of the objective function and constraints. If not provided, differentiations are approximated numerically by either forward finite differencing or central finite differencing. GRG2 provides several alternative ways that can be used to define the search direction. They include the BFGS, the quasi-Newton method, and variations of conjugate gradient methods. The default method is the BFGS method.

MINOS Computer Code

MINOS is a Fortran-based computer code designed to solve large-scale optimization problems. The program solves a linear programming problem by implementing the primal simplex method. When a problem has a nonlinear objective function subject to linear constraints, MINOS uses a reduced gradient algorithm in conjunction with a quasi-Newton algorithm. In case the problem involves nonlinear constraints, the projected Lagrangian algorithm is implemented. Similar to GRG2, MINOS requires that the user provide subroutine FUNOBJ to specify the objective function and its gradient. Also, subroutine FUNCON is to be supplied by the user to input the constraints and as many of their gradients as possible.

GINO Computer Code

GINO (Liebman et al., 1986) is a microcomputer version of GRG2.

GAMS Computer Code

The GAMS (General Algebraic Modeling Systems) family of software (Brooke et al., 1988) consists of three software modules. These are GAMS, which can solve LP problems (Brooke et al., 1988), GAMS/MINOS, which is an adaption of MINOS (Modular In-Core Nonlinear Optimization System) for both linear and nonlinear programming problems (Murtagh and Saunders, 1987), and GAMS/ZOOM, an adaptation of ZOOM (Zero/One Optimization Method) for mixed integer programming problems (Singhal et al., 1987). All three of these software modules are available in PC, workstation, and mainframe versions. GAMS was developed by an economic modeling group at the World Bank in an effort to provide a system structure and programming language in which the conciseness of expression, generality, and portability could be maintained and to use the computer to keep track of as many programming details as possible.

REFERENCES

Brooke, A., Kendrick, D., and Meerhaus, A., *GAMS: A User's Guide*, The Scientific Press, Redwood City, CA, 1988.

Buys, J. D., Dual Algorithms for Constrained Optimization Problems, Ph.D. Thesis, University of Leiden, Netherlands, 1972.

Cooper, L. L. and Cooper, M. W., *Introduction to Dynamic Programming*, Pergamon Press, Elmsford, NY, 1981.

Denardo, E. V., *Dynamic Programming Theory and Applications*, Prentice-Hall, Englewood Cliffs, NJ, 1982.

Dennis, J. E., and Schnable, R. B., *Numerical Methods for Unconstrained Optimization*, Prentice-Hall, Englewood Cliffs, New Jersey, 1983.

Dreyfus, S., and Law, A., *The Art and Theory of Dynamic Programming*, Academic Press, New York, 1977.

Edgar, T. F., and Himmelblau, D. M., *Optimization of Chemical Processes*, McGraw-Hill, New York, 1988.

Fletcher, R., *Practical Methods of Optimization, Vol. 1, Unconstrained Optimization*, John Wiley & Sons, New York, 1980.

Gill, P. E., Murray W., and Wright, M. H., *Practical Optimization*, Academic Press, London, 1981.

Himmelblau, D. M., *Applied Nonlinear Programming*, McGraw-Hill, New York, 1972.

Lasdon, L. S., *Optimization Theory for Large Systems*, Macmillan, New York, 1970.

Lasdon, L. S., Fox, R. L., and Ratner, M. W., Nonlinear optimization using the generalized reduced gradient method, *Revue Francaise d'Automatique, Informatique et Recherche Operationnelle*, Vol. 3, pp. 73–104, November 1974.

Lasdon, L. S., Waren, A. D., Jain, A., and Ratner, M., Design and Testing of a Generalized Reduced Gradient Code for Nonlinear Programming, *ACM Transactions on Mathematical Software*, Vol. 4, pp. 34–50, 1978.

Lasdon, L. S., and Waren, A. D., Generalized Reduced Gradient Software for Linearly and Nonlinearly Constrained Problems, in *Design and Implementation of Optimization Software*, H. J. Greenberg (ed.), Sijthoff and Noordhoff, 1978, pp. 363–397.

Liebman, J. S., Lasdon, L. S., Schrage, L., and Waren, A., *Modeling and Optimization with GINO*, The Scientific Press, Palo Alto, CA, 1986.

Luenberger, D. G., *Introduction to Linear and Nonlinear Programming*, Addison-Wesley, Reading, MA, 1984.

Mantell, J., and Lasdon, L. S., A GRG Algorithm for Econometric Control Problems, *Annals of Economic and Social Management*, Vol. 6, pp. 581–597, 1978.

Mays, L. W., and Tung, Y. K., *Hydrosystems Engineering and Management*, McGraw-Hill, Inc., New York, 1992.

McCormick, G. P., *Nonlinear Programming: Theory, Algorithms, and Applications*, John Wiley & Sons, New York, 1983.

Murtaugh, B. A., and Saunders, M. A., Large-Scale Linearly Constrained Optimization, *Mathematical Programming*, Vol. 14, pp. 41–72, 1978.

Murtaugh, B. A., and Saunders, M. A., MINOS/AUGMENTED User's Manual, *Syst. Optimiz. Lab. Tech. Rep. 80-14*, Department of Operations Research, Stanford University, Stanford, CA, 1980.

Murtaugh, B. A., and Saunder, M. A., "MINOS 5.0 User's Guide," *Syst. Optimiz. Lab. Tech. Rep. 83-20*, 118 pp., Department of Operations Research, Stanford University, Stanford, CA, 1983.

Murtagh, B. A. and Saunders, M. A., MINOS 5.1 User's Guide, Report SOL 83-20R, Stanford University, Dec. 1983, revised 1987.

Powell, M. J. D., Algorithms for Nonlinear Constraints that Use Lagrangian Functions, *Mathematical Programming*, Vol. 14, pp. 224–248, 1978.

Rockafellar, R. T., A Dual Approach to Solving Nonlinear Programming Problems by Unconstrained Optimization, *Mathematical Programming*, Vol. 5, pp. 354–373, 1973a.

Rockafellar, R. T., The Multiplier Method of Hestenes and Powell Applied to Convex Programming, *Journal of Control and Optimization*, Vol. 12, pp. 268–285, 1973b.

Rockafellar, R. T., Augmented Lagrangian Multiplier Functions and Duality in Nonconvex Programming, *Journal of Applied Mathematics*, Vol. 12, pp. 555–562, 1974.

Singhal, J., Marston, R. E., and Morin, T., Fixed Order Branch and Bound Methods for Mixed-Integer Programming: The Zoom System, Working Paper, Management Information Science Department, The University of Arizona, Tucson, 1987.

Chapter 4
Groundwater Management Systems

4.1 PROBLEM IDENTIFICATION

Aquifer simulation models have been used to examine the effects of various groundwater management strategies. Use has primarily been of the "case study" or "what-if" type. The analyst specifies certain quantities and the model predicts the technical and perhaps economic consequences of this choice. The analyst evaluates these consequences and uses his judgment and intuition to specify the next case.

Optimization methods have been used in groundwater management for more than a decade with some success. Most uses focused on explicitly combining simulation and optimization, resulting in so-called simulation-management models. Gorelick (1983) reviewed these models and classified hydraulic management models into two major approaches: embedding and use of a unit response matrix or an "algebraic technological function" (ATF). Embedding incorporates the equations of the simulation model (represented as a set of difference equations) directly into the optimization problem to be solved. This method has limited applications and is mostly used in groundwater hydraulic management, because the optimization problem quickly becomes too large to solve by available algorithms when a large-scale aquifer, especially unconfined, is considered. Previous work based on this approach includes Aguado et al. (1974), Aguado and Remson (1980), Willis and Newman (1977), Aguado et al. (1977), Remson and Gorelick (1980), and Willis and Liu (1984).

The ATF approach generates a unit response matrix by solving the simulation model several times, each with unit pumpage at a single pumping node. Superposition is used to determine the total drawdowns. This yields a smaller optimization problem, but the method has two major limitations. It is exact only for a confined aquifer but has good accuracy for an unconfined aquifer with relatively small drawdowns compared to the aquifer thickness. A drawdown correction method may be used to improve accuracy for an unconfined aquifer with larger drawdowns, but acceptable accuracy can be guaranteed. In addition, the response matrix must be recomputed when exogenous factors such as aquifer boundary conditions or potential well locations change. An alternative is to treat these factors as decision variables and constraints are included in the optimization problem. Work stemming from this approach includes that by Maddock (1972, 1974), Maddock and Haimes (1975), Morel-Seytoux (1975), Morel-Seytoux and Daly (1975), Morel-Seytoux et al. (1980), Illangasakare and Morel-Seytoux (1982), Heidari (1982), and Willis (1984).

Another approach has been to interface a simulation with an optimizer, in which case, the simulator essentially solves the simulator implicitly for the optimizer. Gorelick et al. (1984) applied this method to an aquifer reclamation design to overcome the nonlinearities incurred by the contaminant transport equations. In effect, the dynamic Jacobian matrix, required by the projected Lagrangian method in solving the optimization problem, was determined via forward or central finite differencing, with the contaminant transport simulation used to provide the function values needed in the differencing. This is closely related to the approach described here. We use an analytic rather than differencing approach for computing these same partial derivatives. The possibility of doing this is mentioned in the above reference. However, the hydraulic response was handled by the ATF method.

More recently, groundwater management problems have been looked at using optimal control approaches such as differential dynamic programming (DDP). Jones et al. (1987) began this trend of application as described in Sec. 9.4. Later, Culver and Shoemaker (1992) and others have applied DDP approaches for groundwater reclamation as described in Sec. 9.4. Also refer to Willis and Yeh (1987) and Yeh (1996) for a complete review of groundwater system methodologies.

The work described here attempts to obtain the generality of the hydraulic simulation-management model in combining simulation and optimization to solve the optimal control problem. The overall problem is viewed as one of discrete-time optimal control where variables describing the aquifer system are divided into the system state (head) and control (pumpage). By expressing the head as an implicit function of pumpage, the model constraints are conceptually eliminated, yielding a smaller reduced problem involving only the pumpage variables. Head bounds are incorporated into the objective using an augmented Lagrangian algorithm as described in Chapter 3. This requires the solution of a set of linear difference equations backward in time.

4.2 PROBLEM FORMULATION

4.2.1 Aquifer Model

For non-steady-state heterogeneous anisotropic groundwater flow in saturated media the partial differential equation governing the two-dimensional case is

$$\frac{\partial}{\partial x_i}\left(T_{ij}\frac{\partial h}{\partial x_j}\right) = S\frac{\partial h}{\partial t} + W, \quad i, j = 1, 2 \tag{4.2.1}$$

where T_{ij} is the transmissivity tensor, h is the hydraulic head, W is the volume flux per unit area, S is the storage coefficient, x_i and x_j are the Cartesian coordinates, and t is time. For numerical solution using finite difference methods, the aquifer is divided into T periods which need not be of equal length. For the finite difference grid, see Figure 4.1. The discretization used here leads to the following system of difference equations:

$$\begin{aligned} A_{ij}h_{ij-1,t} &+ B_{ij}h_{i,j+1,t} + C_{ij}h_{i-1j,t} + D_{ij}h_{i+1j,t} \\ &- (A_{ij} + B_{ij} + C_{ij} + D_{ij} + F_{ij} + R_{ij})h_{i,j,t} + F_{ij}h_{i,j,t-1} \\ &- q_{ijt} + R_{ij}\mathrm{RD}_{ij} = 0 \quad \text{for all } i, j; \text{ and } t = 1, \ldots, T \end{aligned} \tag{4.2.2}$$

In the above, h_{ijt} is the head at cell (i, j) at the end of time period t, q_{ijt} is the pumpage (if positive) or recharge (if negative), R_{ij} is the spring water constant, RD_{ij} is the minimum head for spring water to occur, F_{ij} is a coefficient which depends on storativity or specific yield, and A_{ij}, B_{ij}, C_{ij}, and D_{ij} are coefficients which depend on the transmissivity for cells adjacent to (i, j). Coefficients A, B, C, and D can be expressed in terms of aquifer permeability as follows:

$$A_{i,j} = (\mathrm{TH}_{i,j}\Delta x_{j-1} + \mathrm{TH}_{i,j-1}\Delta x_j)\frac{2\Delta y_i \mathrm{PX}_{i,j-1}}{(\Delta x_{j-1} + \Delta x_j)^2} \tag{4.2.3}$$

$$B_{i,j} = (\mathrm{TH}_{i,j}\Delta x_{j+1} + \mathrm{TH}_{i,j+1}\Delta x_j)\frac{2\Delta y_i \mathrm{PX}_{i,j}}{(\Delta x_{j+1} + \Delta x_j)^2} \tag{4.2.4}$$

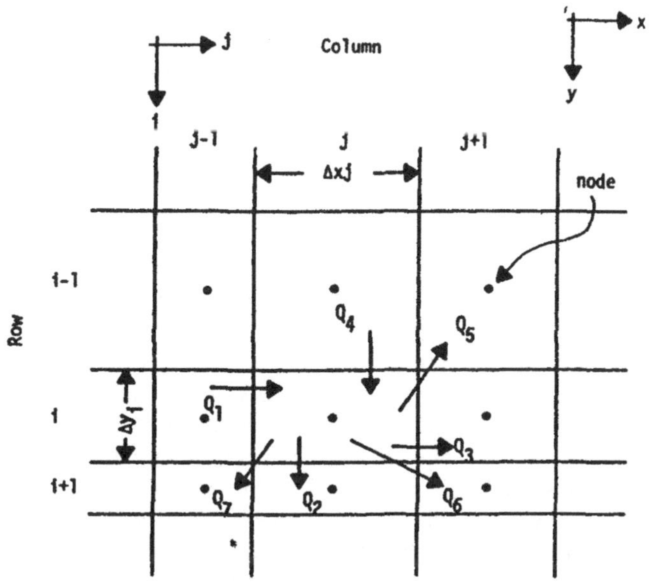

Figure 4.1 Finite difference grid (from TWDB, 1978).

$$C_{i,j} = (\mathrm{TH}_{i,j}\Delta y_{j-1} + \mathrm{TH}_{i-1,j}\Delta y_i)\frac{2\Delta x_j \mathrm{PY}_{i-1,j}}{(\Delta y_{j-1} + \Delta y_i)^2} \tag{4.2.5}$$

$$D_{i,j} = (\mathrm{TH}_{i,j}\Delta y_{j+1} + \mathrm{TH}_{i+1,j}\Delta y_i)\frac{2\Delta x_j \mathrm{PY}_{i,j}}{(\Delta y_{j+1} + \Delta y_i)^2} \tag{4.2.6}$$

where

$\mathrm{PX}_{i,j}$ = aquifer permeability between node (i,j) and $(i, j + 1)$
$\mathrm{PY}_{i,j}$ = aquifer permeability between node (i,j) and $(i + 1, j)$
$\mathrm{TH}_{i,j}$ = aquifer thickness for node (i,j) at time step t
Δx_j, Δy_i = the grid size of the cell (i, j)

Expressions (4.2.3)–(4.2.6) are valid for both artesian and water table conditions; only the thickness terms are defined differently. For a cell (i, j) with water table conditions, the thickness can be computed from

$$\mathrm{TH}_{i,j} = h_{i,j,t} - \mathrm{BOT}_{i,j} \tag{4.2.7}$$

and for artesian conditions

$$\mathrm{TH}_{i,j} = \mathrm{TOP}_{i,j} - \mathrm{BOT}_{i,j} \tag{4.2.8}$$

where $\mathrm{TOP}_{i,j}$ and $\mathrm{BOT}_{i,j}$ are the average elevations at the top and bottom of the aquifer at cell (i, j), respectively. Similarly, the coefficient F is given as

$$F = \frac{S_{i,j}\Delta x_j \Delta y_i}{\Delta t} \tag{4.2.9}$$

where $S_{i,j}$ is either the storage coefficient or the specific yield depending on the condition of the cell (i, j), and Δt is the time step increment. Under water table conditions, the thickness terms defined in Eq. (4.2.7) will cause the system of equations to be nonlinear in terms of the hydraulic head.

The alternating direction implicit (ADI) method is used to solve the system of equations (4.2.2). The method involves iteratively solving the simultaneous equations first, for a given time increment, reducing a large set of the equations down to a number of small sets. This is done by solving the node equations using Gauss elimination of an individual column of the model while all terms related to the node in adjacent columns are held constant. The set of column equations is then implicit in the direction along the column and explicit in the direction orthogonal to the column alignment. The solution of the set of column equations is then a straightforward process of back substitution.

After all column equations have been processed column by column, attention is focused on solving the node equations again by Gauss elimination of an individual row while all items related to adjacent rows are held constant. Finally, after all equations have been solved row by row, an iteration has been completed. The above process is repeated a sufficient number of times to achieve convergence, and this completes the computations for the given time step. The solution is said to converge if the differences between row and column solutions is not greater than the tolerance limit set forth. The computed heads are then used as the initial conditions for the next time step. This total process is repeated for successive time increments with unconditional stability regardless of the size of the time increment. More details on how to rearrange the variables and equations can be found in Prickett and Lonnquist (1971).

The coefficients A_{ij}, B_{ij}, C_{ij}, and D_{ij} are linear functions of the thickness of cell (i, j) and the thickness of one of the adjacent cells. For artesian conditions, this thickness is a known constant, so if cell (i, j) and its neighbors are artesian, the (i, j) equation of (4.2.2) is linear for all t. For water table conditions, the thickness of cell (i, j) is $h_{ijt} - \mathrm{BOT}_{ij}$, where BOT_{ij} is the average elevation of the bottom of the aquifer at cell (i, j). Then (4.2.2) involves products of heads and is nonlinear.

4.2.2 Constraints

Demand Schedule

It is assumed that the flow rates over specified time periods from all wells must either equal specified values or lie within a specified range. If ω is the set of all cells with pumpage, the former restriction is expressed as

$$\sum_{(i,j)\in\omega} q_{ijt} = d_t, \quad t = 1, \ldots, T \tag{4.2.10}$$

the latter one is

$$d_t \leq \sum_{(i,j)\in\omega} q_{ijt} \leq \bar{d}_t, \quad t = 1, \ldots, T \tag{4.2.11}$$

where d_t represents the lower bound on demand for time period t and d_t is the upper bound.

Flow Bounds

The flow bound constraints for recharge and pumpage have the form

$$\underline{q}_{ijt} \leq q_{ijt} \leq \bar{q}_{ijt}, \quad (i, j) \in \omega \tag{4.2.12}$$

where the barred quantities are specified limits on the pumpage or recharge. If $\underline{q}_{ijt}$ is zero, this permits no recharge, whereas a positive lower limit forces pumpage to at least to this level. The expression $\bar{q}_{ijt}$ represents pumpage capacity if positive, whereas $\bar{q}_{ijt} = 0$ and $\underline{q}_{ijt} < 0$ provide for a limited recharge capability.

Head Bounds

The head bounds are expressed as

$$\underline{h}_{ijt} \leq h_{ijt} \leq \bar{h}_{ijt}, \quad (i, j, t) \in S \tag{4.2.13}$$

where S is a subset of cells and time periods where the head is to be controlled. Examples include reducing heads below specified levels in dewatering problems, maintaining heads above certain levels at springs in aquifer management problems, or ensuring that computed heads do not exceed the ground surface for water table conditions.

Groundwater Flow Equations

The difference equations (4.2.2) relating the heads and the well flows in the aquifer are also constraints of the optimization. In the solution approach described in the next section, these equations are used to solve for the heads given the well flows, eliminating the heads and reducing the problem to one involving only the well flows as decision variables. Constraints of this reduction problem will be the demands (4.2.10) or (4.2.11), flow bounds (4.2.12), and head bounds (4.2.13).

4.2.3 Objective Function

Any continuous function of **h** and **q** can be used as an objective function. However, for demonstration purposes, two objective functions are presented here. One is to maximize the sum of the heads at all pumping nodes over all time periods, i.e.,

$$\text{Maximize sum } h = \sum_{(i,j)\in\omega} \sum_{t=1}^{T} h_{ijt} \tag{4.2.14}$$

In conjunction with the demand constraints (4.2.10) or (4.2.11) this objective meets demands while maintaining maximum aquifer potential. The second, used in dewatering problems, is to minimize the pumpage,

$$\text{Minimize sum } q = \sum_{(i,j)\in\omega} \sum_{t=1}^{T} q_{ijt} \tag{4.2.15}$$

4.3 PROBLEM SOLUTION

4.3.1 Overview

The solution methods described here were designed to work with existing aquifer simulation programs. This is a desirable feature in making maximal use of existing technology, and any improvements or changes in the simulation model are automatically incorporated into the optimization scheme.

Aquifer simulators solve for heads and perhaps pollutant concentrations, given certain controllable variables. In the simulator used here, the head is computed, given the well flows. This allows the constraint and objective functions of any aquifer model problems to be viewed as functions of only these controllable variables. As there are relatively few controllable variables, the resulting problem is easier to solve. The major remaining difficulty is to compute first partial derivatives of the objective and constraint functions with respect to the controllable variables. These derivatives can be computed in significantly less time than is required to perform a simulation. Once they are determined, several efficient nonlinear optimization routines are available to solve the problems. These ideas are general and can be applied to any aquifer, modeling both water quantity and quality.

4.3.2 The Reduced Problem

The system of nonlinear difference equations (4.2.2) can be solved for the heads h_{ijt}, given well flows q_{ijt} (and initial and boundary conditions). Let $\mathbf{q}$ be the vector of all well flows in all time periods, and define $h_{ijt}(\mathbf{q})$ as the heads which satisfy these difference equations when the well flows have the values given by $\mathbf{q}$. For purposes of illustration, let the objective function be the sum of the heads at the pumping nodes, $h$sum, given by (4.2.14). Because each head h_{ijt} is a function of $\mathbf{q}$, $h$sum is a function of $\mathbf{q}$ also, expressed as $h\text{sum}(\mathbf{q})$:

$$\overline{h\text{sum}(\mathbf{q})} = \sum_{(i,j)\in\omega} \sum_{t=1}^{T} h_{ijt}(\mathbf{q}) \tag{4.3.1}$$

Similarly, the head bounds (4.2.13) are functions of $\mathbf{q}$ also, rewritten as

$$\underline{h}_{ijt} \le h_{ijt}(\mathbf{q}) \le \bar{h}_{ijt} \quad (i, j, t) \in S \tag{4.3.2}$$

Again, for purposes of illustration, let the demand constraints be equalities as in (4.2.10). These involve only the well flows q_{ijt} and are rewritten here along with the head bounds:

$$\sum_{(i,j)=\omega} q_{ijt} = d_t, \quad t = 1, \ldots, T \tag{4.3.3}$$

$$\underline{q}_{ijt} \le q_{ijt} \le \bar{q}_{ijt}, \quad (i, j) \in \omega, t = 1, \ldots, T \tag{4.3.4}$$

The problem of maximizing $\overline{h\text{sum}}$, (4.3.1), subject to the head bounds (4.3.2), demand constraints (4.3.3), and flow bounds (4.3.4), is the reduced problem. It involves only the well flows and is much smaller than the original problem. Many head variables have been eliminated, as have the aquifer model equations (4.2.2). However, the remaining heads $h_{ijt}(\mathbf{q})$ for $\{(i, j) \in \omega\} \cup \{(i, j, t) \in S\}$ in Eqs. (4.3.1) and (4.3.2) are implicit, possibly nonlinear functions of the well flows, $\mathbf{q}$. The simulation model is used to solve for the implicit function value. Optimization methods require values of the objective and constraint functions and their derivatives with respect to each well flow variable. We will now focus on how these derivatives can be computed efficiently.

4.3.3 Computing the Reduced Gradient

Consider the computation of the gradient of the head sum, $\overline{\nabla h\text{sum}(\mathbf{q})}$, which is the reduced gradient. The function *h*sum is an implicit function of **q** through the groundwater simulation equations (4.2.2). However, the time-staged structure of these equations leads to an efficient procedure for computing $\overline{\nabla h\text{sum}}$. Procedures of this type have been used to compute the reduced gradients of objective functions defined in econometric models, which are also systems of implicit nonlinear difference equations. Details are given by Mantell and Lasdon (1978) and Norman et al. (1982). To apply these results to the problem at hand, some additional notation is needed.

Let $\mathbf{h}_t$ be the vector of all heads at time *t*, with components h_{ijt}, and write the aquifer model difference equations (4.2.2) in vector form as

$$\mathbf{g}_t(\mathbf{h}_t, \mathbf{h}_{t-1}, \mathbf{q}_t) = \mathbf{0}, \quad t = 1, \ldots, T \tag{4.3.5}$$

Also defined are the following matrices of partial derivatives of the model equations with respect to current and lagged heads:

$$\mathbf{B}_t = \frac{\partial \mathbf{g}_t}{\partial \mathbf{h}_t}, \quad t = 1, \ldots, T \tag{4.3.6}$$

$$\mathbf{C}_{t+1,t} = \frac{\partial \mathbf{g}_{t+1}}{\partial \mathbf{h}_t}, \quad t = 1, \ldots, T + 1 \tag{4.3.7}$$

Finally, let $\boldsymbol{\pi}_t$ be a row vector of Lagrange multipliers for the model equations (4.3.5). Each $\boldsymbol{\pi}_t$ has as many components as there are grid blocks in the aquifer discretization. Then the procedure for computing $\overline{\nabla h\text{sum}}$ for a given vector of well flows $\mathbf{q}^+$ is as follows:

Step 1: Solve the simulator equations (4.3.5) forward in time with $\mathbf{q}_t = \mathbf{q}_t^+$, yielding heads $\mathbf{h}_t^+$ for $t = 1, \ldots, T$.

Step 2: Solve the following system of linear difference equations backward in time for the Lagrange multiplier vectors $\boldsymbol{\pi}_t$,

$$\boldsymbol{\pi}_T \mathbf{B}_T = \frac{\partial(h\text{sum})}{\partial \mathbf{h}_T} \tag{4.3.8}$$

$$\boldsymbol{\pi}_t \mathbf{B}_t = \frac{\partial(h\text{sum})}{\partial \mathbf{h}_t} - \boldsymbol{\pi}_{t+1}\mathbf{C}_{t+1,t}, \quad t = T - 1, T - 2, \ldots, 1 \tag{4.3.9}$$

In these equations, the matrices $\mathbf{B}_t$ and $\mathbf{C}_{t+1,t}$ must be evaluated using the well flows and heads obtained in Step 1, and all vectors in Eqs. (4.3.8) and (4.3.9) are row vectors. Then, Eq. (4.3.8) is solved for $\boldsymbol{\pi}_T$ and Eq. (4.3.9) is solved sequentially for $\boldsymbol{\pi}_{T-1}$,

$\pi_{T-2}, \ldots, \pi_1$. Equations (4.3.8) and (4.3.9) are derived from the general reduced gradient equation $\pi\mathbf{B} = \partial f/\partial \mathbf{x}$ in which π is a (row) vector of Lagrange multipliers, $\mathbf{B}$ is the basis matrix, f is the objective function, and $\mathbf{x}$ is the vector of basic variables [see Ladson et al. (1978) or Luenberger (1984) for a derivation]. If all variables $\mathbf{h}_t$ are basic and all $\mathbf{q}_t$ nonbasic, the time-staged structure of Eq. (4.3.5) implies a sequential structure for these Lagrange multiplier equations as well. Refer to Appendix for computation of $\mathbf{B}$.

Step 3: Evaluate the components of $\overline{\nabla h\text{sum}}$ by

$$\frac{\partial(\overline{h\text{sum}})}{\partial q_{ijt}} = \frac{\partial(h\text{sum})}{\partial q_{ijt}} - \pi_t \frac{\partial \mathbf{g}_t}{\partial q_{ijt}} \tag{4.3.10}$$

The most time consuming part of these computations (apart from the groundwater simulation) is computing the partial derivative matrices $\mathbf{B}_t$ and $\mathbf{C}_{t+1,t}$ and solving the linear equations (4.3.8) and (4.3.9). For the difference equations (4.2.2), the structure of the matrices $\mathbf{B}_t$ and $\mathbf{C}_{t+1,t}$ is shown in Figure 4.2. $\mathbf{B}_t$ is a pentadiagonal matrix, whereas $\mathbf{C}_{t+1,t}$ is diagonal, so the right-hand side of Eq. (4.3.9) is easy to compute. $\mathbf{C}_{t+1,t}$ is constant, and $\mathbf{B}_t$ is constant if the entire aquifer is artisian. If some portion has water table conditions, some elements of $\mathbf{B}_t$ are linear functions of head. Hence, for the artesian aquifer, the reduced problem is linear, so the reduced gradient of any problem function [either $h$sum or one of the heads $h_{ijt}(\mathbf{q})$] is constant and need be computed only once. Otherwise, the above computations must be performed each time the well flows are changed.

In addition, because each well flow q_{ijt} appears in only one simulator equation [the one for block (i, j) in period t], $\partial \mathbf{g}_t/\partial q_{ijt}$ is the negative of a unit vector and $\partial(h\text{sum})/\partial q_{ijt}$ is zero, so Eq. (4.3.10) becomes

$$\frac{\partial(\overline{h\text{sum}})}{\partial q_{ijt}} = \pi_{ijt} \tag{4.3.11}$$

Summarizing , while (for water table conditions) the aquifer simulator solves a system of nonlinear difference equations forward in time, yielding the heads, the reduced gradient computation solves the linear system of difference equations (4.3.8) and (4.3.9) backward in time, yielding the Lagrange multipliers π_t. Both systems contain the same number of equations and involve matrices of the same form.

The Lagrange multipliers π_t are more than just an artifice which is useful in computing the reduced gradient. When evaluated at an optimal solution, they supply valuable sensitivity information. For the problem of maximizing $h$sum considered here, the optimal value of π_{ijt} is equal to the

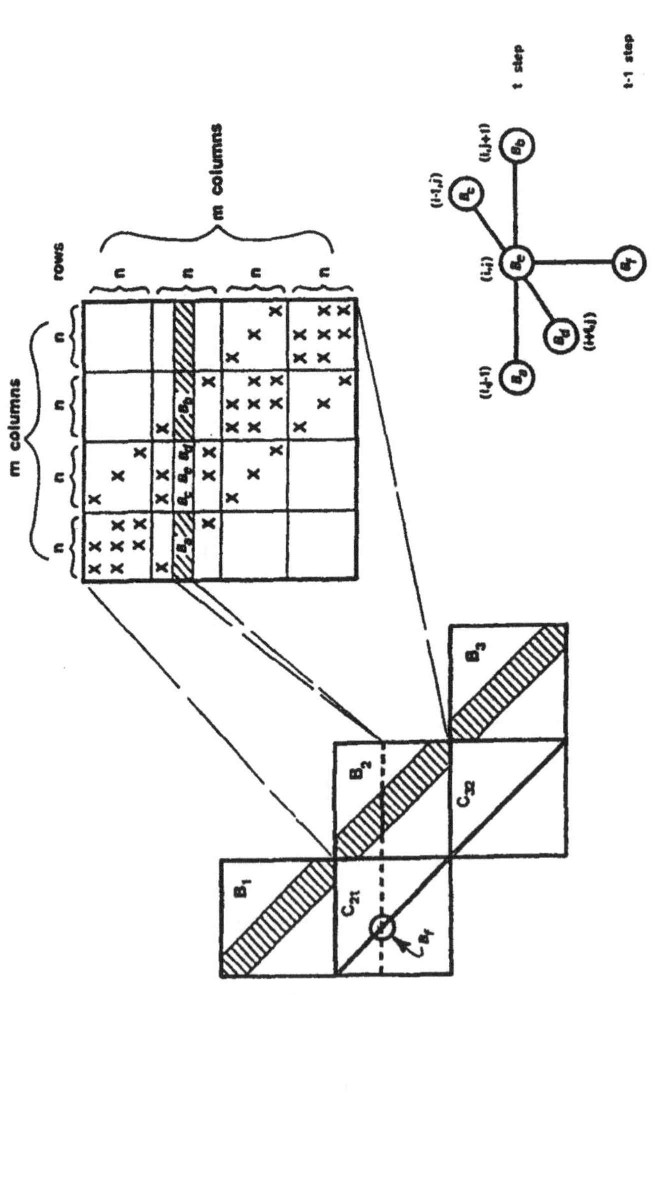

Figure 4.2 Matrix of the constraint coefficients for the $n \times m$ finite difference cells with three time steps (Wanakule, 1984).

change in the optimal $h$sum value caused by an additional thousand gallons of water flowing out of cell (i, j) in period t. This applies whether or not there is a well in cell (i, j). Hence, these multipliers could serve to show where new wells should be located, either for pumping or recharge.

4.3.4 Satisfying the Head Bounds Using an Augmented Lagrangian Function

If a portion of the aquifer has water table conditions, the head bounds (4.3.2) is nonlinear. These constraints involve the implicit functions $h_{ijt}(\mathbf{q})$. Computing the reduced gradient of each of these functions requires performing Steps 2 and 3, previously discussed. Hence, to compute reduced gradients of all head bound constraints, $N_b * T$ systems of linear equations, arising from Eqs. (4.3.8) and (4.3.9), must be solved, where N_b is the number of head bound constraints. Instead, an approach which computes only one reduced gradient, requiring the solution of only T linear systems, was chosen. This approach combines the head bounds and the objective into a penaltylike function called an augmented Lagrangian. The procedure (Chapter 3) is well established in nonlinear programming and is also described by Rockfellar (1973), and Fletcher (1975).

Let

$$c_{ijt}(\mathbf{q}) = \min\{h_{ijt}(\mathbf{q}) - \underline{h}_{ijt}, \bar{h}_{ijt} - h_{ijt}(\mathbf{q})\} \tag{4.3.12}$$

then the head bounds are equivalent to the constraint that $c_{ijt}(\mathbf{q})$ be nonnegative. The appropriate augmented Lagrangian function is

$$\begin{aligned} L(\mathbf{q}, \boldsymbol{\mu}, \sigma) = \overline{h\text{sum}}(\mathbf{q}) + \frac{1}{2}\sigma \sum_{(ijt)\in S} \left[\min\left(0, c_{ijt}(\mathbf{q}) - \frac{\mu_{ijt}}{\sigma}\right)\right]^2 \\ - \frac{1}{2} \sum_{(ijt)\in S} \frac{(\mu_{ijt})^2}{\sigma} \end{aligned} \tag{4.3.13}$$

The parameters μ_{ijt} are Lagrange multipliers for the head bounds, whereas σ is a positive penalty weight. Consider the Lagrangian problem

$$\text{Maximize } L(\mathbf{q}, \boldsymbol{\mu}, \sigma) \tag{4.3.14}$$

subject to constraints (4.3.3) and (4.3.4), where the maximization is over $\mathbf{q}$ and $\boldsymbol{\mu}$ and $\boldsymbol{\sigma}$ are fixed. If a $\boldsymbol{\sigma}$ is larger than some threshold value $\bar{\boldsymbol{\sigma}}$ and $\boldsymbol{\mu}$ is set equal to the optimal multipliers for the head bounds, $\boldsymbol{\mu}^*$, then any optimal solution for this Lagrangian problemsolves the reduced problem

(4.3.1)–(4.3.4). This suggests an algorithm (Fig. 4.2) in which the Lagrangian problem is solved, the parameters μ and σ are adjusted, convergence is checked, and the steps are repeated. The multiplier update rule used is

$$\mu_{ijt}^{+} = \mu_{ijt} - \sigma c_{ijt} \quad \text{if } c_{ijt} \leq \frac{\mu_{ijt}}{\sigma}$$

$$\mu_{ijt}^{+} = 0 \quad \text{if } c_{ijt} > \frac{\mu_{ijt}}{\sigma} \tag{4.3.15}$$

Convergence is tested by checking if the maximum violation of the head bounds is less than a user-supplied tolerance. In general, these violations will be the largest at the start and will diminish as the algorithm proceeds.

If the maximum bound violation has increased over its value at the previous iteration, σ is replaced by 10σ and μ is not updated. If the current largest bound violation is larger than one-fourth of its previous value, σ is replaced by 10σ and the μ are updated. Otherwise, σ is left at its current value when the updating rule (4.3.15) is applied. The algorithm for the methodology is outlined in Figure 4.3.

4.3.5 Solution Using the Code GRG2

The Lagrangian problem (4.3.14) has a nonlinear objective L and linear demand and flow bound constraints. The reduced gradient of L is computed using the previously discussed procedure. To solve the Lagrangian problem (4.3.14), a program called GRG2, described by Ladson et al. (1978), can be used. The algorithm used in this code is of the reduced gradient type, and such methods are particularly effective for linearly constrained problems. GRG2 uses the T demand constraints to eliminate T-dependent well flows in terms of the remaining independent ones. These independent flows are varied by the most efficient algorithm available, the Broyden–Fletcher–Goldfarb–Shanno (BFGS) quasi-Newton method (Fletcher, 1981). This algorithm uses the gradient of the augmented Lagrangian function to estimate the matrix of second partial derivatives of this function and uses this matrix to compute an efficient search direction. A one-dimensional search procedure using quadratic interpolation is used to determine the distance to move along this direction. The procedure is repeated until one of several stopping criteria, described by Ladson et al. (1978) is met. Of course, if there are logical optima distinct from the global optimum, GRG2 cannot guarantee convergence to the global optimum.

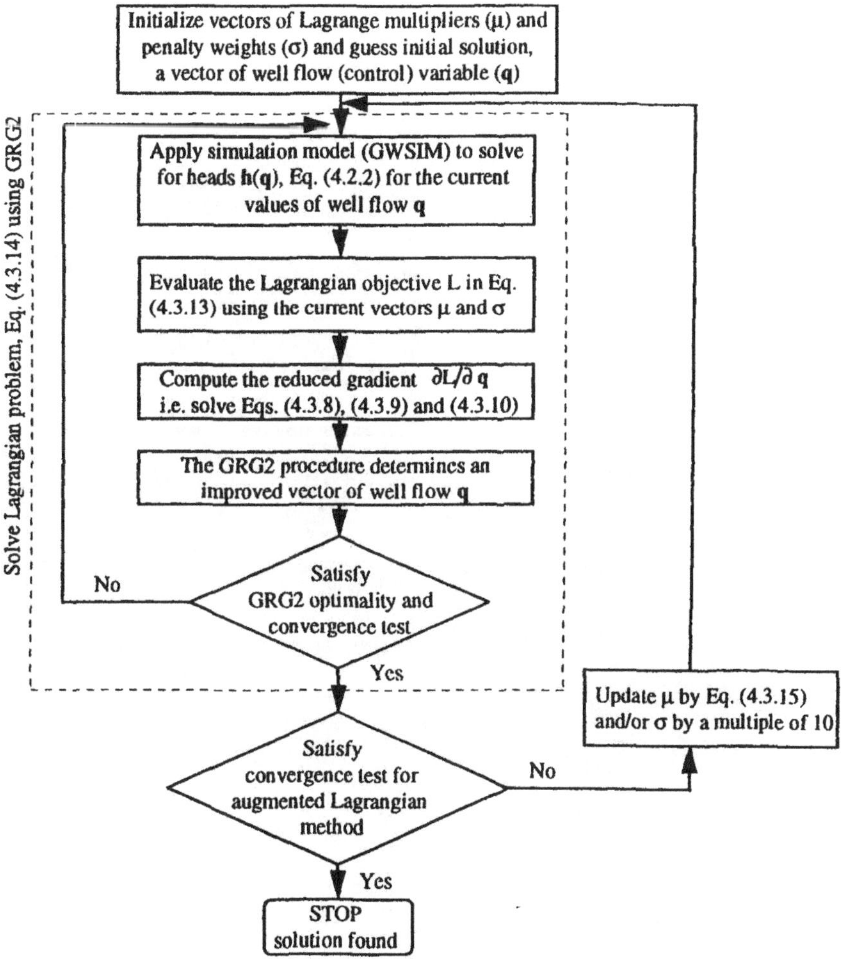

Figure 4.3 Algorithm flowchart (Wanakule, 1984).

The optimization–groundwater simulation system is referred to as GWMAN (see Figure 4.4). It contains GRG2, the generalized reduced gradient model by Ladson et al. (1978), and GWSIM, a groundwater simulation model developed by the Texas Water Development Board (1974). GWSIM is a finite difference simulation model which uses the alternating direction implicit method to solve the finite difference equations.

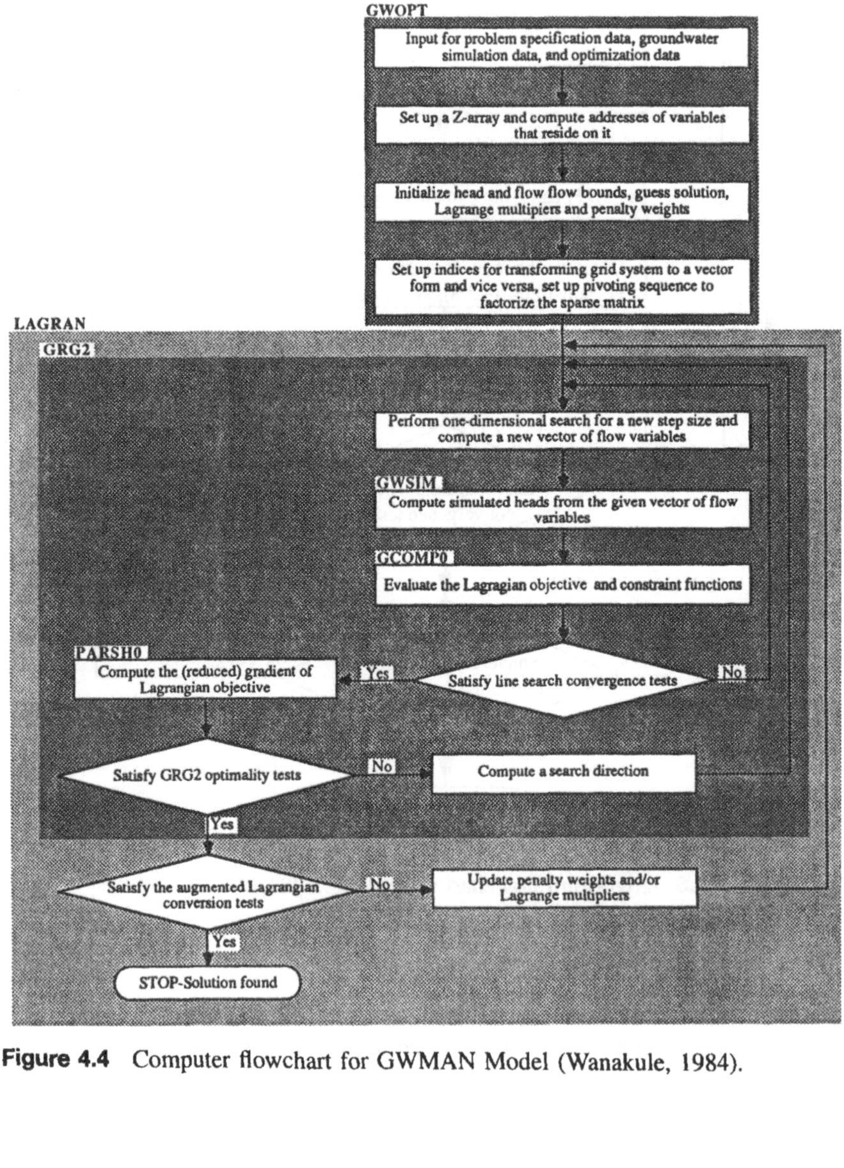

Figure 4.4 Computer flowchart for GWMAN Model (Wanakule, 1984).

4.4 APPLICATION

Wanakule (1984) and Wanakule et al. (1986) developed the model (GWMAN) for determining optimal pumping and recharge for large-scale artesian and/or nonartesian aquifers. The overall problem was viewed as one of discrete-time optimal control, where variables describing the aquifer system were divided into system state (head) and control (pumpage). By expressing head as an implicit function of pumpage, the model constraints were conceptually eliminated, yielding a smaller reduced problem involving only the pumpage variables. Head bounds were incorporated into the objective using an augmented Lagrangian algorithm. The major contribution of their work was an analytic scheme to compute the reduced gradient needed for optimization. This requires the solution of a set of linear difference equations backward in time and has major speed and accuracy advantages over finite differencing.

The following paragraphs describe four groundwater management problems used for comparison.

Problem 1 has a grid system configured as shown in Fig. 4.5a where the grid size is 0.2 miles each side. The bottom elevation is at 150 ft, whereas the thickness in the water table and artesian portions are 100 ft and 50 ft, respectively. Other physical properties are K_x = 600 gal/day/ft^2, K_y = 300 gal/day/ft^2, S_y = 0.1, and S = 0.001. The problem requires maximizing the sum of the heads at the pumping nodes over a period of 5 years, a surrogate objective function for minimizing pumping cost, subject to 2000 acre-ft/yr demand constraints, lower head bounds at 200 ft, and a lower flow bound of 200 acre-ft/yr. The problem has 5, 1-year time intervals with four pumping nodes for each interval.

Problem 2 is the steady-state dewatering example taken from Aguado et al. (1977). The grid system (Fig. 4.5b) consists of 121 rectangular cells of 40 m $\times$ 10 m in size. The problem is to determine minimum total pumpage that will maintain the water level in a rectangular evacuation area located in the center of the homogeneous isotropic unconfined aquifer at 21 m. The bottom of the aquifer elevation is at 0 m and surrounding constant head elevation is at 36 m. The value of hydraulic conductivity is 10.81 m/day.

Problem 3 is a hypothetical example of a hydrocarbon recovery site where the strategy is to create a containment depression near the center of the hydrocarbon plume. The problem is to determine the optimal water pumpage so that the hydrocarbon plume which is floating on the water layer will be confined to the containment area. The finite difference scheme, set up as shown in Figure 4.5c, has a total of 1089 active cells

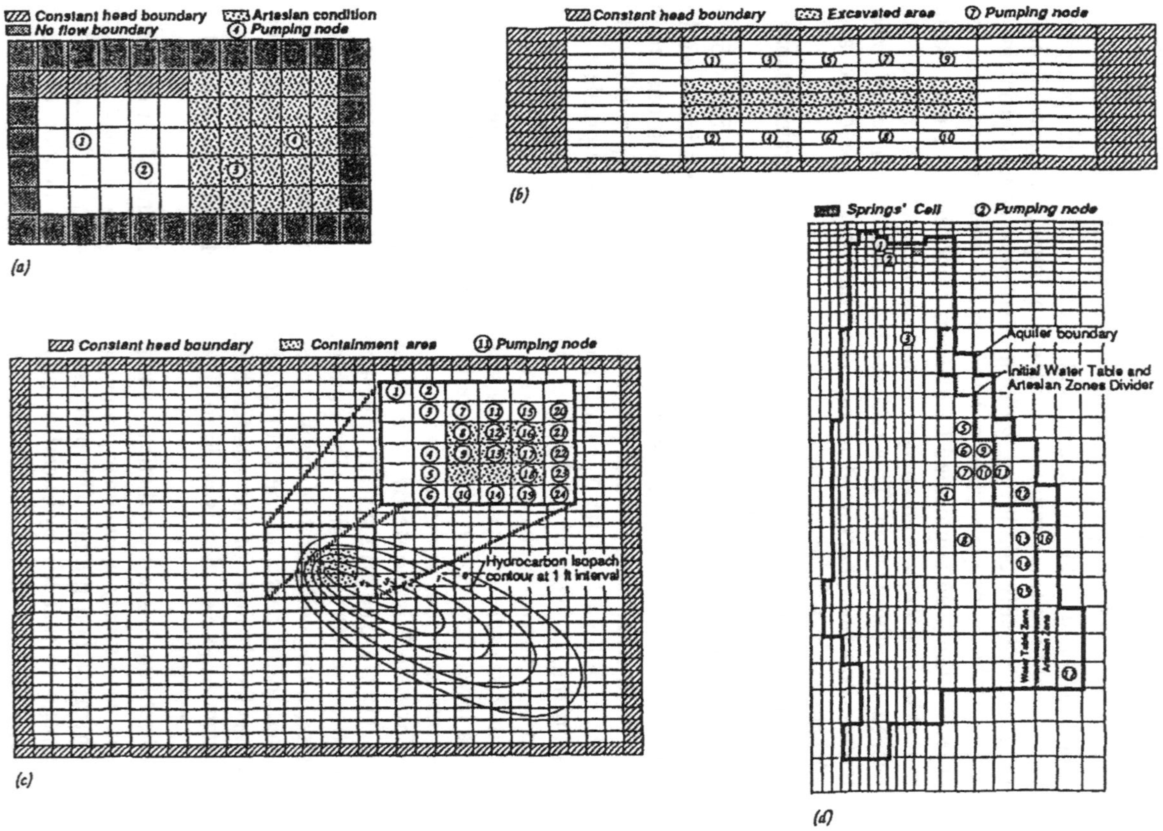

Figure 4.5 Finite difference cells and pumping locations for (a) Problem 1, (b) Problem 2, (c) Problem 3, and (d) Problem 4 (Wanakule, 1984).

whose dimensions are 180 ft × 120 ft. The aquifer is isotropic nonhomogeneous with an average hydraulic conductivity of about 100 gal/day/ft^2.

Problem 4 is a field application to the Barton Springs–Edwards aquifer in Austin, Texas. It is a limestone aquifer where its main recharge openings were created by steep-angle normal faulting across the stream beds. The problem is set up to determine the optimal yields under long-term average recharge conditions subject to maintaining the spring flows at 25 cfs (0.708 m^3/sec). The finite difference grid system contains 330 active cells whose dimensions are varied from 0.379 × 0.283 mi^2 to 0.95 × 1.51 mi^2 (Fig. 4.5d). The total aquifer area includes approximately 150 mi^2. The hydraulic conductivity values vary greatly from 0.1–2.0 ft/day in the outcrop area to 50–1150 ft/day in the eastern side of the aquifer or the confined zone where the main underground flow channels are located. The groundwater flow generally is to the east in the outcrop area and then bends to the north toward Barton Springs.

Table 4.1 compares the results between a VAX and a Mac II. The execution time on the Mac II is about seven times slower, whereas the objective values at optimum obtained from the VAX, in almost all cases, are better than those from the Mac II. The improvement of objective values on the Mac II can be achieved by tightening the convergence limit on the optimizer and/or adjusting the magnitude of penalty weights and the initial estimates of Lagrangian multipliers. This, of course, will increase the number of simulation calls and execution time.

The results clearly indicate the potential for implementing GWMAN on microcomputers. Even though it is slow in execution, the advantages in accessibility and low-cost computing time can compensate for slowness in most medium-sized problems.

APPENDIX: COMPUTATION OF BASIS MATRIX ELEMENTS

This appendix presents the equations for computing elements of the basis matrix and a portion of the Jacobian matrix. Elements of the matrix are the partial derivatives of the groundwater flow system of equations (4.2.2) with respect to the state variable **h**. Each element is evaluated at the current point where **h** and **q** are known. The equations are divided into five groups depending on the aquifer conditions of a cell under consideration. Investigation of Eqs. (4.2.2) reveals that each row of the matrix should contain at most six elements. Represent cells $(i, j-1)$, $(i, j+1)$, $(i-1, j)$, $(i+1, j)$, and (i, j) at time step t by small letters a, b, c, d, e, respectively, and (i, j) at time $t+1$ by f. The configuration of the finite difference scheme

Table 4.1 Comparison of Four Problem Results between the VAX and Mac II Computers

		Problem 1		Problem 2		Problem 3		Problem 4	
		VAX	Mac II	VAX	Mac II	VAX	Mac II	VAX	Mac II
No. of simulation calls		23	41	207	174	243	140	60	44
Exec. time (minutes)		2.13	18.72	3.45	28.52	224.60	1217.28	2.48	17.08
Objective values		4,490	4,490	105,769	105,765	171.99	172.57	35.826	42.940
Pumpage values		acre-ft/yr		m^3/day		cu ft/day		acre-ft/yr	
Pumping node No.	1	1,400	1,400	13,947.0	13,950.0	0.000	0.000	1.673	1.390
	2	200	200	13,658.0	13,655.0	0.000	0.000	1.677	2.688
	3	200	200	8,246.7	8,235.2	0.000	0.000	1.720	2.957
	4	200	200	8,327.9	8,335.3	0.000	0.000	2.134	2.607
	5	1,400	1,400	8,711.3	8,712.1	0.000	0.000	1.759	2.650
	6	200	200	8,698.3	8,701.5	0.000	0.000	1.703	2.468
	7	200	200	8,286.2	8,287.0	10.029	0.000	1.746	2.368
	8	200	200	8,284.7	8,284.8	16.141	15.781	1.791	1.933
	9	1,400	1,400	13,805.0	13,802.0	16.141	16.141	1.723	2.454
	10	200	200	13,804.0	13,802.0	2.523	4.675	1.726	2.316
	11	200	200			0.000	5.898	2.560	3.104
	12	200	200			16.141	16.141	1.788	2.176
	13	1,400	1,400			16.141	16.141	2.067	2.195
	14	200	200			16.141	16.141	3.396	3.384
	15	200	200			1.970	12.576	3.396	3.266
	16	200	200			16.141	16.141	2.047	2.177
	17	1,400	1,400			16.141	16.141	2.920	2.809
	18	200	200			16.141	16.141		
	19	200	200			16.141	16.141		
	20	200	200			0.000	0.000		
	21					0.000	0.000		
	22					12.203	3.381		
	23					0.000	1.131		
	24					0.000	0.000		

for the system of Eqs. (4.2.2) can be viewed as shown in Figure 4.6. The figure also shows the positions of the elements in the structured matrix.

Case I

When all six nodes are under water table conditions, all terms in Eqs. (4.2.2) are nonlinear. The elements of the matrix in a row can be computed from the following expressions:

$$B_a = [(h_e - \text{BOT}_e)\Delta x_a + (2h_a - \text{BOT}_a)\Delta x_e - h_e\Delta x_e] \frac{2\Delta y_e \text{PX}_a}{(\Delta x_a + \Delta x_e)^2} \tag{4.A.1}$$

$$B_b = [(h_e - \text{BOT}_e)\Delta x_b + (2h_b - \text{BOT}_b)\Delta x_e - h_e\Delta x_e] \frac{2\Delta y_e \text{PX}_e}{(\Delta x_b + \Delta x_e)^2} \tag{4.A.2}$$

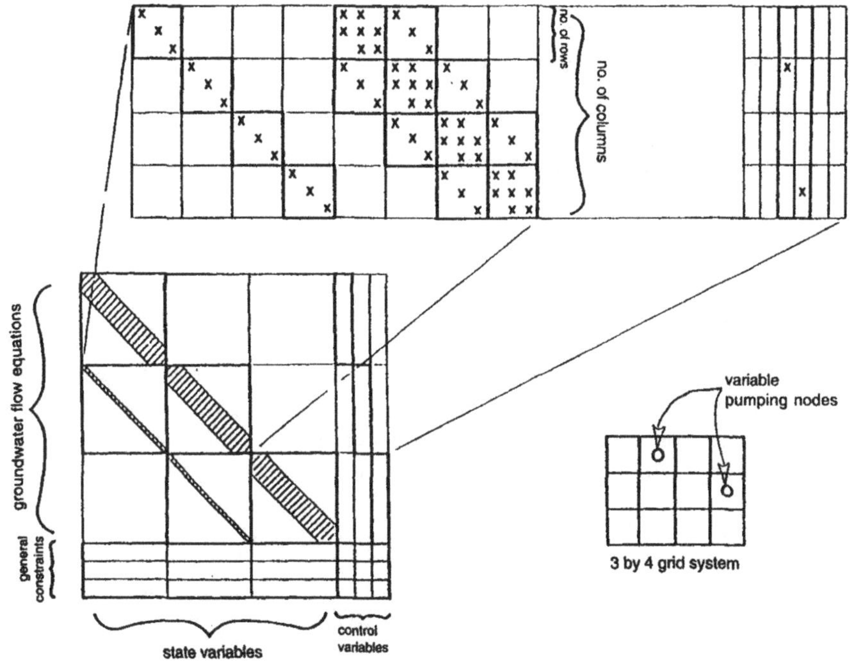

Figure 4.6 Basis matrix structure for groundwater flow system of equations for three rows by four columns and three time steps (Wanakule, 1984).

$$B_c = [(h_e - \text{BOT}_e)\Delta y_c + (2h_c - \text{BOT}_c)\Delta y_e - h_e\Delta y_e] \frac{2\Delta x_e \text{PY}_c}{(\Delta y_c + \Delta y_e)^2} \tag{4.A.3}$$

$$B_d = [h_e - \text{BOT}_e)\Delta y_d + (2h_d - \text{BOT}_d)\Delta y_e - h_e\Delta y_e] \frac{2\Delta x_e \text{PY}_c}{(\Delta y_d + \Delta y_e)^2} \tag{4.A.4}$$

$$B_e(a) = [h_a\Delta x_a - (2h_e - \text{BOT}_e)\Delta x_a + (h_a - \text{BOT}_a)\Delta x_e] \frac{2\Delta y_e \text{PX}_a}{(\Delta x_a + \Delta x_e)^2} \tag{4.A.5}$$

$$B_e(b) = [h_b\Delta x_b - (2h_e - \text{BOT}_e)\Delta x_b + (h_b - \text{BOT}_b)\Delta x_e] \frac{2\Delta y_e \text{PX}_e}{(\Delta x_b + \Delta x_e)^2} \tag{4.A.6}$$

$$B_e(c) = [h_c\Delta y_c - (2h_e - \text{BOT}_e)\Delta y_c - (h_b - \text{BOT}_b)\Delta y_e] \frac{2\Delta x_e \text{PY}_c}{(\Delta y_c + \Delta y_e)^2} \tag{4.A.7}$$

$$B_e(d) = [h_d\Delta y_d - (2h_e - \text{BOT}_e)\Delta y_d - (h_b - \text{BOT}_b)\Delta y_e] \frac{2\Delta x_e \text{PY}_e}{(\Delta y_d + \Delta y_e)^2} \tag{4.A.8}$$

$$B_e = B_e(a) + B_e(b) + B_e(c) + B_e(d) - B_f - R_e \tag{4.A.9}$$

$$B_f = \frac{S_e\Delta x_e\Delta y_e}{\Delta t} \tag{4.A.10}$$

Case II

When the center node e is under water table conditions but some of the neighboring nodes are artesian, Eqs. (4.2.2) are partially nonlinear. The equations are the same as in Case I, except that the elements corresponding to the artesian nodes are replaced by the following expressions:

$$B_a = [h_e - \text{BOT}_e)\Delta x_a + (\text{TOP}_a - \text{BOT}_a)\Delta x_e] \frac{2\Delta y_e \text{PX}_a}{(\Delta x_a + \Delta x_e)^2} \tag{4.A.11}$$

$$B_b = [(h_e - \text{BOT}_e)\Delta x_b + (\text{TOP}_b - \text{BOT}_b)\Delta x_e] \frac{2\Delta y_e \text{PX}_e}{(\Delta x_b + \Delta x_e)^2} \tag{4.A.12}$$

$$B_c = [(h_e - \text{BOT}_e)\Delta y_c + (\text{TOP}_c - \text{BOT}_c)\Delta y_e] \frac{2\Delta x_e \text{PY}_c}{(\Delta y_c + \Delta y_e)^2} \tag{4.A.13}$$

$$B_d = [(h_e - \mathrm{BOT}_e)\Delta y_d + (\mathrm{TOP}_d - \mathrm{BOT}_d)\Delta y_e] \frac{2\Delta x_e \mathrm{PY}_c}{(\Delta y_d + \Delta y_e)^2} \quad (4.A.14)$$

$$B_e(a) = [h_a\Delta x_a - (2h_e - \mathrm{BOT}_e)\Delta x_a - (\mathrm{TOP}_a - \mathrm{BOT}_a)\Delta x_e] \frac{2\Delta y_e \mathrm{PX}_a}{(\Delta x_a + \Delta x_e)^2} \quad (4.A.15)$$

$$B_e(b) = [h_b\Delta x_b - (2h_e) - \mathrm{BOT}_e)\Delta x_b - (\mathrm{TOP}_b - \mathrm{BOT}_b)\Delta x_e] \frac{2\Delta y_e \mathrm{PX}_e}{(\Delta x_b + \Delta x_e)^2} \quad (4.A.16)$$

$$B_e(c) = [h_c\Delta y_c - (2h_e - \mathrm{BOT}_e)\Delta y_c - (\mathrm{TOP}_b - \mathrm{BOT}_b)\Delta y_e] \frac{2\Delta x_e \mathrm{PY}_c}{(\Delta y_c + \Delta y_e)^2} \quad (4.A.17)$$

$$B_e(d) = [h_d\Delta y_d - (2h_e - \mathrm{BOT}_e)\Delta y_d - (\mathrm{TOP}_b - \mathrm{BOT}_b)\Delta y_e] \frac{2\Delta x_e \mathrm{PY}_e}{(\Delta y_d + \Delta y_e)^2} \quad (4.A.18)$$

Case III

When all six nodes are artesian, Eqs. (4.2.2) are linear and their elements in a row can be computed as follows:

$$B_a = [(\mathrm{TOP}_e - \mathrm{BOT}_e)\Delta x_a + (\mathrm{TOP}_a - \mathrm{BOT}_a)\Delta x_e] \frac{2\Delta y_e \mathrm{PX}_a}{(\Delta x_a + \Delta x_e)^2} = -B_e(a) \quad (4.A.19)$$

$$B_b = [(\mathrm{TOP}_e - \mathrm{BOT}_e)\Delta x_b + (\mathrm{TOP}_b - \mathrm{BOT}_b)\Delta x_e] \frac{2\Delta y_e \mathrm{PX}_e}{(\Delta x_b + \Delta x_e)^2} = -B_e(b) \quad (4.A.20)$$

$$B_c = [(\mathrm{TOP}_e - \mathrm{BOT}_e)\Delta y_c - (\mathrm{TOP}_c - \mathrm{BOT}_c)\Delta y_e] \frac{2\Delta x_e \mathrm{PY}_c}{(\Delta y_c + \Delta y_e)^2} = -B_e(c) \quad (4.A.21)$$

$$B_d = [(\mathrm{TOP}_e - \mathrm{BOT}_e)\Delta y_d - (\mathrm{TOP}_d - \mathrm{BOT}_d)\Delta y_e] \frac{2\Delta x_e \mathrm{PY}_e}{(\Delta y_d + \Delta y_e)^2} = -B_e(d) \quad (4.A.22)$$

$$B_e = B_e(a) + B_e(b) + B_e(c) + B_e(d) - B_f - R_e \quad (4.A.23)$$

$$B_f = \frac{S_e \Delta x_e \Delta y_e}{\Delta t} \tag{4.A.24}$$

Case IV

When the center node e is artesian but some neighboring nodes are under water table conditions, the nonlinear terms appear corresponding to the water table nodes. The computation is the same as Case III except the elements at the water table nodes are substituted by the following:

$$B_a = [(\text{TOP}_e - \text{BOT}_e)\Delta x_a + (2h_a - \text{BOT}_a)\Delta x_e - h_e \Delta x_e] \frac{2\Delta y_e \text{PX}_a}{(\Delta x_a + \Delta x_e)^2} \tag{4.A.25}$$

$$B_b = [(\text{TOP}_e - \text{BOT}_e)\Delta x_b + (2h_b - \text{BOT}_b)\Delta x_e - h_e \Delta x_e] \frac{2\Delta y_e \text{PX}_e}{(\Delta x_b + \Delta x_e)^2} \tag{4.A.26}$$

$$B_c = [(\text{TOP}_e - \text{BOT}_e)\Delta y_c + (2h_c - \text{BOT}_c)\Delta y_e - h_e \Delta y_e] \frac{2\Delta x_e \text{PY}_c}{(\Delta y_c + \Delta y_e)^2} \tag{4.A.27}$$

$$B_d = [(\text{TOP}_e - \text{BOT}_e)\Delta y_d + (2h_d - \text{BOT}_d)\Delta y_e - h_e \Delta y_e] \frac{2\Delta x_e \text{PY}_e}{(\Delta y_d + \Delta y_e)^2} \tag{4.A.28}$$

$$B_e(a) = -[(\text{TOP}_e - \text{BOT}_e)\Delta x_a + (h_a - \text{BOT}_a)\Delta x_e] \frac{2\Delta y_e \text{PX}_a}{(\Delta x_a + \Delta x_e)^2} \tag{4.A.29}$$

$$B_e(b) = -[(\text{TOP}_e - \text{BOT}_e)\Delta x_b + (h_b - \text{BOT}_b)\Delta x_e] \frac{2\Delta y_e \text{PX}_e}{(\Delta x_b + \Delta x_e)^2} \tag{4.A.30}$$

$$B_e(c) = -[(\text{TOP}_e - \text{BOT}_e)\Delta y_c + (h_c - \text{BOT}_c)\Delta y_e] \frac{2\Delta x_e \text{PY}_c}{(\Delta y_c + \Delta y_e)^2} \tag{4.A.31}$$

$$B_e(d) = -[(\text{TOP}_e - \text{BOT}_e)\Delta y_d + (h_d - \text{BOT}_d)\Delta y_e] \frac{2\Delta x_e \text{PY}_e}{(\Delta y_d + \Delta y_e)^2} \tag{4.A.32}$$

Case V

When the middle node e is a constant head cell, the matrix elements in the corresponding row are constant as follows:

$$B_a = B_b = B_c = B_d = B_f = 0; \qquad B_e = 1 \tag{4.A.33}$$

This is a consequence of taking derivatives of the equation

$$h_e - H = 0 \tag{4.A.34}$$

in place of the groundwater flow equations (4.2.1), where H is the constant head elevation. It follows that the elements B_a, B_b, B_c, and B_d are also zeros in all four cases above if they represent a constant head node.

REFERENCES

Aguado, E. and Remson, I., Groundwater Management with Fixed Charges, *Journal of Water Resources Planning and Management*, Vol. 106, No. 2, pp. 375–382, 1980.

Aguado, E., Remson, I., Pikul, M. F., and Thomas, W. A., Optimum Pumping for Aquifer Dewatering, *Journal of Hydraulics*, Vol. 100, No. 7, pp. 860–877, 1974.

Aguado, E., Sitar, N., and Remson, I., Sensitivity Analysis in Aquifer Studies, *Water Resources Research*, Vol. 13, No. 4, pp. 733–737, 1977.

Culver, T. and Shoemaker, C., Dynamic Optimal Control for Groundwater Remediation with Flexible Management Periods, *Water Resources Research*, Vol. 28, No. 3, pp. 629–641, 1992.

Fletcher, R., An Ideal Penalty Function for Constrained Optimization, *Journal of Mathematics and its Applications*, Vol. 15, pp. 319–342, 1975.

Fletcher, R., *Practical Methods of Optimization*, Vol. 1, John Wiley & Sons, New York, 1981.

Gorelick, S. M., A Review of Distributed Parameter Groundwater Management Modeling Methods, *Water Resources Research*, Vol. 19, No. 2, pp. 305–319, 1983.

Gorelick, S. M., Voss, C. I., Gill, P. E., Murray, W., Saunders, M. A., and Wright, M. M., Aquifer Reclamation Design: The Use of Contaminant Transport Simulation Combined with Nonlinear Programming, *Water Resources Research*, Vol. 20, No. 4, pp. 415–427, 1984.

Guyton, W. F. and Associates, Geohydrology of Comal, San Marcos, and Hueco Springs, *Report 234*, Texas Department of Water Resources, June 1979.

Heidari, M., Application of Linear Systems Theory and Linear Programming to Groundwater Management in Kansas, *Water Resources Bulletin*, Vol. 18, pp. 1003–1012, 1982.

Illangasekare, T. H., and Morel-Seytoux, H. J., Stream Aquifer Influence Coefficients for Simulation and Management, *Water Resources Research*, Vol. 18, No. 1, pp. 168–176, 1982.

Jones, L. C., Willis, R., and Yeh, W. W., Optimal Control of Nonlinear Groundwater Hydraulics Using Differential Dynamic Programming, *Water Resources Research*, Vol. 23, No. 11, pp. 2097–2217, 1987.

Klemt, W. B., Knowles, T. R., Elder, G. R., and Sieh, T. W., Groundwater Resources and Model Application for the Edwards (Balcones Fault Zone) Aquifer in the San Antonio Region, Texas, *Report 239*, Texas Department of Water Resources, October 1979.

Knowles, T., *GWSIM III—Groundwater Simulation Program, Program Document and User's Manual*, Texas Department of Water Resources, Austin, 1981.

Lasdon, L. S., Warren, A. D., Jain, A., and Ratner, M., Design and Testing of a Generalized Reduced Gradient Code for Nonlinear Programming, *ACM Transactions on Mathematical Software*, Vol. 4, pp. 34–50, 1978.

Luenberger, D. G., *Introduction to Linear and Nonlinear Programming*, Addison-Wesley, Reading, Mass., 1984.

Maddock, T., III, Algebraic Technological Function for a Simulation Model, *Water Resources Research*, Vol. 8, No. 1, 129–134, 1972.

Maddock, T., III, Nonlinear Technological Functions for Aquifers Whose Transmissivities Vary with Drawdown, *Water Resources Research*, Vol. 10, No. 4, pp. 877–881, 1974.

Maddock, T., III and Haimes, Y. Y., A Tax System for Groundwater Management, *Water Resources Research*, Vol. 11, No. 1, 7–14, 1975.

Mantell, J. and Lasdon, L. S., A GRG Algorithm for Econometric Control Problems, *Annals of Economic and Social Management*, Vol. 6, pp. 581–597, 1978.

Morel-Seytoux, H. J., A Simple Case of Conjunctive Surface-Ground Water Management, *Ground Water*, Vol. 13, pp. 505–515, 1975.

Morel-Seytoux, H. J., and Daly, C. J., A Discrete Kernel Generator for Stream-Aquifer Studies, *Water Resources Research*, Vol. 11, pp. 253–260, 1975.

Morel-Seytoux, H. J., Peters, G., Young, R., and Illangasekare, T., Groundwater Modeling for Management, International Symposium on Water Resource Systems, Water Resources Development and Training Center, University of Roorkee, Roorkee, India, 1980.

Norman, A. L., Lasdon, L. S., and Hsin, J. K., A Comparison of Methods for Solving and Optimizing a Large Nonlinear Econometric Model, Discussion paper, Center for Economic Research, University of Texas, Austin, 1982.

Prickett, T. A. and Lonnquist, C. G., Selected Digital Computer Techniques for Groundwater Resource Evaluation, *Bulletin 55*, Illinois State Water Survey, Urbana, Ill., 1971.

Remson, I. and Gorelick, S. M., Management Models Incorporating Groundwater Variables, in *Operation Research in Agriculture and Water Resources*, D. Yaron, and C. S. Tapiero (eds.), North-Holland, Amsterdam, 1980.

Rockafellar, R. T., A Dual Approach to Solving Nonlinear Programming Problems by Unconstrained Optimization, *Mathematical Programming*, Vol. 5, pp. 354–373, 1973.

Texas Water Development Board (TWDB), GWSIM—Groundwater Simulation Program, Program Document and User's Manual, *UM S7405,* Austin, Texas, 1974.

Wanakule, N., A Model for Determining Optimal Pumping and Recharge of Large-Scale Aquifers, Ph.D. dissertation, University of Texas, Austin, 1984.

Wanakule, N., Mays, W. L., and Lasdon, L. S., Optimal Management of Large-Scale Aquifer: Methodology and Application, *Water Resources Research,* Vol. 22, No. 4, pp. 447–466, 1986

Willis, R., A Unified Approach to Regional Groundwater Management, in *Groundwater Hydraulics,* J. S. Rosenshein and G. D. Bennett (eds.), American Geophysical Union, Washington, DC, 1984.

Willis, R. and Liu, P., Optimization Model for Ground-Water Planning, *Journal of Water Resources Planning and Management,* Vol. 110, No. 3, pp. 333–347, 1984.

Willis, R. and Newman, B. A., Management Model for Groundwater Development, *Journal of Water Resources Planning and Management,* Vol. 103, No. 1, pp. 159–171, 1977.

Willis, R., and Yeh, W. W-G., Groundwater Systems Planning and Management, Prentice-Hall, Englewood Cliffs, N.J., 1987.

Yeh, W-G., Groundwater Systems, in *Water Resources Handbook,* L. W. Mays (ed.), McGraw-Hill, New York, 1996.

Chapter 5
Real-Time Operation of River-Reservoir Systems

5.1 PROBLEM IDENTIFICATION

Real-time operation of multireservoir systems involves various hydrologic, hydraulic, operational, technical, and institutional considerations. For efficient operation, a monitoring system is essential that provides the reservoir operator with the flows and water levels at various points in the river system including upstream extremities, tributaries, and major creeks as well as reservoir levels, and precipitation data for the watersheds whose outputs (runoff from rainfall) are not gauged. A flow routing procedure is needed to predict the impacts of observed and/or predicted inflow hydrographs on the downstream parts of the river system. A reservoir operation policy or a methodology is another component which reflects the flood control objectives of the system, the operational and institutional constraints on flood operations, and other system-related considerations. An integral part of these components is the reservoir operation model that predicts the results of a given operation policy for forecasted flood hydrographs.

Flood forecasting, in general, and real-time flood forecasting, in particular, have always been an important problem in operational hydrology, especially when the operation of storage reservoirs is involved. The forecasting problem, as in most hydrological problems, can be viewed as a system with inputs and outputs. The system output is related to its causative input through a process, either linear or nonlinear. In the reservoir management problem, the system is the river system that includes the main river and its tributaries, catchments, and natural and man-made structures on the path of the flood waters. The system inputs are inflow hydrographs at the upstream ends of the river system, and runoff from the rainfall (and snowmelt, where applicable) in the intervening catchments. The system outputs are flow rates and/or water levels at control points of the river system. The operations involved are the operations of the reservoir(s) in order to control flood waters. The term "forecasting" refers to the prediction of the discharges and water surface elevations at various points of a river system as a result of the observed portion of the flood hydrograph.

Multireservoir operation can be characterized by the integrated operation of multiple facilities on river systems for multiple objectives. Flood control is one of the major purposes of many reservoirs in the United States. Many reservoirs were built several years ago and operation policies were established. However, many of these reservoirs cannot be operated in the manner that they were initially intended to be operated. One of the major reasons is the uncontrolled urbanization into the floodplains of the rivers and reservoirs. Other reasons are due to inadequate spillways for passing floods, legal constraints, and reduced downstream conveyance capacities.

Many of the reservoir systems are characterized by conditions that result in significant backwater conditions due to gate operation, tributary flows, hurricane surge flows, conditions, and flow constrictions in the rivers. These conditions cannot be described by the use of hydrologic routing methods and, as a result, must be described by more accurate hydraulic routing models such as DWOPER (Fread, 1982), which is based on a finite difference solution of the Saint-Venant equations. Also, flows through reservoirs having considerable length are not properly predicted by the simple hydrologic methods, particularly when the inflow hydrograph is a flash flood, that is, has a short time base.

There have been many reservoir operation models reported in the literature but only a few have been directed at reservoir operations under flooding conditions. Jamieson and Wilkinson (1972) developed a dynamic programming (DP) model for flood control with forecasted inflows being the inputs to the model. Windsor (1973) employed a recursive linear programming procedure for the operation of flood control systems, using the Muskingum method for channel routing and the mass balance equation for reservoir computations.

The U.S. Army Corps of Engineers (1973a, 1973b, 1979) developed HEC-5 and HEC-5C for reservoir operation for flood control, where releases are selected by applying a fixed set of heuristic rules and priorities that are patterned after typical operation studies. These models are based on hydrologic routing techniques and provide no optimal strategy for operation. One application of these models was to the Kanawha River Basin (U.S. Army Corps of Engineers, 1983) which contributes flow to the Ohio River at Pt. Pleasant, West Virginia. Figure 5.1 illustrates observed and

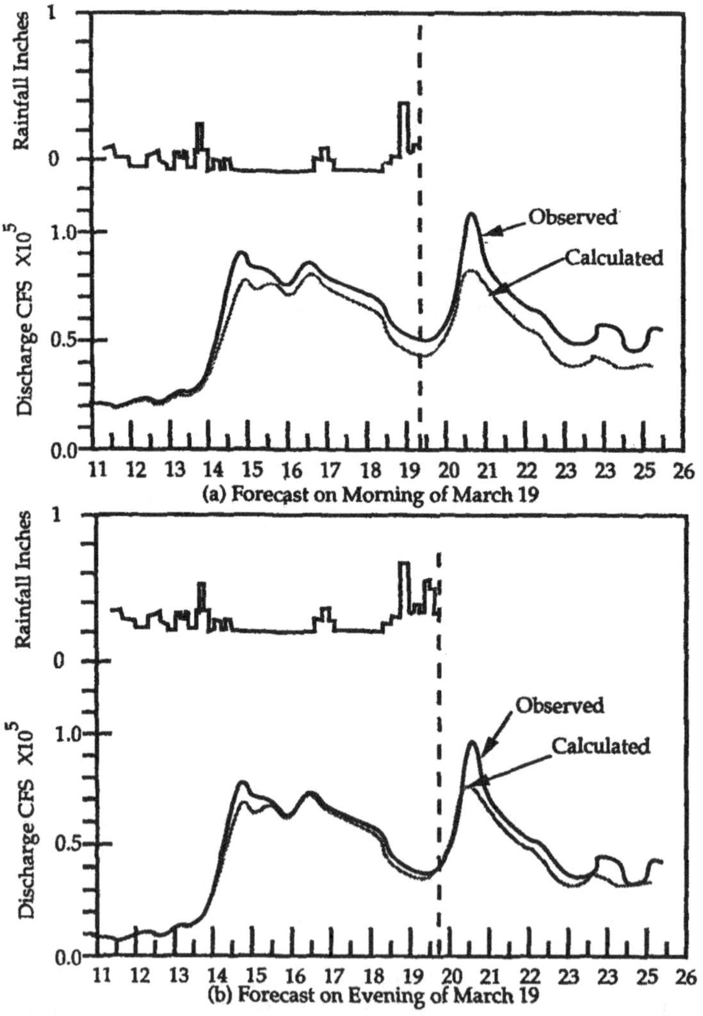

Figure 5.1 Observed and forecasted hydrographs at Kanawha Falls, resulting forecast of the March 1967 Flood Event (from U.S. Army Corps of Engineers, 1983)

forecasted hydrographs at Kanawha Falls for the March 1967 event. The vertical dashed line represents the time of the forecast.

The Tennessee Valley Authority (1974) developed an incremental dynamic programming and successive approximations technique for real-time operations, with flood control and hydropower generation being the objectives. Can and Houck (1984) developed a goal programming model for the hourly operations of a multireservoir system and applied it to the Green River basin in Indiana. The model objective is defined by a hierarchy of goals, with the best policy being a predetermined rule curve.

Wasimi and Kitanidis (1983) developed an optimization model for the daily operations of a multireservoir system during floods which combines linear quadratic Gaussian optimization and a state space mathematical model for flow forecasting. Yazicigil (1980) developed a linear programming (LP) optimization model for the daily real-time operations of the Green River basin in Indiana, a system of four multipurpose reservoirs. The model inputs are deterministic. The objective of operation is to follow a set of target states, deviations from which are penalized. The channel routing is performed using a linear routing procedure similar to the Muskingum method, called multi-input linear routing. The reservoir calculations are based on mass-balance equations which take into account precipitation input.

The flood forecasting model for the Lower Colorado River–Highland Lakes system in Texas developed by Unver et al. (1987) was developed for a real-time framework to make decisions on reservoir operations during flooding. This model is an integrated computer program with components for flood routing, rainfall-runoff modeling, and graphical display, and is controlled by interactive software. Input to the model includes automated real-time precipitation and stream flow data from various locations in the watershed.

The real-time reservoir operation problem involves the operation of a reservoir system by making decisions on reservoir releases as information becomes available, with relatively short time intervals which may vary between several minutes and several hours. A new methodology is presented for operating the reservoir system under flooding conditions that incorporates (a) a simulation model that adequately simulates the hydraulics of the system for a given flood hydrograph and a set of operating decisions and (b) a systematic way that will improve the trial decisions made previously and generate a set of operating decisions that would cause the least damage to the protected areas.

5.2 PROBLEM FORMULATION

The optimization problem for the operation of multireservoir systems under flooding conditions can be stated as

1. Objective:

$$\text{Minimize } z = f(\mathbf{h}, \mathbf{Q}) \tag{5.2.1}$$

2. Constraints:

   (a) Hydraulic constraints defined by the Saint-Venant equations for one-dimensional gradually varied unsteady flow and other relationships such as upstream, downstream, and internal boundary conditions and initial conditions that describe the flow in the different components of the river-reservoir system,

$$\mathbf{G}(\mathbf{h}, \mathbf{Q}, \mathbf{r}) = \mathbf{0} \tag{5.2.2}$$

   (b) Bounds on discharges defined by minimum and maximum allowable reservoir releases and flow rates at specific locations,

$$\underline{\mathbf{Q}} \leq \mathbf{Q} \leq \overline{\mathbf{Q}} \tag{5.2.3}$$

   (c) Bounds on elevations defined by minimum and maximum allowable water surface elevations at specified locations (including reservoir levels),

$$\underline{\mathbf{h}} \leq \mathbf{h} \leq \overline{\mathbf{h}} \tag{5.2.4}$$

   (d) Physical and operational bounds on gate operations,

$$\mathbf{0} \leq \underline{\mathbf{r}} \leq \mathbf{r} \leq \overline{\mathbf{r}} \leq \mathbf{1} \tag{5.2.5}$$

   (e) Other constraints such as operating rules, target storages, storage capacities, and so forth.

$$\mathbf{W}(\mathbf{r}) \leq \mathbf{0} \tag{5.2.6}$$

The objective z is defined by minimizing the total flood damage or deviations from the target levels or water surface elevations in flood areas or spills from reservoirs or maximizing storage in reservoirs. The variables $\mathbf{h}$ and $\mathbf{Q}$ are respectively the water surface elevations and the discharge at the computational points, and $\mathbf{r}$ is the gate setting, all given in matrix form to consider the time and space dimensions of the problem. Bars above and below a variable denote the upper and lower bounds for that variable, respectively.

5.2.1 Simulator Equations

The governing equations for one-dimensional unsteady flow are the Saint-Venant equations defined in conservation form are as follows:

Continuity:

$$\frac{\partial Q}{\partial x} + \frac{\partial (A + A_o)}{\partial t} - q = 0 \tag{5.2.7}$$

Momentum:

$$\frac{\partial Q}{\partial t} + \frac{\partial (\beta Q^2/A)}{\partial x} + gA\left(\frac{\partial h}{\partial x} + S_f + S_e\right) - \beta q v_x + W_f B = 0 \tag{5.2.8}$$

where

x = longitudinal distance along the channel or river
t = time
A = cross-sectional area of flow
A_o = cross-sectional area of off-channel dead storage (contributes to continuity, but not momentum)
q = lateral inflow per unit length along the channel
h = water surface elevation
v_x = velocity of lateral flow in the direction of channel flow
S_f = friction slope
S_e = eddy loss slope
B = width of the channel at the water surface
W_f = wind shear force
β = momentum correction factor
g = acceleration due to gravity

Weighted four-point finite difference approximations are used for dynamic routing with the Saint-Venant equations. The spatial derivatives $\partial Q/\partial x$ and $\partial h/\partial x$ are estimated between adjacent time lines,

$$\frac{\partial Q}{\partial x} = \theta \frac{Q_{i+1}^{j+1} - Q_i^{j+1}}{\Delta x_i} + (1 - \theta)\frac{Q_{i+1}^j - Q_i^j}{\Delta x_i} \tag{5.2.9}$$

$$\frac{\partial h}{\partial x} = \theta \frac{h_{i+1}^{j+1} - h_i^{j+1}}{\Delta x_i} + (1 - \theta)\frac{h_{i+1}^j - h_i^j}{\Delta x_i} \tag{5.2.10}$$

and the time derivatives are estimated using

$$\frac{\partial (A + A_o)}{\partial t} = \frac{(A + A_o)_i^{j+1} + (A + A_o)_{i+1}^{j+1} - (A + A_o)_i^j - (A + A_o)_{i+1}^j}{2\Delta t_j} \tag{5.2.11}$$

$$\frac{\partial Q}{\partial t} = \frac{Q_i^{j+1} + Q_{i+1}^{j+1} - Q_i^j - Q_{i+1}^j}{2\Delta t_j} \tag{5.2.12}$$

The nonderivative terms, such as q and A, are estimated between adjacent time lines using

$$\begin{aligned} q &= \theta \frac{q_i^{j+1} + q_{i+1}^{j+1}}{2} + (1 - \theta) \frac{q_i^j + q_{i+1}^j}{2} \\ &= \theta \bar{q}_i^{j+1} + (1 - \theta)\bar{q}_i^j \end{aligned} \tag{5.2.13}$$

$$\begin{aligned} A &= \theta \frac{A_i^{j+1} + A_{i+1}^{j+1}}{2} + (1 - \theta) \frac{A_i^j + A_{i+1}^j}{2} \\ &= \theta \bar{A}_i^{j+1} + (1 - \theta)\bar{A}_i^j \end{aligned} \tag{5.2.14}$$

where $\bar{q}_i$ and $\bar{A}_i$ indicate the lateral flow and cross-sectional area averaged over the reach Δx_i, respectively.

The finite difference form of the continuity equation is produced by substituting Eqs. (5.2.9), (5.2.11), and (5.2.13) into Eq. (5.2.7) and rearranging to obtain

$$\begin{aligned} &\theta(Q_{i+1}^{j+1} - Q_i^{j+1} - \bar{q}_i^{j+1}\,\Delta x_i) + (1 - \theta)(Q_{i+1}^j - Q_i^j - \bar{q}_i^j \Delta x_i) \\ &+ \frac{\Delta x_i}{2\Delta t_j}\,[(A + A_o)_i^{j+1} + (A + A_o)_{i+1}^{j+1} \\ &- (A + A_o)_i^j - (A + A_o)_{i+1}^j] = 0 \end{aligned} \tag{5.2.15}$$

Similarly, the momentum equation in finite difference form is

$$\begin{aligned} &\frac{\Delta x_i}{2\Delta t_j}(Q_i^{j+1} + Q_{i+1}^{j+1} - Q_i^j - Q_{i+1}^j) \\ &+ \theta \left\{ \left(\frac{\beta Q^2}{A}\right)_{i+1}^{j+1} - \left(\frac{\beta Q^2}{A}\right)_i^{j+1} \right. \\ &+ g\bar{A}_i^{j+1}[h_{i+1}^{j+1} - h_i^{j+1} + (\bar{S}_f)_i^{j+1}\,\Delta x_i + (\bar{S}_e)_i^{j+1}\,\Delta x_i] \\ &\left. - (\overline{\beta q v_x})_i^{j+1}\,\Delta x_i + (\overline{W_f B})_i^{j+1}\,\Delta x_i \right\} \\ &+ (1 - \theta) \left\{ \left(\frac{\beta Q^2}{A}\right)_{i+1}^{j} - \left(\frac{\beta Q^2}{A}\right)_i^{j} \right. \\ &+ g\bar{A}_i^j[h_{i+1}^j - h_i^j + (\bar{S}_f)_i^j\,\Delta x_i + (\bar{S}_e)_i^j\,\Delta x_i] \\ &\left. - (\overline{\beta q v_x})_i^j\,\Delta x_i + (\overline{W_f B})_i^j\,\Delta x_i \right\} = 0 \end{aligned} \tag{5.2.16}$$

where the average values (marked with an overbar) over a reach are defined as

$$\overline{\beta}_i = \frac{\beta_i + \beta_{i+1}}{2} \tag{5.2.17}$$

$$\overline{A}_i = \frac{A_i + A_{i+1}}{2} \tag{5.2.18}$$

$$\overline{B}_i = \frac{B_i + B_{i+1}}{2} \tag{5.2.19}$$

$$\overline{Q}_i = \frac{Q_i + Q_{i+1}}{2} \tag{5.2.20}$$

Also,

$$\overline{R}_i = \frac{\overline{A}_i}{\overline{B}_i} \tag{5.2.21}$$

for use in Manning's equation. Manning's equation may be solved for S_f and written in the form shown below, where the term $|Q|Q$ has magnitude Q^2 and sign positive or negative depending on whether the flow is downstream or upstream, respectively:

$$(\overline{S}_f)_i = \frac{\overline{n}_i^2|\overline{Q}_i|\overline{Q}_i}{2.208\overline{A}_i^2\overline{R}_i^{4/3}} \tag{5.2.22}$$

The minor head losses arising from contraction and expansion of the channel are proportional to the difference between the squares of the downstream and upstream velocities, with a contraction/expansion loss coefficient K_e:

$$(\overline{S}_e)_i = \frac{(K_e)_i}{2g\Delta x_i}\left[\left(\frac{Q}{A}\right)_{i+1}^2 - \left(\frac{Q}{A}\right)_i^2\right] \tag{5.2.23}$$

The velocity of the wind relative to the water surface, V_r, is defined by

$$(\overline{V}_r)_i = \left(\frac{\overline{Q}_i}{\overline{A}_i}\right) - (\overline{V}_w)_i \cos \omega \tag{5.2.24}$$

where ω is the angle between the wind and the water directions. The wind shear factor is then given by

$$(\overline{W}_f)_i = (C_w)_i|(\overline{V}_r)_i|(\overline{V}_r)_i \tag{5.2.25}$$

where C_w is the friction drag coefficient.

The terms having superscript j in Eqs. (5.2.15) and (5.2.16) are known either from initial conditions or from a solution of the Saint-Venant equations for a previous time line. The terms g, Δx_i, β_i, K_e, C_w, and V_w are known and must be specified independently of the solution. The unknown terms are Q_i^{j+1}, Q_{i+1}^{j+1}, h_i^{j+1}, h_{i+1}^{j+1}, A_i^{j+1}, A_{i+1}^{j+1}, B_i^{j+1}, and B_{i+1}^{j+1}. However, all the terms can be expressed as functions of the unknowns, Q_i^{j+1}, Q_{i+1}^{j+1}, h_i^{j+1}, and h_{i+1}^{j+1}, so there are actually four unknowns. The unknowns are raised to powers other than unity, so Eqs. (5.2.15) and (5.2.16) are nonlinear equations.

The continuity and momentum equations are considered at each of the $N - 1$ rectangular grids shown in Figure 5.2, between the upstream boundary at $i = 1$ and the downstream boundary at $i = N$. This yields $2N - 2$ equations. There are two unknowns at each of the N grid points (Q and h), so there are $2N$ unknowns in all. The two additional equations required to complete the solution are supplied by the upstream and downstream boundary conditions. The upstream boundary condition is usually specified as a known inflow hydrograph, whereas the downstream boundary condi-

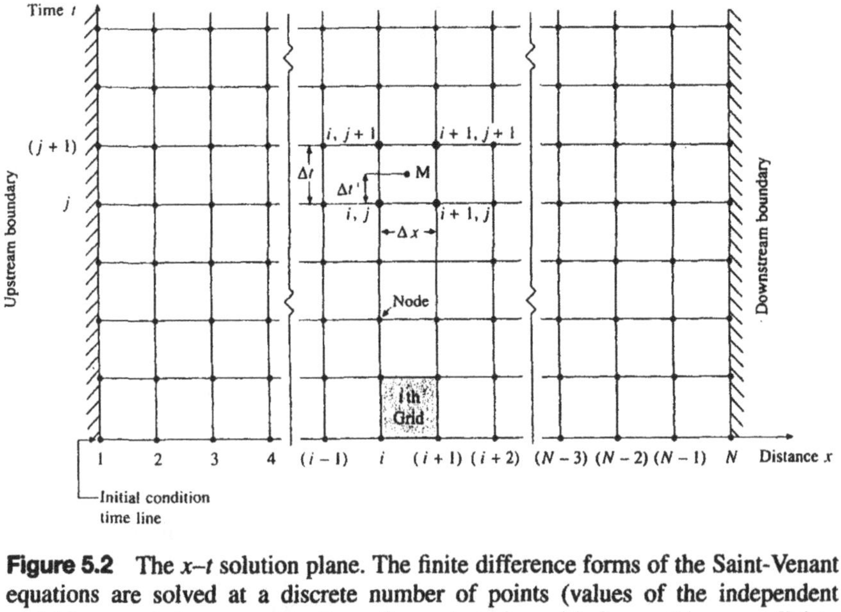

Figure 5.2 The x–t solution plane. The finite difference forms of the Saint-Venant equations are solved at a discrete number of points (values of the independent variables x and t) arranged to form the rectangular grid shown. Lines parallel to the time axis represent locations along the channel, and those parallel to the distance axis represent times (Fread, 1974).

tion can be specified as a known stage hydrograph, a known discharge hydrograph, or a known relationship between stage and discharge, such as a rating curve.

5.2.2 Constraints

The constraints of the model can be divided into two groups: the hydraulic constraints and the operational constraints. The hydraulic constraints are equality constraints consisting of the equations that describe the flow in the system. These are (a) the Saint-Venant equations for all the computational reaches except internal boundary reaches, (b) the relationship to describe the upstream and downstream boundary conditions in addition to the Saint-Venant equations for the extremities, and (c) the internal boundary conditions, including the continuity equation and a flow relationship.

Internal boundary conditions cannot be described by the Saint-Venant equations such as critical flow resulting from flow over a spillway or waterfall. The operational constraints are basically greater-than or less-than type constraints that define variable bounds, operational targets, structural limitations, capacities, and so forth. Options for the operator to set or limit the values of certain variables are also classified under this category. The solution methodology used in this study separately solves the hydraulic and operational constraints. The hydraulic constraints are solved implicitly by the simulation model, DWOPER, whereas the operational constraints are solved by the optimization model, GRG2. The DWOPER model performs the unsteady flow computations.

Bound constraints are used to impose operational or optimization-related requirements. Non-negativity constraints on discharges are not used because discharges are allowed to take on negative values in order to be able to realistically represent the reverse-flow phenomena (backwater effects) due to a rising lake or due to large tributary inflows into a lake. Non-negativity of water surface elevations is always satisfied because the system hydraulics are solved implicitly by the simulation model, DWOPER. The lower limits on elevations and discharges can be used to impose water quality considerations, minimum required reservoir releases, and other policy requirements. The upper bounds on elevations and discharges can be used to set the maximum allowable levels (values beyond are either catastrophic or physically impossible) such as the overtopping elevations for major structures, spillway capacities, and so forth. When the objective function, Eq. (5.2.23) or (5.2.24), is used, the damaging elevations and/or discharges must be given to the model through the constraints, as neither objective function has any terms to control them.

The third model variable, gate openings, are allowed to vary between 0 and 1, which corresponds to 0% and 100% opening of the total available gate area, respectively. The upper and lower bounds on the model variables are expressed mathematically as

$$\underline{Q}_i^j \leq Q_i^j \leq \overline{Q}_i^j, \quad \forall\, i, j \tag{5.2.26}$$

$$\underline{h}_i^j \leq h_i^j \leq \overline{h}_i^j, \quad \forall\, i, j \tag{5.2.27}$$

$$0 \leq \underline{r}_i^j \leq r_i^j \leq \overline{r}_i^j \leq 0.1, \quad \forall\, i, j, i \in l_r \tag{5.2.28}$$

where variables with a bar above them denote upper limits, those with a bar below them denote lower limits; i and j are respectively the time and location index, and l_r is the set containing the reservoir locations. Q, h, and r denote the discharge, water surface elevation, and gate opening, respectively.

The bounds on gate settings are intended primarily to reflect the physical limitations on gate operations as well as to enable the operator to prescribe any portion(s) of the operation for any reservoir(s). Operational constraints other than bounds can be imposed for various purposes. The maximum allowable rates of change of gate openings, for instance, for a given reservoir, can be specified through this formulation, as a time-dependent constraint. This particular formulation may be very useful, especially for cases where sharp changes in gate operations (i.e., sudden openings and closures), are not desirable or physically impossible. It is handled by setting an upper bound to the change in the percentage of gate opening from one time step to the next. This constraint can also be used to model another important aspect to gate operations for very short time intervals (i.e., the gradual settings that have to be followed when opening or closing a gate). For this case, the gate cannot be opened (or closed) by more than a certain percentage during a given time interval. This can be expressed in mathematical terms as follows:

$$-r_{ci}^j \leq r_i^{j+1} - r_i^j \leq r_{oi}, \quad i \in l_r \tag{5.2.29}$$

where r_o and r_c are the maximum allowable (or possible) percentages by which to open and close the gate. This constraint can be used to model manually operated gates, for example, for all or a portion of the time intervals. The same constraint can be used, for example, to incorporate an operational rule that ties the operations of a reservoir to those of the upstream reservoir such as a multisite constraint.

5.2.3 Objective Functions

The model can be based on any of a number of objective functions reflecting various approaches to real-time reservoir operation for flood control.

The first objective function is based on minimizing total flood damages which are defined as a function of water surface elevations in flood-prone areas. A damage–elevation relationship is provided to the model for each location where flood damage potential exists. The overall damage to be minimized is the summation of the total damages at each location. The mathematical expression for this objective function is

$$\min z = \sum_i \sum_j c_i h_i^j, \quad i \in I_c, j \in T \tag{5.2.30}$$

where z is the objective function value, i is the location index, I_c is the set that contains flood control locations, j is the time index, T is the time horizon, and c is the unit flood damage defined as a function of the water surface elevation, h_i^j. The unit flood damage, c, is expressed in terms of the water surface elevation at flood control locations. It must be noted that, unlike the more common approach to damage functions (e.g., Windsor, 1973), the damage is not a function of the maximum water surface elevation for any given location but rather a function of all elevations that are individually damaging. This approach was chosen to keep all water surface elevations in the nondamaging range individually and when this is not possible, to minimize the number of times a damaging elevation occurs. The total damage cost, however, may not have a real meaning in dollar value due to the nature of this formulation.

The second objective function is basically the same as the first one except that flood damages are expressed in terms of discharges instead of water surface elevations, given as

$$\min z = \sum_i \sum_j c_i' Q_i^j, \quad i \in I_p, j \in T \tag{5.2.31}$$

where c' is the unit flood damage as a function of discharge, Q_j. The unit flood damages, c', are expressed in terms of the discharge at the flood control locations. This objective function is provided for cases where it is more convenient to express damages in terms of flow rates for certain locations, or the available data is in this form. It must be noted that this objective function would normally be used for natural channels as the damages in lakes are almost always a function of flood stages.

The third objective function is a combination of the first two for cases where both discharges and water surface elevations are used to define the flood damages given as

$$\min z = \sum_{i \in I_c} \sum_j c_i h_i^j + \sum_{i \in I_p} \sum_j c_i' Q_i^j, \quad j \in T \tag{5.2.32}$$

where l_c is the set that contains locations where damage is a function of water surface elevation and l_c is the set that contains locations where damage is a function of discharge. The myopic nature of short-term operation is usually handled by constraints that represent the end-of-the-period, or medium-term targets or goals. For example, the possibility of ending up with an empty reservoir is usually prevented by defining a lower limit for the water surface elevation of the headwater location for time step T. An alternative to this is given by the fourth objective function. The objective of operation is defined as the maximization of the total reservoir storages while keeping the water stages and/or flow rates within nondamaging ranges through the constraint set. The fourth objective function is

$$\max z = \sum_i \sum_j Q_i^j, \quad j \in T \tag{5.2.33}$$

where all terms are as defined earlier.

Zoning is another very common approach used in modeling the real-time operation objectives (e.g., Yazicigil, 1982; Can and Houck, 1984; Wasimi and Kitanidis, 1983). In order to use this approach, operation targets (or ideal levels) are defined prior to operation and deviations from these are penalized through a penalty function. Zones are identified for different levels (deviations) and a unit penalty (or a penalty coefficient) is assigned to each, almost always in such a way that the resulting function is convex. Although the solution methodology presented in the next section has provisions for violated bounds on discharges and water surface elevations, a penalty-type objective function is presented here as the sixth objective function, for cases where data are already available or the reservoir operator opts to use a penalty function. The mathematical expression for the sixth objective is

$$\min z = \sum_i \sum_j c_i h_i^j + \sum_i \sum_j c_i' Q_i^j, \quad i \in l_s, j \in T \tag{5.2.34}$$

where l_s is the set that contains locations for which a target is specified and c and c' are the unit penalties associated with water surface elevation, h, and discharge, Q. It must be noted that water surface elevations in this formulation replace the deviations used in most penalty functions. However, this is justified by the fact that the inclusion of the target into the objective function contributes a constant to the objective value, which does not affect the optimization, within the given range of unit penalties. Different unit penalties for different locations are used to reflect the relative importance of each location.

5.3 PROBLEM SOLUTION

5.3.1 Overview

The optimization problem stated above is a large mathematical programming problem for most real-world situations. In modeling a river system, computational points are used to discretize the river channels and reservoirs. Each computational point, for each time step of the operation, contributes two flow variables (water surface elevation and discharge) and two hydraulic constraints (the Saint-Venant equations or other flow relations) to the problem. In addition, each reservoir contributes another variable (the setting of the equivalent gate) per time step. The external boundaries each contribute an additional hydraulic relationship. Thus, a typical 24-h-operation horizon with 1-hr time steps for a river system with 5 reservoirs and 150 computational points would give rise to a problem with more than 7200 flow equations (twice the product of the number of time steps and computational nodes) and over 7200 flow variables. This is beyond the capacity of existing nonlinear programming codes. A logical approach in solving a problem this large would be to reduce its size. Traditionally, the problem size has been reduced by replacing the unsteady flow equations by more simplistic relationships. In this work, a different approach is taken to alleviate the dimensionality problem. The optimum control model presented here leads to an efficient algorithm to solve the optimization problem without sacrificing the hydraulic model accuracy.

The basic idea is to solve the hydraulic constraints (Saint-Venant equations) using an unsteady flow routing model such as the U.S. National Weather Service Dynamic Wave Operational (DWOPER) model. For each iteration of the optimization model, the simulator (DWOPER) solves for the water surface elevations, h, and the flow rates, Q, given the gate operations which are the control variables. This allows the constraints and the objective function of the reservoir optimization problem to be viewed as a function of only the controllable variables. As there are relatively few controllable variables, the resulting reduced problem is easier to solve. The major remaining difficulty is to compute the first partial derivatives of the objective and constraint functions with respect to the controllable variables. Once the derivatives are determined, several efficient nonlinear optimization routines could be used to solve the reduced optimization problem.

5.3.2 The Reduced Problem

The operations problem [Eqs. (5.2.1)–(5.2.6)], referred to as the general operations model (GOM) has certain characteristics that can be used in reducing it to a smaller problem. The GOM has the general structure of a discrete-time control problem with three basic groups of constraints: hydraulic constraints, those concerning the state of the system; and bound and operational constraints, those describing the system controls. The GOM yields an efficient solution algorithm when the state variables (discharges and water surface elevations) and the control variables (gate settings) are treated separately, in a coordinated manner. The hydraulic constraints [Eq. (5.2.2)] can be solved sequentially forward in time for water surface elevations, **h**, and the flow rates, **Q**, by using the DWOPER simulation model, once the gate settings, **r**, are specified. The general optimal control approach to the real-time reservoir operation problem is shown in Figure 5.3. Through this simulator-optimizer formulation, the problem is solved efficiently by incorporating the simulation model into a procedure when a set of gate operations, **r** (control vector), is chosen; the simulation model is run subject to the selected control vector, to solve the hydraulic constraint set, **g**, for the elevations and discharges (state vector). Then the objective function is evaluated, the bound constraints are checked for any violations, and the procedure is repeated with an updated set of gate op-

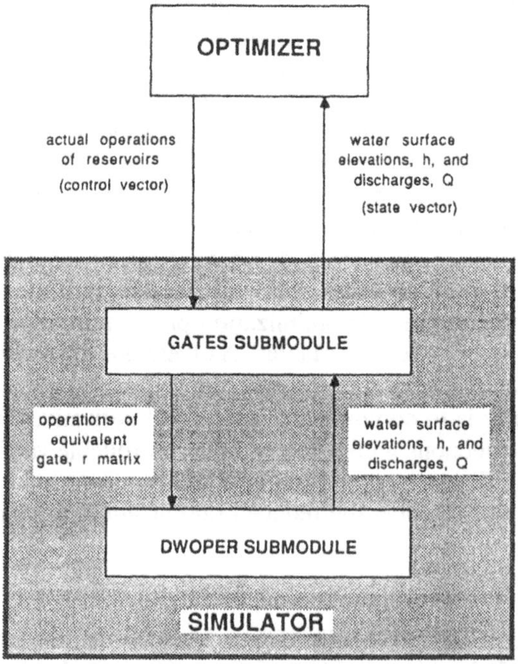

Figure 5.3 Optimal control approach to operations problem (Unver, 1987).

erations until a convergence criterion is satisfied and no bound constraints are violated.

It must be noted that the optimization is performed only on the gate settings in this procedure. The new optimization problem, called the reduced operations model (ROM), has $N_r * T$ variables compared to the $(2N * T + N_r * T)$ variables of the GOM, where N, T, and N_r are the total number of computational points, time steps, and reservoirs, respectively. The number of constraint equations has also been reduced by the same amount $(2N * T)$ with the elimination of the hydraulic constraints, **g**. The transformation of the operations problem is shown in Figure 5.4, along with the problem size at each step of the transformation for an example

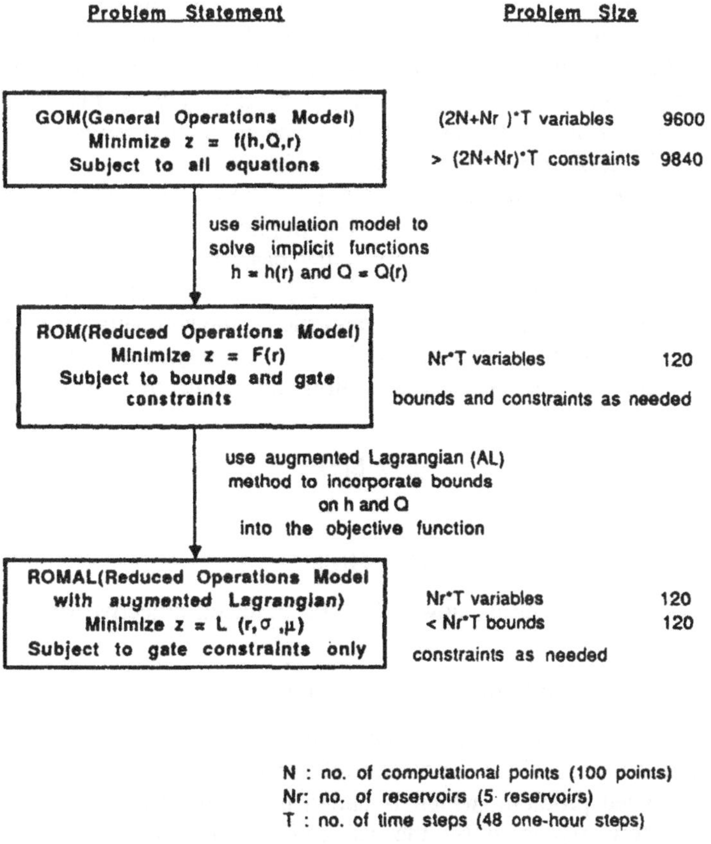

Figure 5.4 Transformation of operations problem (Unver, 1987).

system. The problem size for an example with 100 computational points, 5 reservoirs, and 48 time steps is drastically reduced, from over 9000 variables and constraints to 120 variables and 120 bound constraints because of the simulator-optimizer formulation.

The hydraulic constraint set, **g**, has a special staircase banded structure that can be exploited to construct an efficient overall algorithm. The model presented herein combines the simulation model, DWOPER, and the optimization model, GRG2, within the framework of an optimum control formulation. The transformation of the original problem into the reduced one is similar to the generalized reduced gradient approach, which is also used to solve the reduced (transformed) problem.

The original problem, GOM, can be converted into a reduced problem as suggested by the implicit function theorem (Luenberger, 1984). The implicit function theorem states that if some of the problem variables can be solved in terms of the remaining variables, then a reduced problem can be devised which can be manipulated more easily. The approach is applied to the problem given by Eqs. (5.2.1)–(5.2.6) in such a way that the hydraulic constraints [Eq. (5.2.2)] are handled separately by the simulator and the other constraints by the optimizer. The simulation model computes the values of the state variables, **h** and **Q**, for given values of the control variables **r** and the optimization model seeks the optimal values of **r** that will minimize the objective function. The implicit function theorem states that **h**(**r**) and **Q**(**r**) exist if and only if the basic matrix [the Jacobian of the system of equations given by Eq. (5.2.2)] is nonsingular. This condition is always satisfied when a solution is possible, as the simulator (DWOPER) uses the same matrix for the finite difference unsteady flow computations.

Expressing the water surface elevation and discharge as a function of the control variable, **r**,

$$\mathbf{h} = \mathbf{h}(\mathbf{r}) \tag{5.3.1}$$

and

$$\mathbf{Q} = \mathbf{Q}(\mathbf{r}), \tag{5.3.2}$$

then the objective function, now called the reduced objective function is expressed as

$$\text{Minimize } z = F(\mathbf{r}) = f[\mathbf{h}(\mathbf{r}), \mathbf{Q}(\mathbf{r})] \tag{5.3.3}$$

The objective function can be evaluated once the state variables, **h** and **Q**, are computed for the given set of control variables, **r**.

The reduced problem, which is called the reduced operations model (ROM), is now expressed by the reduced objective function, equation (5.3.3), subject to Eqs. (5.2.3)–(5.2.6). The ROM is much smaller in size than the GOM, with the simulator determining the implicit functions $\mathbf{h}(\mathbf{r})$ and $\mathbf{Q}(\mathbf{r})$ by performing the unsteady flow computations thus eliminating the constraint matrix $\mathbf{g}$ that describes the hydraulics.

In solving the ROM by a nonlinear programming algorithm, the Jacobian of the matrix $\mathbf{g}(\mathbf{h}, \mathbf{Q}, \mathbf{r})$ will be required as well as the gradients of the functions $F(\mathbf{r})$, $\mathbf{h}(\mathbf{r})$, and $\mathbf{Q}(\mathbf{r})$, which are also called the reduced gradients. The Jacobian matrix is defined as

$$\mathbf{J}(\mathbf{h}, \mathbf{Q}, \mathbf{r}) = \left[\frac{\partial \mathbf{g}}{\partial \mathbf{h}}, \frac{\partial \mathbf{g}}{\partial \mathbf{Q}}, \frac{\partial \mathbf{g}}{\partial \mathbf{r}}\right] = [\mathbf{B}, \mathbf{C}] \tag{5.3.4}$$

or

$$\mathbf{J}(\mathbf{y}, \mathbf{r}) = \left[\frac{\partial \mathbf{g}}{\partial \mathbf{y}}, \frac{\partial \mathbf{g}}{\partial \mathbf{r}}\right] = [\mathbf{B}, \mathbf{C}] \tag{5.3.5}$$

where $\mathbf{y}$ denotes the state variables $(\mathbf{h}, \mathbf{Q})$ and $\mathbf{B}$ is the basis matrix. The basis matrix of the optimal control problem is the same as the Jacobian matrix used in the Newton–Raphson solution procedure in the simulation model (DWOPER). Thus, the two elements of the Jacobian matrix $\mathbf{J}$ are available (with the basis $\mathbf{B}$ explicitly computed, and terms in $\mathbf{C}$ already available) after a simulation run. The basis matrix is a banded sparse matrix with at most four nonzero elements in each row around the matrix's main diagonal.

The reduced gradients can be calculated by applying the two-step scheme used by Mantell and Lasdon (1978) and also by Wanakule et al. (1986). Letting $\mathbf{B}_t = \partial \mathbf{g}_t / \partial \mathbf{y}_t$ denote the basis matrix for time step t, the following scheme is adapted for the ROM:

(i) Solve the system of finite difference equations for the last time step T to find the values of the Lagrange multipliers $\boldsymbol{\pi}_T$

$$\boldsymbol{\pi}_T \mathbf{B}_T = \frac{\partial f}{\partial \mathbf{y}_T} \tag{5.3.6}$$

then solve for the $\boldsymbol{\pi}_t$ backward in time

$$\boldsymbol{\pi}_t \mathbf{B}_t = \frac{\partial f}{\partial \mathbf{y}_t} - \boldsymbol{\pi}_{t+1} \left(\frac{\partial \mathbf{g}_{t+1}}{\partial \mathbf{y}_t}\right) \quad \text{for } t = T-1, T-2, \ldots, 2, 1 \tag{5.3.7}$$

(ii) Calculate the value of the reduced gradient

$$\frac{\partial F}{\partial \mathbf{r}_t} = \frac{\partial f}{\partial \mathbf{r}_t} - \boldsymbol{\pi}_t \left(\frac{\partial \mathbf{g}_t}{\partial \mathbf{r}_t}\right) \quad \text{for } t = 1, 2, \ldots, T \tag{5.3.8}$$

The Lagrange multipliers, π_t, can be used in a sensitivity analysis as they show the effect of a small change in the corresponding term in the objective value.

5.3.3 Solution of Reduced Problem

The reduced problem, ROM, can be solved by a nonlinear programming algorithm. As the reduced problem still contains bound-type constraints on the state variables **h** and **Q**, the algorithm adopted should have provisions to assure the feasibility of the simulation model solutions for the state variables. An augmented Lagrangian (AL) algorithm that incorporates the bounds on the state variables into the objective function is used for this purpose. An application of this type can be found in Hsin (1980), where the bounds on the state variables are violated until the solution converges. The reduced problem with AL terms is

$$\min L_A(\mathbf{r}, \boldsymbol{\mu}, \boldsymbol{\sigma}) = F(\mathbf{r}) + 0.5 \sum_i \sigma_i \min\left[0, \left(b_i - \frac{\mu_i}{\sigma_i}\right)\right]^2 + 0.5 \sum_i \frac{\mu_i^2}{\sigma_i} \tag{5.3.9}$$

where i denotes the constraint set which is formed of the bounds on the state variables (i.e., the water surface elevations and discharges) and σ_i and μ_i are respectively the penalty weight and the Lagrange multiplier associated with the ith bound. The term b_i is the violation term defined as

$$b_i = \min[(y_i - \underline{y}_i), (\bar{y}_i - y_i)] \tag{5.3.10}$$

The constraints of the new problem are the bounds on the control variables and the operating constraints.

A reduced gradient approach is adopted to solve the reduced problem with AL terms. This new problem, which will be referred to as the reduced operations model with augmented Lagrangian (ROMAL) can be expressed as

$$\text{Minimize } L_A(\mathbf{r}, \boldsymbol{\sigma}, \boldsymbol{\mu}) \tag{5.3.11}$$

subject to Eqs. (5.2.5) and (5.2.6).

The solutions to this is a two-step procedure with an inner and an outer problem that must be solved. The objective function of this inner–outer problem combination is

$$z = \min_{\boldsymbol{\sigma}, \boldsymbol{\mu}} \left[\min_{\mathbf{r} \in \mathbf{S}} L_A(\mathbf{r}, \boldsymbol{\sigma}, \boldsymbol{\mu})\right] \tag{5.3.12}$$

where **r** is selected from **S**, the set of feasible gate settings defined by Eq.

(5.2.5). The inner problem involves the optimization of the augmented Lagrangian objective by using GRG2 to determine optimal values of **r** while keeping **μ** and **σ** fixed. Then, the outer problem is iterated by updating the values of **μ** and **σ** for the next solution run of the inner problem. The overall optimization is attained when **μ** and **σ** need no further updating, within a given tolerance level. The updating formula used for **μ** is

$$\mu_i^{k+1} = \begin{cases} \mu_i^{(k)} - \sigma_i b_i & \text{if } c_i < \dfrac{\mu_i}{\sigma_i} \\ 0 & \text{otherwise} \end{cases} \tag{5.3.13}$$

where k is the number of the current iteration. The value of **σ** is normally adjusted once during early iterations and then kept constant (Powell, 1978).

In applying the generalized reduced gradient approach to the ROMAL formulation, the gradient of the new objective function is evaluated as

$$\nabla L_A(\mathbf{r}, \boldsymbol{\mu}, \boldsymbol{\sigma}) = \frac{\partial L_A}{\partial r_i} - \boldsymbol{\pi}\left(\frac{\partial \mathbf{g}}{\partial r_i}\right) \quad \text{for all } i = 1 \text{ to } 2N \tag{5.3.14}$$

The solution of the inner problem (i.e., finding the optimal **r** for fixed **μ** and **σ**) is accomplished by GRG2 (Lasdon and Waren, 1983), which is based on the generalized reduced gradient technique. The basic steps of the optimal control algorithm are shown in Figure 5.5 and a diagram of the algorithm is shown in Figure 5.6.

5.4 APPLICATION

Unver and Mays (1990) developed a model for the real-time optimal flood control operation of reservoir systems. This model combined the GRG2 and the U.S. National Weather Service DWOPER codes. The resulting model was applied to Lake Travis on the Lower Colorado River in Texas. Lake Travis is one of the seven reservoirs of the Highland Lake (Fig. 5.7) system located on the Colorado River in central Texas near Austin with a total contributing drainage area of 27,352 miles^2. A major tributary is the Pedernales River, which has a watershed area of approximately 1280 miles^2. Lake Travis (Fig. 5.8) is about 64 miles long with a designated flood control capacity of 3,223,000 acre-ft. Mansfield Dam originally was built primarily for flood control and hydroelectric production. Of the seven reservoirs in the lake chain, Lake Travis is the only reservoir with designated flood control stage. Development in the floodplain of the Highland Lakes has caused severe problems in operation of the reservoir under flooding conditions.

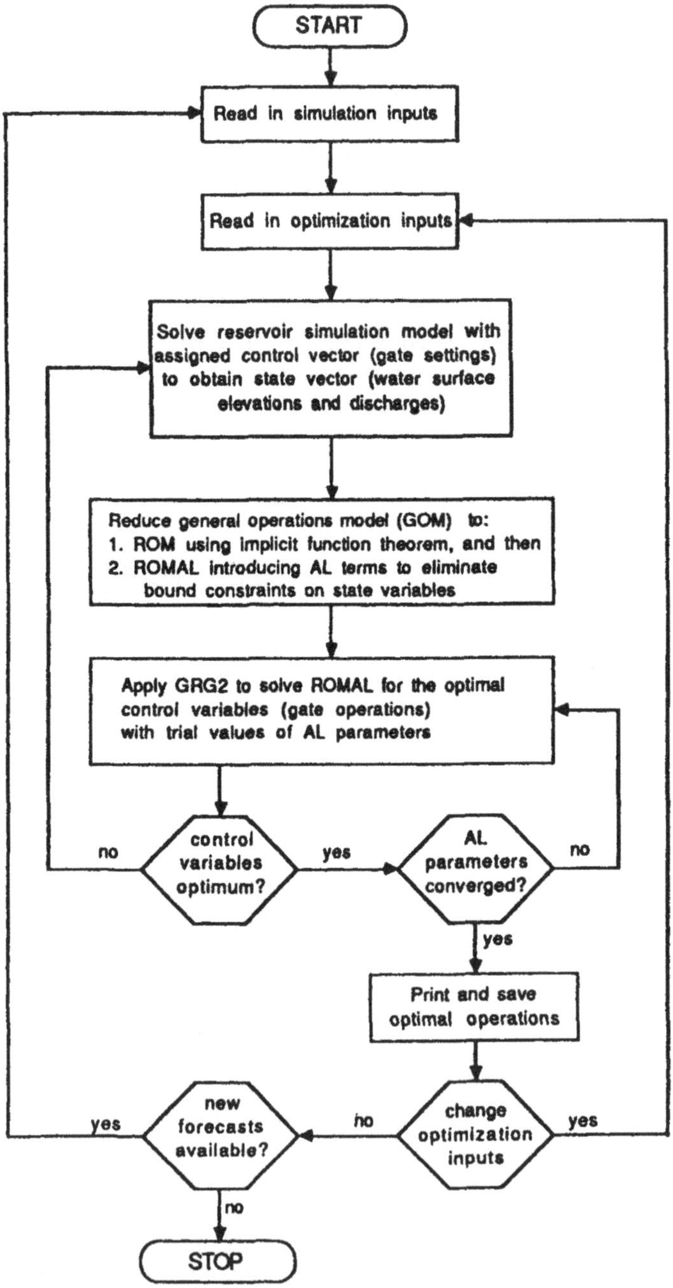

Figure 5.5 Basic steps of optimal control algorithm (Unver, 1987).

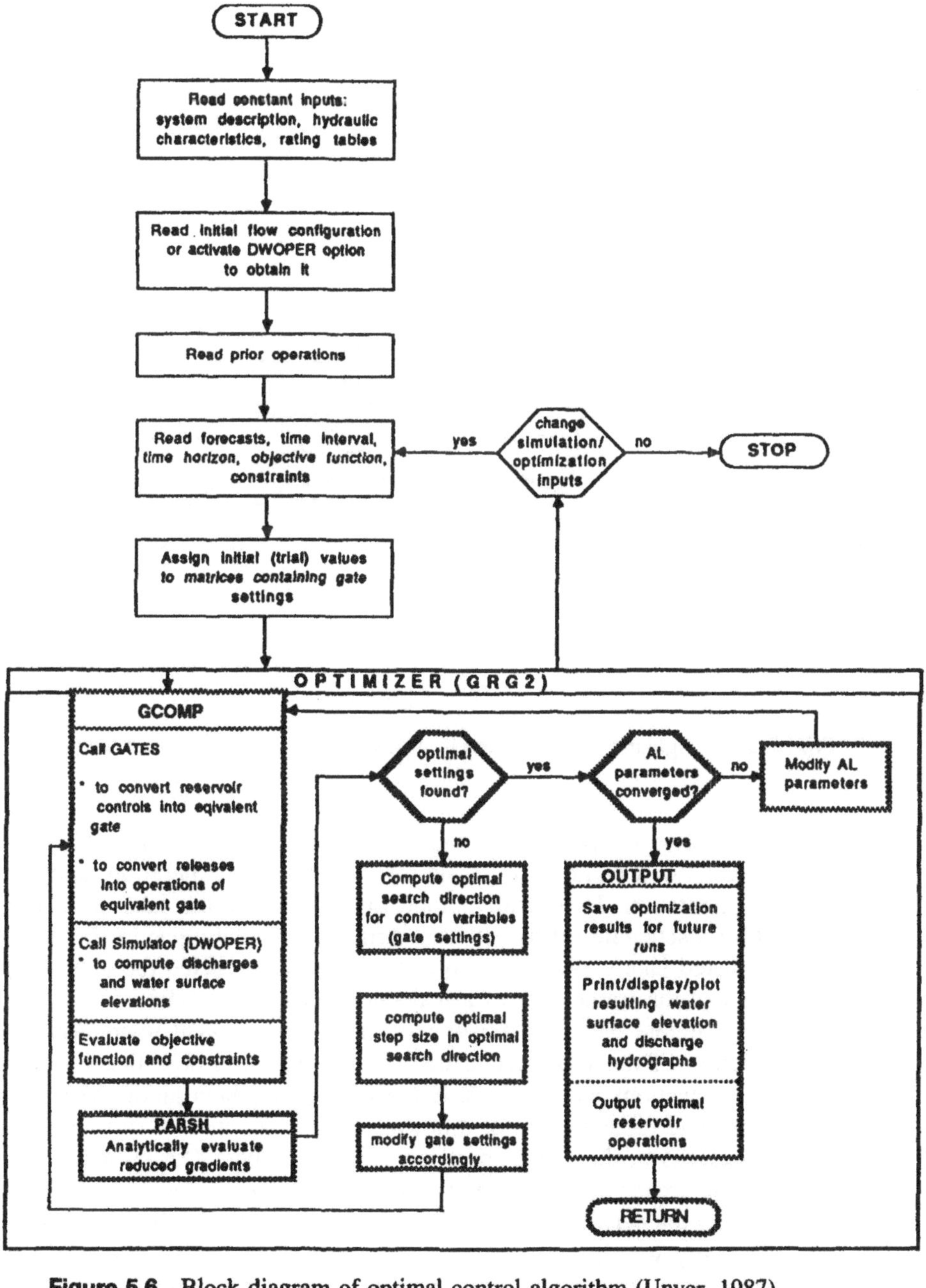

Figure 5.6 Block diagram of optimal control algorithm (Unver, 1987).

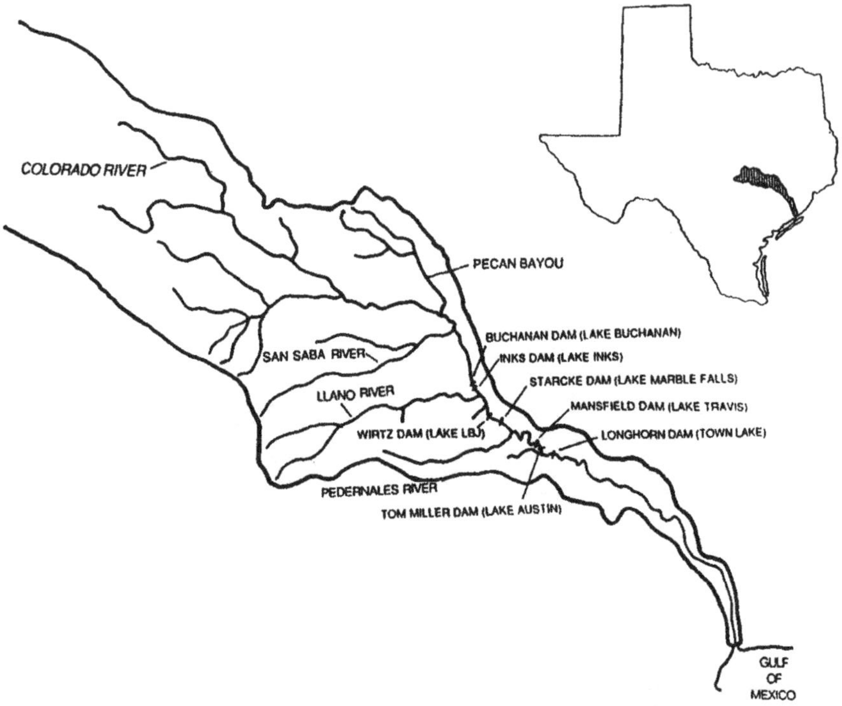

Figure 5.7 Highland Lakes system in Texas: Lower Colorado River Basin.

Lake Travis is operated by the Lower Colorado River Authority for purposes of flood control, water supply, hydropower generation, and low flow augmentation. Historically, flood operations for the lake were prescribed by a schedule set by the U.S. Army Corps of Engineers, based on the forecasted lake elevations. Unver et al. (1987) modeled the lake using DWOPER in the framework of a real-time simulation model.

The available data and existing operational and legal restrictions on the operation of the lake dictate that the optimization objective consider both lake elevations and releases. Extensive development along the shores of the lake and the existence of a downstream urban area (Austin) are two major concerns in the flood operation of Lake Travis. The critical elevations in the existing flood operations schedule were used to set up an elevation versus penalty weight table for the lake area. Similarly, a release versus penalty weight relationship was established to model the damages resulting from excessive releases from the lake. The optimization model

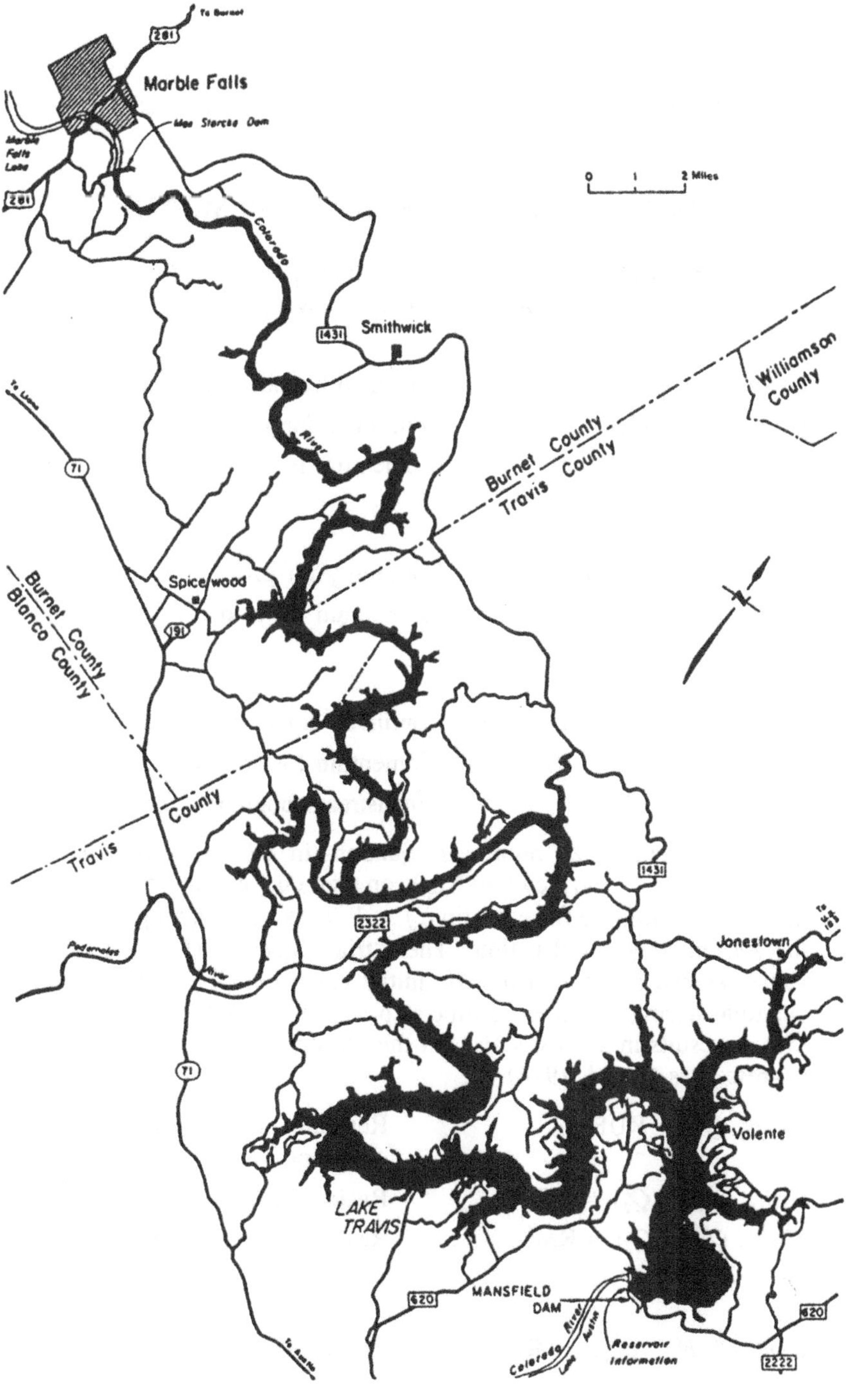

Figure 5.8 Highland Lakes system in Texas: Lake Travis (Texas Water Development Board, 1971).

was run to determine the optimal hourly operation of Lake Travis for the June 1981 flood event. See Unver and Mays (1990) for details.

APPENDIX A: SOLUTION OF SIMULATION MODEL

The system of nonlinear equations can be expressed in functional form in terms of the unknowns **h** and **Q** at time level $j + 1$, as follows (Fread, 1974):

$$
\begin{aligned}
&\mathrm{UB}(h_1, Q_1) = 0 && \text{Upstream boundary conditions}\\
&C_1(h_1, Q_1, h_2, Q_2) = 0 && \text{Continuity for grid 1}\\
&M_1(h_1, Q_1, h_2, Q_2) = 0 && \text{Momentum for grid 1}\\
&\quad\vdots\\
&C_i(h_i, Q_i, h_{i+1}, Q_{i+1}) = 0 && \text{Continuity for grid } i\\
&M_i(h_i, Q_i, h_{i+1}, Q_{i+1}) = 0 && \text{Momentum for grid } i\\
&\quad\vdots\\
&C_{N-1}(h_{N-1}, Q_{N-1}, h_N, Q_N) = 0 && \text{Continuity for grid } N-1\\
&M_{N-1}(h_{N-1}, Q_{N-1}, h_N, Q_N) = 0 && \text{Momentum for grid } N-1\\
&\mathrm{DB}(h_N, Q_N) = 0 && \text{Downstream boundary condition}
\end{aligned}
\tag{5.A.1}
$$

This system of $2N$ nonlinear equations in $2N$ unknowns is solved for each time step by the Newton–Raphson method (see Figure 5.9). The computational procedure for each time $j + 1$ starts by assigning trial values to the $2N$ unknowns at that time. These trial values of **Q** and **h** can be the values known at time j from the initial condition (if $j = 1$) or from calculations during the previous time step. Using the trial values in the system (5.A.1) results in $2N$ residuals. For the kth iteration, these residuals can be expressed as (Fread, 1974)

$$
\begin{aligned}
&\mathrm{UB}(h_1^k, Q_1^k) = \mathrm{RUB}^k && \text{Residual for upstream boundary condition}\\
&C_1(h_1^k, Q_1^k, h_2^k, Q_2^k) = \mathrm{RC}_1^k && \text{Residual for continuity at grid 1}\\
&M_1(h_1^k, Q_1^k\ h_2^k, Q_2^k) = \mathrm{RM}_1^k && \text{Residual for momentum at grid 1}\\
&\quad\vdots\\
&C_i(h_i^k, Q_i^k, h_{i+1}^k, Q_{i+1}^k) = \mathrm{RC}_i^k && \text{Residual for continuity at grid } i
\end{aligned}
\tag{5.A.2}
$$

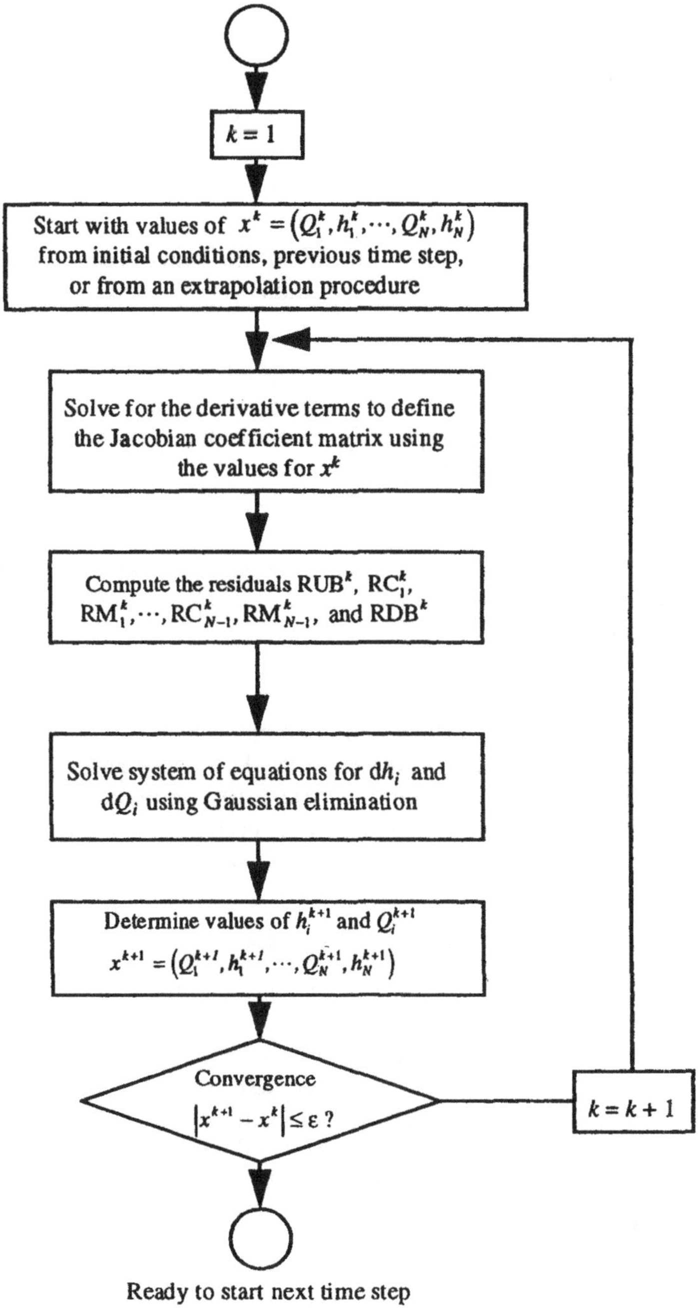

Figure 5.9 Procedure for solving a system of difference equations for each time step using the Newton–Raphson method.

$M_i(h_i^k, Q_i^k, h_{i+1}^k, Q_{i+1}^k) = \mathrm{RM}_i^k$	Residual for momentum at grid i
$\vdots$	
$C_{N-1}(h_{N-1}^k, Q_{N-1}^k, h_N^k, Q_N^k) = \mathrm{RC}_{N-1}^k$	Residual for continuity at grid $N-1$
$M_{N-1}(h_{N-1}^k, Q_{N-1}^k, h_N^k, Q_N^k) = \mathrm{RM}_{N-1}^k$	Residual for momentum at grid $N-1$
$\mathrm{DB}(h_N^k, Q_N^k) = \mathrm{RDB}_N^k$	Residual for downstream boundary condition

The solution is approached by finding values of the unknowns **Q** and **h** so that the residuals are forced to zero or very close to zero.

The Newton–Raphson method is an iterative technique for solving a system of nonlinear algebraic equations. Consider the system of equations (5.A.2) denoted in vector form as

$$\mathbf{f}(\mathbf{x}) = \mathbf{0} \tag{5.A.3}$$

where $\mathbf{x} = (Q_1, h_1, Q_2, h_2, \ldots, Q_N, h_N)$ is the vector of unknown quantities and for iteration k, $\mathbf{x}^k = (Q_1^k, h_1^k, Q_2^k, h_2^k, \ldots, Q_N^k, h_N^k)$. The nonlinear system can be linearized to

$$f(\mathbf{x}^{k+1}) \approx f(\mathbf{x}^k) + J(\mathbf{x}^k)(\mathbf{x}^{k+1} - \mathbf{x}^k) \tag{5.A.4}$$

where $\mathbf{J}(\mathbf{x}^k)$ is the Jacobian, which is a coefficient matrix made up of the first partial derivatives of $\mathbf{f}(\mathbf{x})$ evaluated at $\mathbf{x}^k$. The right-hand side of Eq. (5.A.4) is the linear vector function of $\bar{\mathbf{x}}^k$. Basically, an iterative procedure is used to determine $\mathbf{x}^{k+1}$ that forces the residual error $\mathbf{f}(\bar{\mathbf{x}}^{k+1})$ in Eq. (5.A.4) to zero. This can be accomplished by setting $\mathbf{f}(\bar{\mathbf{x}}^{k+1}) = \mathbf{0}$ and rearranging Eq. (5.A.4) to read

$$\mathbf{J}(\mathbf{x}^k)(\mathbf{x}^{k+1} - \mathbf{x}^k) = -\mathbf{f}(\mathbf{x}^k) \tag{5.A.5}$$

This system is solved for $(\mathbf{x}^{k+1} - \mathbf{x}^k) = \Delta\mathbf{x}^k$, and the improved estimate of the solution, $\mathbf{x}^{k+1}$, is determined knowing $\Delta\mathbf{x}^k$. The process is repeated until $\mathbf{x}^{k+1} - \mathbf{x}^k$ is smaller than some specified tolerance.

The system of linear equations represented by Eq. (5.A.5) involves $\mathbf{J}(\mathbf{x}^k)$, the Jacobian of the set of Eqs. (5.A.1) with respect to **h** and **Q**, and $-\mathbf{f}(\mathbf{x}^k)$, the vector of the negatives of the residuals in Eq. (5.A.2). The resulting system of equations is (Fread, 1974)

$$\frac{\partial \mathrm{UB}}{\partial h_1} dh_1 + \frac{\partial \mathrm{UB}}{\partial Q_1} dQ_1 = -\mathrm{RUB}^k$$

$$
\begin{aligned}
&\frac{\partial C_1}{\partial h_1} dh_1 + \frac{\partial C_1}{\partial Q_1} dQ_1 + \frac{\partial C_1}{\partial h_2} dh_2 + \frac{\partial C_1}{\partial Q_2} dQ_2 = -\mathrm{RC}_1^k \\
&\frac{\partial M_1}{\partial h_1} dh_1 + \frac{\partial M_1}{\partial Q_1} dQ_1 + \frac{\partial M_1}{\partial h_2} dh_2 + \frac{\partial M_1}{\partial Q_2} dQ_2 = -\mathrm{RM}_1^k \\
&\vdots \\
&\frac{\partial C_i}{\partial h_i} dh_i + \frac{\partial C_i}{\partial Q_i} dQ_i + \frac{\partial C_i}{\partial h_{i+1}} dh_{i+1} + \frac{\partial C_i}{\partial Q_{i+1}} dQ_{i+1} = -\mathrm{RC}_i^k \\
&\frac{\partial M_i}{\partial h_i} dh_i + \frac{\partial M_i}{\partial Q_i} dQ_i + \frac{\partial M_i}{\partial h_{i+1}} dh_{i+1} + \frac{\partial M_i}{\partial Q_{i+1}} dQ_{i+1} = -\mathrm{RM}_i^k \\
&\vdots \\
&\frac{\partial C_{N-1}}{\partial h_{N-1}} dh_{N-1} + \frac{\partial C_{N-1}}{\partial Q_{N-1}} dQ_{N-1} + \frac{\partial C_{N-1}}{\partial h_N} dh_N + \frac{\partial C_{N-1}}{\partial Q_N} dQ_N = -\mathrm{RC}_{N-1}^k \\
&\frac{\partial M_{N-1}}{\partial h_{N-1}} dh_{N-1} + \frac{\partial M_{N-1}}{\partial Q_{N-1}} dQ_{N-1} + \frac{\partial M_{N-1}}{\partial h_N} dh_N + \frac{\partial M_{N-1}}{\partial Q_N} dQ_N = -\mathrm{RM}_{N-1}^k \\
&\frac{\partial \mathrm{DB}}{\partial h_N} dh_N + \frac{\partial \mathrm{DB}}{\partial Q_N} dQ_N = -\mathrm{RDB}^k
\end{aligned}
\tag{5.A.6}
$$

APPENDIX B: COMPUTATION OF BASIS MATRIX ELEMENTS

For any given time step and a given set of reservoir gate settings, the system of equations has a banded structure, as shown in Figure 5.10, with at most four elements around the main diagonal. This structure is exploited in the simulation model by an efficient solution algorithm. The system of equations (5.A.1) constitutes the **g** matrix of the optimization model when written for all time steps. Each equation in the matrix has at most four nonzero terms that belong to the current time step, and another four nonzero terms that belong to the previous time step. The **g** matrix thus has a special banded staircase structure as shown in Figure 5.11, which can be exploited. The partial derivatives of the **g** matrix with respect to the problem variables have to be computed to be used in the optimization model. The mathematical expressions for the partial derivatives are given in the next two sections. Section 5.B.1 gives the partial derivatives with respect to the variables at the current time step (i.e., time step $j + 1$), and Section 5.B.2 gives the expressions with respect to the previous time step.

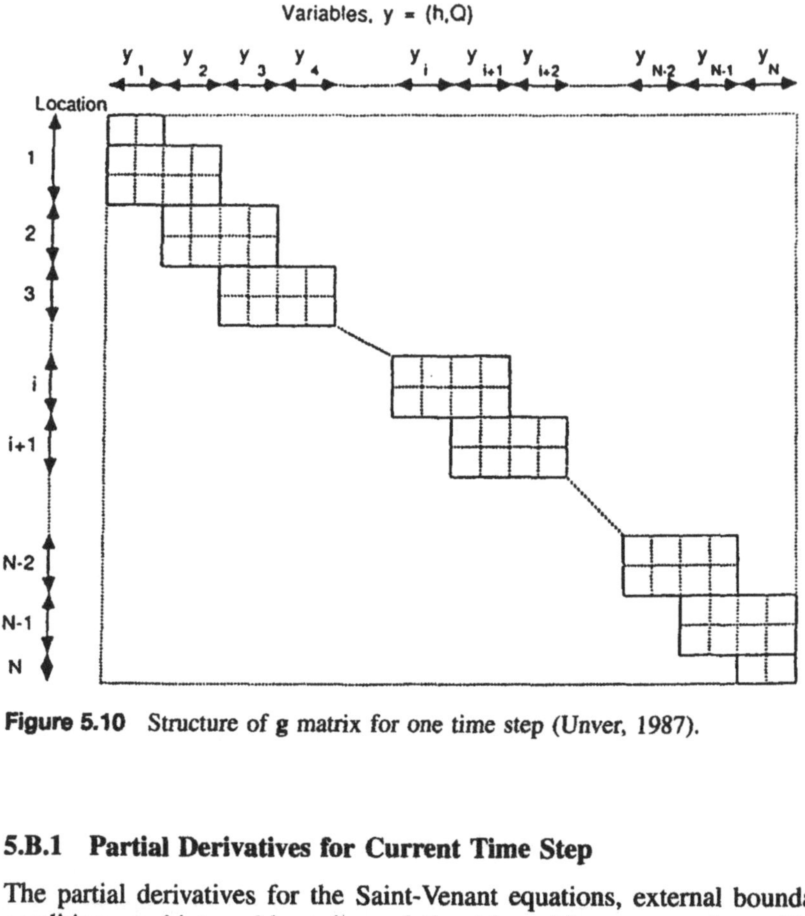

Figure 5.10 Structure of **g** matrix for one time step (Unver, 1987).

5.B.1 Partial Derivatives for Current Time Step

The partial derivatives for the Saint-Venant equations, external boundary conditions, and internal boundary relationships with respect to the problem variables water surface elevations **h**, discharges **Q**, and gate setting **r** at the current time step are given below under separate headings. In the following, the superscripts denote the time step, and the subscripts denote the location of the variables.

5.B.2 Partial Derivatives for Saint-Venant Equations (Fread, 1974)

$$\frac{\partial C_i}{\partial h_i} = \frac{\Delta x_i}{2\Delta t_j}(B + B_o)_i^{j+1} \tag{5.B.1}$$

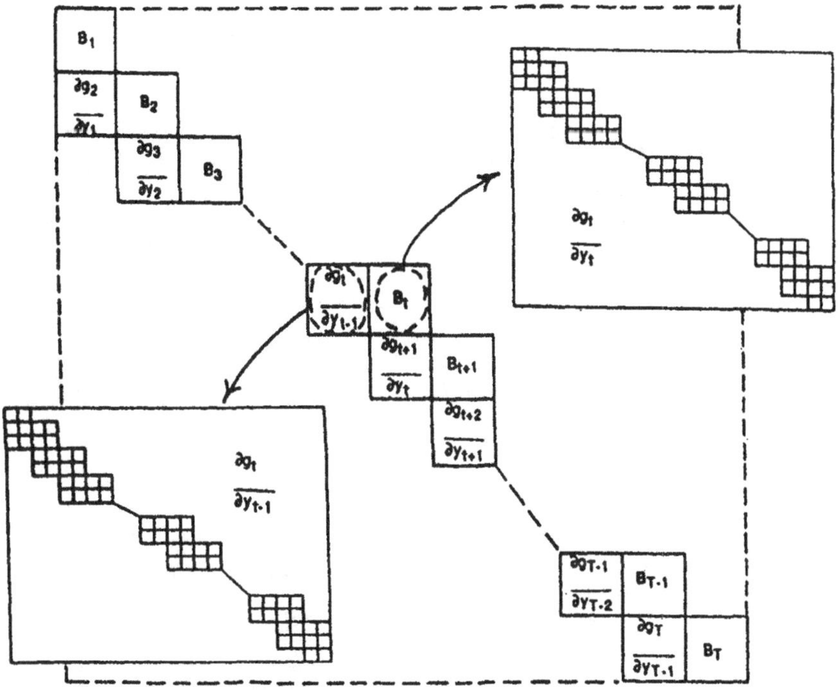

Figure 5.11 Matrix of partial derivatives with respect to $\mathbf{y} = (\mathbf{h}, \mathbf{Q})$ for T time steps (Unver, 1987).

$$\frac{\partial C_i}{\partial Q_i} = -\theta \tag{5.B.2}$$

$$\frac{\partial C_i}{\partial h_{i+1}} = \frac{\Delta x_i}{2\Delta t_j} (B + B_o)_{i+1}^{j+1} \tag{5.B.3}$$

$$\frac{\partial C_i}{\partial Q_{i+1}} = \theta \tag{5.B.4}$$

$$\frac{\partial M_i}{\partial h_i} = \theta \left\{ \left(\frac{\beta Q^2 B}{A^2} \right)_i^{j+1} + g\bar{A}_i^{j+1} \left[-1 + \left(\frac{\partial \bar{S}_f}{\partial h} \right)_i^{j+1} \Delta x_i + \left(\frac{\partial \bar{S}_e}{\partial h} \right)_i^{j+1} \Delta x_i \right] + \frac{gB_i^{j+1}}{2} \left[h_{i+1}^{j+1} - h_i^{j+1} + \bar{S}_{f_i}^{j+1} \Delta x_i + (\bar{S}_e)_i^{j+1} \Delta x_i \right] + \frac{1}{2} \left(\overline{W}_f \frac{dB}{dh} \right)_i^{j+1} \Delta x_i \right\} \tag{5.B.5}$$

$$\frac{\partial M}{\partial Q_i} = \left(\frac{\Delta x_i}{2\Delta t_j}\right) + \theta\left\{-2\left(\frac{\beta Q}{A}\right)_i^{j+1} + g\bar{A}_i^{j+1}\left[\left(\frac{\partial \bar{S}_f}{\partial Q}\right)_i^{j+1}\Delta x_i + \left(\frac{\partial \bar{S}_e}{\partial Q}\right)_i^{j+1}\Delta x_i\right]\right\} \tag{5.B.6}$$

$$\begin{aligned}\frac{\partial M}{\partial h_{i+1}} = \theta\Bigg\{&-\left(\frac{\beta Q^2 B}{A^2}\right)_{i+1}^{j+1} + g\bar{A}_i^{j+1}\left(1 + \left(\frac{\partial \bar{S}_f}{\partial h}\right)_i^{j+1}\Delta x_i + \left(\frac{\partial \bar{S}_e}{\partial h}\right)_{i+1}^{j+1}\Delta x_i\right)\\ &+ \frac{gB_{i+1}^{j+1}}{2}(h_{i+1}^{j+1} - h_i^{j+1} + \bar{S}_{f_i}^{j+1}\Delta x_i + \bar{S}_{e_i}^{j+1}\Delta x_i) + \left(\frac{\bar{W}_f B}{2}\right)_i^{j+1}\Delta x_i\Bigg\}\end{aligned} \tag{5.B.7}$$

$$\frac{\partial M}{\partial Q_{i+1}} = \left(\frac{\Delta x_i}{2\Delta t_j}\right) + \theta\left\{2\left(\frac{\beta Q}{A}\right)_{i+1}^{j+1} + g\bar{A}_i^{j+1}\left[\left(\frac{\partial \bar{S}_f}{\partial Q}\right)_{i+1}^{j+1}\Delta x_i + \left(\frac{\partial \bar{S}_e}{\partial Q}\right)_{i+1}^{j+1}\Delta x_i\right]\right\} \tag{5.B.8}$$

in which

$$\frac{\partial \bar{S}_f}{\partial h_i} = 2\bar{S}_{f_i}\left(\frac{d\bar{n}/dh_i}{\bar{n}_i} - \frac{5B_i}{6\bar{A}} + \frac{dB_i/dh_i}{3\bar{B}}\right) \tag{5.B.9}$$

$$\frac{\partial \bar{S}_f}{\partial h_{i+1}} = 2\bar{S}_{f_i}\left(\frac{d\bar{n}/dh_{i+1}}{\bar{n}_i} - \frac{5B_{i+1}}{6\bar{A}} + \frac{dB_{i+1}/dh_{i+1}}{3B_{i+1}}\right) \tag{5.B.10}$$

$$\frac{\partial \bar{S}_f}{\partial Q_i} = 2\bar{S}_{f_i}\left(\frac{d\bar{n}/dQ_i}{\bar{n}_i} + \frac{1}{2Q_i}\right) \tag{5.B.11}$$

$$\frac{\partial \bar{S}_f}{\partial Q_{i+1}} = 2\bar{S}_{f_i}\left(\frac{d\bar{n}/dQ_{i+1}}{\bar{n}_i} + \frac{1}{2Q}\right) \tag{5.B.12}$$

$$\frac{\partial \bar{S}_e}{\partial h_i} = \left(\frac{2\bar{S}_{e_i}B_iV_i^2}{A_i(V_{i+1}^2 - V_i^2)}\right) \tag{5.B.13}$$

$$\frac{\partial \bar{S}_e}{\partial h_{i+1}} = \left(\frac{-2\bar{S}_{e_i}B_{i+1}V_{i+1}^2}{A_{i+1}(V_{i+1}^2 - V_i^2)}\right) \tag{5.B.14}$$

$$\frac{\partial \bar{S}_e}{\partial Q_i} = \left(\frac{-2\bar{S}_{e_i}V_i}{(V_{i+1}^2 - V_i^2)A_i}\right) \tag{5.B.15}$$

$$\frac{\partial \bar{S}_e}{\partial Q_{i+1}} = \left(\frac{2\bar{S}_{e_i}V_{i+1}}{(V_{i+1}^2 - V_i^2)A_{i+1}}\right) \tag{5.B.16}$$

$$\frac{dB_i}{dh_i} = \frac{\Delta B_i}{\Delta h_i} \tag{5.B.17}$$

$$V_i = \frac{Q_i}{A_i} \tag{5.B.18}$$

$$V_{i+1} = \frac{Q_{i+1}}{A_{i+1}} \tag{5.B.19}$$

$$\frac{d\bar{n}}{dh_i} = \frac{\Delta\bar{n}}{\Delta\hat{h}}\frac{d\hat{h}}{dh_i} \tag{5.B.20}$$

$$\frac{d\bar{n}}{dQ_i} = \frac{\Delta\bar{n}}{\Delta\hat{Q}}\frac{d\hat{Q}}{dQ_i} \tag{5.B.21}$$

$$\hat{h} = \frac{h_m + h_{m+1}}{2} \tag{5.B.22}$$

$$\hat{Q} = \frac{Q_m + Q_{m+1}}{2} \tag{5.B.23}$$

and

$$\left.\begin{aligned} \frac{d\hat{h}}{dh_i} &= 0 \quad \text{if } i \neq m \\ \frac{d\hat{h}}{dh_i} &= \frac{1}{2} \quad \text{if } i = m \end{aligned}\right\} \tag{5.B.24}$$

$$\left.\begin{aligned} \frac{d\hat{Q}}{dQ_i} &= 0 \quad \text{if } i \neq m \\ \frac{d\hat{Q}}{dQ_i} &= \frac{1}{2} \quad \text{if } i = m \end{aligned}\right\} \tag{5.B.25}$$

REFERENCES

Can, E. K. and Houck, M. H., Real-Time Reservoir Operations by Goal Programming, *Journal of Water Resources Planning and Management*, Vol. 110, pp. 297–309, 1984.

Fread, D. L., Numerical Properties of Implicit Four-Point Finite Difference Equations of Unsteady Flow, NOAA Technical Memorandum NWS HYDRO-18, National Weather Service, NOAA, U.S. Department of Commerce, Silver Spring, MD, 1974.

Fread, D. L., *National Weather Service Operational Dynamic Wave Model*, Hydrologic Research Laboratory, U.S. National Weather Service, Silver Spring, MD, 1982.

Hsin, J. K., The Optimal Control of Deterministic Econometric Planning Models, Ph.D. dissertation, The University of Texas, Austin, 1980.

Jamieson, D. G. and Wilkinson, J. C., River Dee Research Program, 3, a Short-Term Control Strategy for Multipurpose Reservoir Systems, *Water Resources Research* Vol. 8, pp. 911–920, 1972.

Lasdon, L. S. and Waren, A. D., *GRG2 User's Guide*, The University of Texas, Austin, 1983.

Luenberger, D. G., *Introduction to Linear and Nonlinear Programming*, Addison-Wesley, Menlo Park, 1984.

Mantell, J., and Lasdon, L. S., A GRG Algorithm for Econometric Control Problems, *Annals of Economic and Social Mangement*, Vol. 6, p. 581–597, 1978.

Powell, M. J. D., Algorithms for Nonlinear Constraints That Use Lagrangian Functions, *Mathmatical Programming*, Vol. 14, No. 2, 1978.

Tennessee Valley Authority, Development of a Comprehensive TVA Water Resource Management Program, Technical Report, Div. of Water Cont. Plan., Tenn. Valley Auth., Knoxville, 1974.

Texas Water Development Board, Engineering Data on Dams and Reservoirs in Texas, Part III, Report No. 126, Texas Water Development Board, Austin, 1971.

U.S. Army Corps of Engineers, Hydrologic Engineering Center, HEC-5, Simulation of Flood Control and Conservation Systems, U.S. Army Corps. of Engineers, Davis, CA, 1973a.

U.S. Army Corps of Engineers, Hydrologic Engineering Center, HEC-5C, A Simulation Model for System Formulation and Evaluation, U.S. Army Corps of Engineers, Davis, CA, 1973b.

U. S. Army Corps of Engineers, Hydrologic Engineering Center, HEC-5, Simulation of Flood Control and Conservation Systems, Davis, CA, 1979.

U.S. Army Corps of Engineers, Hydrologic Engineering Center, Real-Time Flood-casting and Reservoir Control for the Kanauha, Special Projects Memorandum, No. 83-10, Davis, CA, Sep. 1983.

Unver, O. L., Simulation and Optimization for Real-Time Operation for Multi-reservoir Systems Under Flooding Conditions, Ph.D. Dissertation, University of Texas at Austin, 1987.

Unver, O., Mays, L. W., and Lansey, K., Real-Time Flood Management Model for Highland Lake system, *Journal of Water Resource Planning and Management*, Vol. 5, No. 3, pp. 620–638, 1987.

Unver, O. L., and Mays, L. W., Model for Real-Time Optimal Flood Control Operation of a Reservoir System, *Water Resources Management*, Vol. 4, Kluwer Academic Publishers, Netherlands, pp. 21–46, 1990.

Wanakule, N., Mays, L. W., and Lasdon, L., Optimal Management of Large-Scale Aquifers: Methodology and Applications, *Water Resources Research*, Vol. 22, No. 2, 447–466, 1986.

Wasimi, S. A. and Kitanidis, P. K., Real-Time Forecasting and Daily Operation of a Multireservoir System During Floods by Linear Quadratic Gaussian Control, *Water Resources Research*, Vol. 19, pp. 1511–1522, 1983.

Windsor, J. S., Optimization Model for the Operation of Flood Control Systems, *Water Resources Research*, Vol. 9, pp. 1219–1226, 1973.

Yazicigil, H., Optimal Operation of a Reservoir System Using Forecasts, Ph.D. Dissertation, Purdue University, West Lafayette, IN, 1980.

Chapter 6
Water Distribution System Operation

6.1 PROBLEM IDENTIFICATION

A methodology based on solving a large-scale nonlinear programming (NLP) problem is presented for the optimal operation of pumping stations in water distribution systems. Optimal operation refers to the scheduling of pump operation that results in the minimum operating cost for a given set of operating conditions. The methodology is based on an optimal control framework which interfaces a nonlinear optimization model with a hydraulic simulation model. The objective function is to minimize pumping cost over a planning horizon, and the constraint set includes system constraints, which account for the hydraulics involved in a water distribution system, bound constraints on decision variables, and other constraints that may reflect operator preferences or system limitations.

There are a myriad of reasons why pumping stations operate inefficiently (Ormsbee, 1991): (1) the pumps were incorrectly selected; (2) the pumps have worn out; (3) there is limited capacity in the transmission or distribution system; (4) there is limited storage capacity; (5) there is inefficient operation of pressure (hydropneumatic) tanks; (6) they have inadequate or inaccurate telemetry equipment; (7) there is an inability to automatically or remotely control pumps and valves; (8) there is a penalty due to time-of-day or seasonal energy pricing; (9) there is a lack of understanding of demand or capacity power charges; (10) there is operator error; and (11) they have suboptimal control strategies. The optimal control problem for a water distribution system is complicated by the fact that the mathematical problem can be very large in the number of constraints, many of which are nonlinear, and the large number of decision variables that are nonlinear. This is complicated even further by the fact that the controls (pumps on and off) are discrete. Several approaches using dynamic programming (DP) have been proposed (Solanas and Vergés, 1974; Solanas and Montolio, 1987; Cohen, 1982; Joalland and Cohen, 1980; Carpentier and Cohen, 1985; Coulbeck and Orr, 1985; Sabet and Helweg, 1985; Zessler and Shamir, 1985; Ormsbee et al., 1987). All of these DP approaches suffer from the curse of dimensionality limiting the size of problems (number of pumps, storage facilities, and size of network) that can be considered; as a result, the DP approaches are only applicable to very small systems. Other previous techniques (Fallside and Perry, 1975; Coulbeck and Sterling, 1978) that were not based on dynamic programming were also not very successful. Chase and Ormsbee (1989) proposed a nonlinear programming approach based on using a nonlinear programming optimizer and a hydraulic simulator to solve the hydraulic constraints of the optimizer. Brion (1990) and Brion and Mays (1991) presented a methodology to solve the problem as a discrete time optimal control problem building on the work by Mays and Lansey (1989), Chase and Ormsbee (1989), and Duan et al. (1990). This methodology is presented in this chapter.

6.2 PROBLEM FORMULATION

Consider a water distribution system composed of J nodes, M pipes, P pumps, K primary loops, F fixed grade nodes, and S storage tanks. The mathematical statement of the optimal pump operation problem considering T time periods is to minimize the energy cost, Z_p, given as

$$\text{Min } Z_p = \text{Minimize} \sum_t^T \sum_p^P \frac{1}{\text{EFF}_{pt}} \text{UC}_t \frac{0.746\ \gamma Q_{pt} H_{pt}}{550} D_{pt} \qquad (6.2.1)$$

where EFF_{pt} is the efficiency of pump p in time period t, UC_t is the unit pumping cost ($/kwh) during time period t, γ is the specific weight of water (lb/ft^3); and D_{pt} is the length of time pump p operates during time period t (hr).

The constraints that have to be satisfied at all time periods include the conservation of mass at nodes,

$$\sum_i (q_{i,j})_t = Q_{jt}, \quad j = 1, \ldots, J, \; t = 1, \ldots, T \qquad (6.2.2)$$

where $(q_{i,j})_t$ is the flow rate in the pipe connecting nodes i and j during time step t and Q_{jt} is the external demand at node j during time period t. This constraint, which is linear in $(q_{i,j})_t$, assumes that the fluid is incompressible and is written for each node j in the network.

The conservation of energy for primary loops is

$$\sum_{i,j \in k} h_{kt} - \sum_{p \in k} H_{pt} = 0, \quad k = 1, \ldots, K, t = 1, \ldots, T \qquad (6.2.3)$$

where h_{kt} is the head loss in the pipe connecting nodes i and j contained in primary loop k at time t and H_{pt} is the pumping head delivered by pump p in primary loop k at time t.

The conservation of energy for paths between two points of known total grade (fixed grade nodes) is

$$\sum_{i,j \in f} h_{ft} - \sum_{p \in f} H_{pt} = \Delta E_f, \quad f = 1, \ldots, F - 1, \quad t = 1, \ldots, T \qquad (6.2.4)$$

where h_{ft} is the head loss in the pipe connecting nodes i and j contained in path f at time t, H_{pt} is the pumping head delivered by pump p in path f at time t, and ΔE_f is the difference in total grade, expressed as elevation plus gauge pressure, between two fixed grade nodes (FGNs) located at both ends of path f. This constraint, which is nonlinear in pipe flow rate, is written for $F - 1$ paths, where F is the number of fixed grade nodes in the network for all time periods. Constraint equation (6.2.4) is a special case of constraint equation (6.2.3). In fact, energy conservation by (6.2.3) and (6.2.4) apply to primary loops and pseudo-loops (i.e., independent path equations), respectively. The total number of the above hydraulic constraints, all equality constraints in this case, is $J + K + F - 1$. The total number of unknowns, the M pipe flows, is the same as the number of equations.

The pump operation problem is, inherently, an extended period simulation problem. For this type of analysis, water levels, E_{st}, in storage tanks for the current time period are functions of water levels from the previous time period, which can be expressed as

$$E_{st} = f(E_{st-1}), \quad s = 1, \ldots, S, \quad t = 1, \ldots, T \qquad (6.2.5)$$

This relationship involves the flow rate in the pipe connected to each tank evaluated at the previous time period.

The lower and upper bounds on the length of time pump P operates, D_{pt}, within each time period are given as

$$\Delta_{t\,\min} \leq D_{pt} \leq \Delta_{t\,\max}, \quad p = 1, \ldots, P, \quad t = 1, \ldots, T \qquad (6.2.6)$$

where $\Delta_{t\,\min}$ can be zero in order to simulate pump line closing and $\Delta_{t\,\max}$ is the length of one time period. This constraint limits the operating time of a pump within a given time period t. D_{pt} is a non-negative number which cannot exceed the total length of a time period. The smaller the time period used, the more closely continuous pump operation is approximated. D_{pt} appears implicitly in Eq. (6.2.5).

The pressure head bounds on nodal heads are

$$\underline{H}_{jt} \leq H_{jt} \leq \overline{H}_{jt}, \quad j = 1, \ldots, J, \quad t = 1, \ldots, T \qquad (6.2.7)$$

where $\underline{H}_{jt}$ and $\overline{H}_{jt}$ are the lower and upper bounds, respectively, on the pressure head, H_{jt}, at each node j at time t. No universally accepted values for either bound exists. Normally, the minimum desired pressure at the demand nodes fall in the range of 20–40 psi. This may be true during average loading conditions but may be significantly lowered during emergency situations such as when a fire starts. The upper bounds, on the other hand, are fixed by structural limits on the pipes. They depend on the type of material used as well as the age of the pipe. This constraint is extended to handle bounds on storage capacities in tanks expressed in terms of water surface elevations.

The bounds on the tank water surface elevations are

$$\underline{E}_{st} \leq E_{st} \leq \overline{E}_{st}, \quad s = 1, \ldots, S, \quad t = 1, \ldots, T \qquad (6.2.8)$$

where $\underline{E}_{st}$ and $\overline{E}_{st}$ are the lower and upper bounds, respectively, on the water surface elevation, E_{st}, for each storage tank s during time t. These storage bounds can be imposed for all time periods. Normally, these bounds correspond to physical limits of the tank. During the last time period, a tighter bound is usually placed on all tanks whereby all tank levels are preferred to revert back to the level at the beginning of the first time period. This is evident from a practical point of view because at the end of the night rate period, which usually has the cheapest rate and is the start of the simulation, it is preferred that storage tanks be full. Cohen (1982) stated that optimizing the operation of a network over a limited horizon, say 24 hr, has no meaning without the requirement of some periodicity in operation. A simple way to do this is to constrain all final states or tank levels to be the same as the initial states within some tolerance. In the methodology, it was decided that the final water surface elevation in each tank be approximately the same as its initial level and that its lower and upper bounds be expressed as functions of the initial water surface elevation.

The above formulation results in a large-scale nonlinear programming problem where the $(q_{i,j})_t$, H_{jt}, E_{st}, and D_{pt} are the decision variables. Chase and Ormsbee (1989) also used similar decision variables in their nonlinear formulation. Additional bound constraints on system characteristics, such as the pump/pipe flow rate at each time period and total energy consumption, can be imposed. Prespecified operating rules, such as limits on the number of times a pump can be turned on and off during the entire planning horizon, can also be considered. However, a mixed integer–nonlinear programming formulation would result, one that requires an optimization technique different from the one presented in this study. Thus, these constraints were not implemented.

6.3 PROBLEM SOLUTION

6.3.1 Overview

The problem is formulated in an optimal control framework where an optimal solution to the problem is arrived at by interfacing a hydraulic simulation code with a nonlinear optimization code. The hydraulic simulation model is used to implicitly solve the hydraulic constraints that define the flow phenomena each time the optimizer needs to evaluate these constraints (see Appendix A). A general formulation of the problem is stated as follows:

$$\text{Minimize energy costs} = f(\mathbf{H}, \mathbf{Q}, \mathbf{D}) \tag{6.3.1}$$

subject to

(a) Conservation of flow and energy constraints and pump operation, see Eqs. (6.2.2)–(6.2.5) and Appendix A for more detail

$$\mathbf{G}\,(\mathbf{H}, \mathbf{Q}, \mathbf{D}, \mathbf{E}) = \mathbf{0} \tag{6.3.2}$$

(b) Bounds on pump operation time, Eq. (6.2.6)

$$\mathbf{0} \leq \mathbf{D} \leq \boldsymbol{\Delta}_{t\ \max} \tag{6.3.3}$$

(c) Nodal pressure head bands, Eq. (6.2.7)

$$\underline{\mathbf{H}} \leq \mathbf{H} \leq \overline{\mathbf{H}} \tag{6.3.4}$$

(d) Storage bounds and final tank levels, Eq. (6.2.8)

$$\underline{\mathbf{E}} \leq \mathbf{E} \leq \overline{\mathbf{E}} \tag{6.3.5}$$

6.3.2 The Reduced Problem

First, the decision variables are partitioned into two sets such that one set can be expressed in terms of the other. Let D_{pt} be the set of "control" or independent variables, and H_{jt} and E_{st} form the set of dependent or "state" variables. The justification is based on the implicit function theorem (Luenberger, 1984), which states (dropping subscripts for brevity): if $\mathbf{H}(\mathbf{D}^*)$ and $\mathbf{E}(\mathbf{D}^*)$ solve the hydraulic constraint equations for $\mathbf{D} = \mathbf{D}^*$ and the basis matrix of the equations is nonsingular, then **H(D)** and **E(D)** exist in the neighborhood of **D***. Thus, for the given set of **D**, there is always a solution of **H** and **E** which satisfies the hydraulic equations, implying that **H** and **E** can be written in terms of **D**, or **H(D)** and **E(D)**. Similarly, the objective function can be written in terms of the control variables and is referred to as the reduced objective function F:

$$F(D_{pt}) = f(H_{jt}(D_{pt}), E_{sT}(D_{pt})) \tag{6.3.6}$$

By implicitly expressing the state variables in terms of the control variables, a smaller nonlinear optimization problem can be solved explicitly by an NLP code while delegating the burden of satisfying the hydraulic constraint equations [thus establishing the implicit functions **H(D)** and **E(D)**] through an hydraulic simulation code. The hydraulic simulation code KYPIPE by Wood (1980) for water distribution networks not only solves the hydraulic constraint equations but likewise satisfies the storage bound constraints from period $t = 1$ to $t = T - 1$. This added incentive is reflected in the discussion that follows. The reduced problem takes the form

$$\text{Min } Z_{\text{RP}} = \text{Min } F(H_{jt}(D_{pt}), E_{sT}(D_{pt}), D_{pt}) \tag{6.3.7}$$

subject to upper and lower bounds on D_{pt}, E_{sT}, and H_{jt}, where E_{st} and H_{jt} are written in terms of D_{pt} as $E_{st}(D_{pt})$ and $H_{jt}(D_{pt})$, and the reduced objective function F is expressed as a function of D_{pt}. The hydraulic simulator (KYPIPE by Wood, 1980) satisfies the set of hydraulic constraint equations, including the storage bound constraints except for the final time period, E_{sT}. It calculates the implicit functions $E_{sT}(D_{pt})$ and $H_{jt}(D_{pt})$. Figure 6.1 shows the linkage between the optimization and simulation codes.

In mathematical programming, the control and state variables may also be referred to as nonbasic and basic variables, respectively. Improvements in the objective function of the nonlinear programming problem is attained by a systematic variation of the nonbasic variables. NLP codes restrict the step size by which the nonbasic variables change so as not to violate their bounds. In an optimal control formulation, the determination of step size of the control (nonbasic) variables does not take ino consideration the values of the (basic) state variables. If the bounds on these state variables are violated, more iterations would be required to obtain a feasible solution.

As mentioned earlier, the procedure initially reduces the problem size by expressing the pressure heads, H_{jt}, and tank levels, E_{sT}, as functions of pump durations, D_{pt}. A penalty function method offers a further reduction in problem size. In general, the method incorporates the upper and lower state bounds into the objective function in the form of penalty terms. Specifically, the simple penalty function method approximates the optimal so-

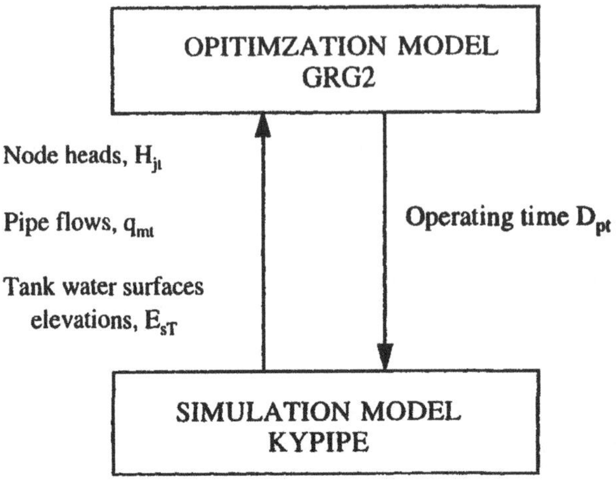

Figure 6.1 Optimization-simulation model linkage (Brion, 1990).

lution to the problem from exterior points. However, it has ill-conditioning effects when the penalty weights become excessively large; that is, the problem terminates before finding the real local optimizer (Bazaraa and Shetty, 1979). A variant of the general penalty method, the augmented Lagrangian method (Hsin, 1980), is used to formulate the optimal control problem. The mathematical derivation of the equations using the augmented Lagrangian method is given in Chapter 3.

Each state bound constraint is converted to the form of a penalty term and is added to the original objective function. The head bound penalty terms and final storage bound penalty terms are added to the original objective function f to develop the augmented Lagrangian function, AL,

$$\text{Min AL}\,(H_{jt}, E_{sT}, D_{pt}, \mu_{jt}, \eta_{st}, \sigma_{jt}, \beta_{st}) \tag{6.3.8}$$
$$= f(H_{jt}, E_{sT}, D_{pt}) + \frac{1}{2}\sum_i \sigma_i \left\{\min\left[0, b_i - \frac{\mu_i}{\sigma_i}\right]\right\}^2 - \frac{1}{2}\sum_i \frac{\mu_i^2}{\sigma_i}$$

The index i is a one-dimensional index representation of the double index (j, t) for head bound penalty terms and double index (s, t) for storage bound penalty terms; σ_i and μ_i are penalty eights and Lagrange multipliers for the ith penalty term, respectively. Furthermore, b_i is the bound constraint violation term which is negative if a bound constraint violation indeed occurs. Bound constraints, which are sets of upper and lower limit constraints, are incorporated into the objective function as a single penalty term. At any given time, only one of the two bounds may be violated. State bound constraints are now considered in the determination of the step size used in the search for the optimal solution.

Combining the two approaches presented so far, the original problem is recast into the reduced problem with the augmented Lagrangian formulation, given by

$$\text{Min } L(D_{pt}, \mu_{jt}, \eta_s, \sigma_{jt}, \beta_s) = f(H_{jt}, E_{sT}, D_{pt})$$
$$+ \frac{1}{2}\sum_{tj} \sigma_{jt} \left\{\min\left[0, b_{jt} - \frac{\mu_{jt}}{\sigma_{jt}}\right]\right\}^2 - \frac{1}{2}\sum_{tj}\frac{\mu_{jt}^2}{\sigma_{jt}}$$
$$+ \frac{1}{2}\sum_{s} \beta_s \left\{\min\left[0, c_s - \frac{\eta_s}{\beta_s}\right]\right\}^2 - \frac{1}{2}\sum_{s}\frac{\eta_s^2}{\beta_s} \quad (6.3.9)$$

subject to

$$0 \le D_{pt} \le \Delta_t, \quad p = 1, \ldots, P, t = 1, \ldots, T \quad (6.3.10)$$

Lagrange multipliers, μ_{jt}, and penalty weights, σ_{jt}, are associated with the head bound penalty terms; η_s and β_s are the Lagrange multipliers and penalty weights, respectively, associated with the final storage bound penalty terms. Head bound violations, b_{jt}, and final storage bound violations, c_s, are defined as

$$b_{jt} = \min(\underline{b}_{jt}, \overline{b}_{jt}) \quad \text{with } \underline{b}_{jt} = H_{jt} - \underline{H}_{jt}, \overline{b}_{jt} = \overline{H}_{jt} - H_{jt} \quad (6.3.11)$$

$$c_s = \min(\underline{c}_s, \overline{c}_s) \quad \text{with } \underline{c}_s = E_{sT} - \underline{E}_{sT}, \overline{c}_s = \overline{E}_{sT} - E_{sT} \quad (6.3.12)$$

Again, the penalty terms associated with the storage bound constraints for $t = 1$ to T were not considered because they are implicitly satisfied by the hydraulic simulation code (KYPIPE).

6.3.3 Solution of the Reduced Problem

The reduced problem, Eqs. (6.3.9)–(6.3.10), is the final formulation to be solved by the nonlinear (NL) optimizer, such as the generalized reduced gradient code GRG2 by Lasdon and Waren (1986). The set of decision variables is narrowed down to include only the control variables. The upper and lower bounds imposed on the control variables are simply handled by the NLP algorithm in order to approach or maintain feasibility during its search of an optimal solution to the problem. The reduction steps of the solution process are summarized in Figure 6.2. It should be noted that in this figure, the variables are evaluated only within an inner level optimization as explained next.

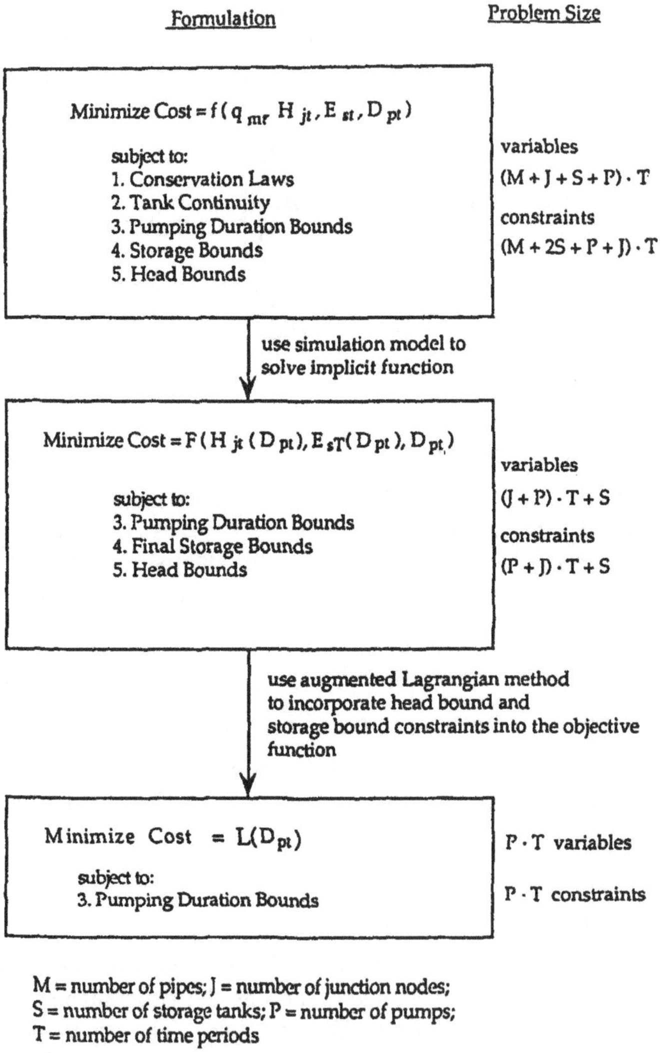

Figure 6.2 Transformation of optimal pump operation problem (Brion, 1990).

Based on the above formulation, the proposed solution methodology can be summarized as follows (also see Figure 6.3). The method is a two-level optimization where the final set of variables are partitioned into the control variables, D_{pt}, and the augmented Lagrangian variables μ, η, σ, and β, and has the objective function

$$\text{Min } Z_{\text{RPAL}} = \text{Min}\{\text{Min } L(D_{pt}, \mu_{jt}, \eta_{st}, \sigma_{jt}, \beta_{st})\} \qquad (6.3.13)$$

Initially, the penalty weights and Lagrange multipliers are fixed and the optimizer is used to solve the inner level minimization for D_{pt}. Given these optimal values of D_{pt}, the outer level minimization involving μ, η, σ, and β is then carried out. If a convergence criterion is not met or if an iteration limit is yet to be reached, the outer level variables are updated and passed inside the inner level minimization, and the procedure is repeated. The inner level or loop is a nonlinear optimization subproblem which takes different forms based on the current values of the penalty weights and Lagrange multipliers. This loop is solved repeatedly by the generalized reduced gradient code GRG2.

The outer level or loop is the master problem and is solved by a heuristic based on Fletcher's (1975) algorithm. Using updating formulas (Brion, 1990), the values of the Lagrange multipliers and penalty weights are revised. The types of updating formulas vary and the one that follows the steepest descent criteria has the form

$$\begin{aligned} \mu_i^{m+1} &= \mu_i^m - \sigma_i c_i \quad && \text{if } c_i \le \frac{\mu_i}{\sigma_i} \\ &= 0 && \text{if } c_i > \frac{\mu_i}{\sigma_i} \end{aligned} \qquad (6.3.14)$$

where m is the outer loop iteration index. Convergence is checked at each iteration by evaluating a convergence factor r and testing its value against a preset convergence limit. At the very beginning, r is set to a large absolute number and is then updated by the formula

$$r = \max\left|\min\left\{\sigma_i, \frac{\mu_i}{\sigma_i}\right\}\right| \qquad (6.3.15)$$

If r is less than or equal to the preset convergence limit, then the augmented Lagrangian method is said to have converged; otherwise, the outer loop undergoes another iteration.

6.3.4 Computation of Reduced Gradients

The state of a water distribution system can be defined by a specification of all pipe flow rates or all nodal heads at any given time of the day. This

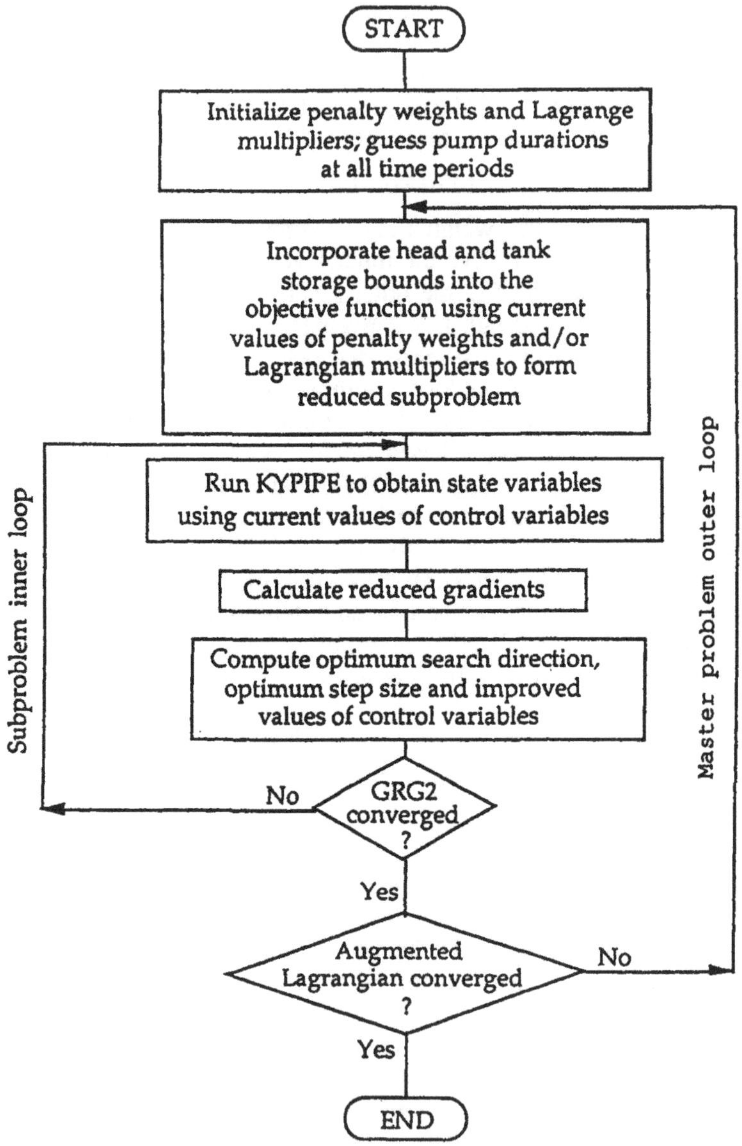

Figure 6.3 Flow chart of overall optimization model (Brion, 1990).

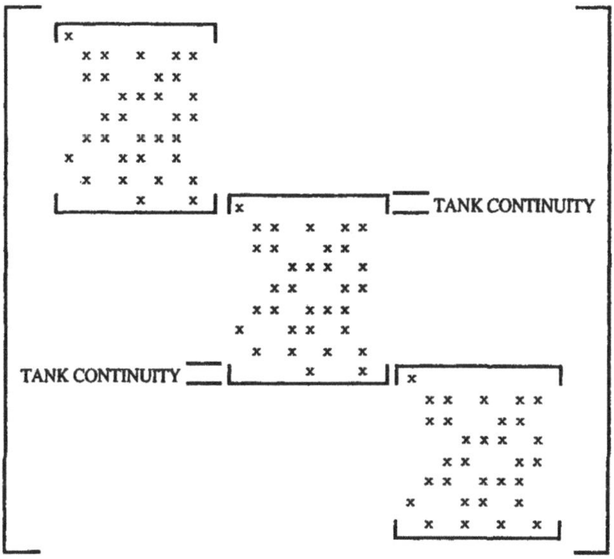

Figure 6.4 Matrix structure for time sequence of demands (Brion, 1990).

is evident by the use of the Hazen–Williams equation which ensures that the pipe flow rates determine nodal pressures and vice versa. Furthermore, the solution of the system hydraulic equations would represent system response to a steady-state simulation for independent loading conditions or demand patterns. For an extended perod simulation where a series of demand patterns make up the daily cycle loads, tank continuity equations link the submatrices representing the different sets of system equations for each demand pattern, as shown in Figure 6.4. Thus, for a time sequence of demands, the state space would comprise the entire set of nodal pressure heads plus tank levels. The state is said to be defined at the junction nodes and at the storage tanks. The gradient computations are performed using **D** as the vector of control variables and **H** as the contiguous vector of state variables. The water level at a fixed grade node is not part of the state space because its value is constant whatever the demand pattern. No system equation is required at this location in order to perform a pipe network simulation.

The derivatives of the reduced objective function with respect to the control variables are called the reduced gradients. Functions $F(\mathbf{D})$ and $\mathbf{H}(\mathbf{D})$ are implicit functions and all the gradients cannot be directly calculated. All are differentials; however, these functions cannot be evaluated in closed form. The two-step procedure of Mantell and Lasdon (1978) can be used to compute the reduced gradients. First, by applying the chain rule,

$$\frac{\partial F}{\partial \mathbf{D}} = \frac{\partial f}{\partial \mathbf{D}} + \left(\frac{\partial f}{\partial \mathbf{H}}\right)^T \frac{\partial \mathbf{H}}{\partial \mathbf{D}} \tag{6.3.16}$$

Only the matrix $\partial \mathbf{H}/\partial \mathbf{D}$ cannot be directly determined given the objective function. In order to evaluate this, the hydraulic constraints, or the **G** equations, are used. Taking the derivative of the flow constraint equations with respect to **D**,

$$\frac{\partial \mathbf{G}}{\partial \mathbf{D}} = \frac{\partial \mathbf{g}}{\partial \mathbf{D}} + \left(\frac{\partial \mathbf{g}}{\partial \mathbf{H}}\right)^T \frac{\partial \mathbf{H}}{\partial \mathbf{D}} = \mathbf{0} \tag{6.3.17}$$

where

$$\frac{\partial \mathbf{H}}{\partial \mathbf{D}} = -\left(\frac{\partial \mathbf{g}}{\partial \mathbf{H}}\right)^{-1} \frac{\partial \mathbf{g}}{\partial \mathbf{D}} \tag{6.3.18}$$

Substituting Eq. (6.3.18) into Eq. (6.3.17)

$$\frac{\partial F}{\partial \mathbf{D}} = \frac{\partial f}{\partial \mathbf{D}} - \left[\left(\frac{\partial f}{\partial \mathbf{H}}\right)^T \left(\frac{\partial \mathbf{g}}{\partial \mathbf{H}}\right)^{-1} \frac{\partial \mathbf{g}}{\partial \mathbf{D}}\right] \tag{6.3.19}$$

and defining

$$\boldsymbol{\pi}^T = \left(\frac{\partial f}{\partial \mathbf{H}}\right)^T \left(\frac{\partial \mathbf{g}}{\partial \mathbf{H}}\right)^{-1} \tag{6.3.20}$$

from which $\boldsymbol{\pi}$, the Lagrange multipliers, can be calculated by solving the system of linear equations:

$$\left(\frac{\partial \mathbf{g}}{\partial \mathbf{H}}\right)^T \boldsymbol{\pi} = \frac{\partial f}{\partial \mathbf{H}} \tag{6.3.21}$$

In this procedure, the derivatives of the network equations and objective function are needed with respect to each variable, **D** and **H**. Using these gradients, the Lagrange multipliers can be computed by Eq. (6.3.21). With these multipliers and the remaining known gradients, the derivative of the reduced objective can be calculated from Eq. (6.3.19). The Lagrange multiplier $\boldsymbol{\pi}$ has a useful physical meaning. At the optimum, it defines the change in the objective function due to a small change in the duration of pumping at the corresponding pump locations.

Although the flow equations are solved by a hydraulic simulator in order to obtain **q** and **H**, the head equations and tank continuity equations are equally complete and accurate representations of the state of the system. The loop equations are easier to solve at the expense of requiring analysis of the geometry of the network. The node equations, on the other hand, are more manageable for gradient computation because they basically represent the system node connectivity matrix at each time step, with the tank continuity equations as links between adjacent time steps. This is shown in Figure 6.4. To calculate the reduced gradients of the objective function, four terms, $\partial f/\partial \mathbf{H}$, $\partial f/\partial \mathbf{D}$, $\partial \mathbf{g}/\partial \mathbf{H}$, and $\partial \mathbf{g}/\partial \mathbf{D}$, are to be computed. The first two terms are actually the derivatives of the objective function; that is, the cost function with respect to the state and control variables, respectively. The last two terms, taken together, form the Jacobian matrix of the **G** system equations; that is, the first partial differential of the matrix $\mathbf{g}(\mathbf{H}, \mathbf{D})$ with respect to the vector $(\mathbf{H}, \mathbf{D})$. In compact form,

$$\mathbf{J}(\mathbf{H}, \mathbf{D}) = \left[\frac{\partial \mathbf{g}}{\partial \mathbf{H}}, \frac{\partial \mathbf{g}}{\partial \mathbf{D}}\right] = [\mathbf{B}, \mathbf{C}] \tag{6.3.22}$$

where the basis matrix $\mathbf{B}$ $(= \partial \mathbf{g}/\partial \mathbf{H})$ is assumed to be nonsingular at all points and is comprised of diagonal block matrices; the right partition $\mathbf{C}$ $(= \partial \mathbf{g}/\partial \mathbf{D})$ is defined only at the lower portions of each diagonal block corresponding to the tank continuity equations. The elements of $\mathbf{B}$ are nonlinear functions of $\mathbf{H}$; the elements of $\mathbf{C}$ are linear in terms of $\mathbf{D}$. For a time sequence of demands, the submatrices of $\mathbf{B}$, which represent separate demand patterns, are symmetrical and are very sparse. This special structure is lost when tank continuity equations link these submatrices to form the whole time sequence of demands (Fig. 6.4). The Lagrange multipliers are computed for the entire sequence of demands which involves the solution to a sparse matrix $\partial \mathbf{g}/\partial \mathbf{H}$. Sparse matrix solvers, such as the one used in KYPIPE, are available, but a more efficient way, based on optimal control theory, considers each demand pattern individually rather than as a whole.

The calculation of $\boldsymbol{\pi}$ starts at the final time step T by solving the system of equations

$$\left(\frac{\partial \mathbf{g}_T}{\partial \mathbf{H}_t}\right)^T \boldsymbol{\pi}_T = \left(\frac{\partial \mathbf{f}}{\partial \mathbf{H}}\right)_T \tag{6.3.23}$$

Multipliers for time steps $T - 1$ to 1 are solved backward in succession by

$$\left(\frac{\partial \mathbf{g}_t}{\partial \mathbf{H}_t}\right)^T \boldsymbol{\pi}_T = \left(\frac{\partial \mathbf{f}}{\partial \mathbf{H}}\right)_t - \boldsymbol{\pi}_{t+1} \frac{\partial \mathbf{g}_{t+1}}{\partial \mathbf{H}_t} \quad \text{for } t = T - 1, \ldots, 1 \tag{6.3.24}$$

Finally, the computation of the reduced gradients is done by appropriate substitution of the Lagrange multipliers in

$$\left(\frac{\partial F}{\partial \mathbf{D}}\right)_t = \left(\frac{\partial f}{\partial \mathbf{D}}\right)_t - \pi_t \frac{\partial \mathbf{g}_t}{\partial \mathbf{D}_t} \quad \text{for } t = 1, \ldots, T \tag{6.3.25}$$

Appendix 6.B shows how to compute each entry in the Jacobian matrix $[\partial \mathbf{g}/\partial \mathbf{H}, \partial \mathbf{g}/\partial \mathbf{D}]$, and the partial derivatives of the objective function with respect to the state and control variables, $\partial \mathbf{AL}/\partial \mathbf{H}$ and $\partial \mathbf{AL}/\partial \mathbf{D}$, respectively. The derivatives of the augmented Lagrangian (penalty) terms in Eq. (6.3.9), $\partial \mathbf{AL}/\partial \mathbf{H}$, are with reference to Eq. (6.3.16); the augmented Lagrangian terms, Λ_A, are

$$\Lambda_A(\mathbf{H}, \boldsymbol{\mu}, \boldsymbol{\sigma}) = \sum_i \begin{cases} -\mu_i c_i + \dfrac{1}{2}\sigma_i c_i^2 & \text{if } c_i \leq \dfrac{\mu_i}{\sigma_i} \\ -\dfrac{1}{2}\dfrac{\mu_i^2}{\sigma_i} & \text{otherwise} \end{cases} \tag{6.3.26}$$

where **H** represents the whole vector of states, nodal heads plus tank levels, and c_i is defined by (6.3.12). The derivatives of these terms with respect to the state variables become

$$\frac{\partial \Lambda_A}{\partial \mathbf{H}} = \begin{cases} 0 & \text{if } c_i > \dfrac{\mu_i}{\sigma_i} \\ -\mu_i + \sigma_i c_i & \text{if } c_i \leq \dfrac{\mu_i}{\sigma_i} \text{ and } c_i = H_i - \underline{H}_i \\ \mu_i - \sigma_i c_i & \text{if } c_i \leq \dfrac{\mu_i}{\sigma_i} \text{ and } c_i = \overline{H}_i - H_i \end{cases} \tag{6.3.27}$$

6.4 APPLICATION

Brion (1990) developed a computer code, PMPOPR, that interfaces GRG2 and KYPIPE for determining the optimal operation of pumping stations in water distribution systems. A very extensive system of software was developed in order to combine the augmented Lagrangian algorithm into a single cohesive computer code. Brion (1990) and Brion and Mays (1991) presented an application of the model to a pressure zone in Austin, Texas for a typical 24-hr day. This pressure zone consists of 126 pipes, 98 nodes, 5 pressure watchpoints, 3 pumps, 1 storage tank, and 12, 2-hr time periods.

Various computer runs of the model showed savings in pumping costs ranging from 5.2% to 17.3% over the actual operating costs for the day.

In the future, water distribution systems may be operated using optimal control systems consisting of the components shown in Figure 6.5. Such systems will be able to provide operators with an optimal operating policy for pump stations in a water distribution system. The optimal control sys-

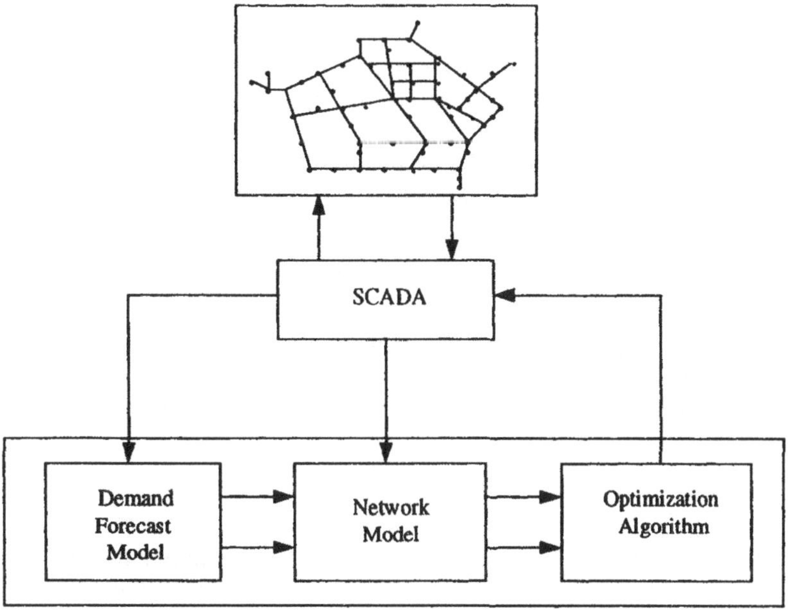

Figure 6.5 Optimal Control System (Ormsbee, 1991).

tem can be directly integrated with a SCADA (Supervisory Control and Data Acquisition) system to provide the link between the optimal control software and the system operator.

APPENDIX A: SIMULATION MODEL

6.A.1 Simulator Equations

One approach in steady-state analysis of pipe networks would be the generation and solution of mass continuity and energy conservation equations in terms of discharge in each pipe section. The resulting equations are referred to as loop equations, as opposed to node equations wherein the same set of equations are expressed in terms of total grades at junction nodes. At any rate, the solution to both formulations should yield flow rate in each pipe and pressure head at each node. It has been shown that the loop equations have superior convergence characteristics. These equations are used in the simulation model KYPIPE (Wood, 1980).

The mathematical relationship among a number of pipes, primary loops, junction nodes, and fixed grade nodes for all pipe systems is

$$M = N + \text{LP} + F_{\max} - 1 \tag{6.A.1}$$

where M is the number of pipe sections, N is the number of junction nodes, LP is the number of primary loops, and $F_{\max}$ is the number of fixed grade nodes.

For each junction node, a continuity relationship can be written wherein flow into the junction equals flow out of the junction:

$$\left(\sum q_{\text{in}}\right)_i - \left(\sum q_{\text{out}}\right)_i = (q_{\text{ext}})_i \quad i = 1, 2, \ldots, N \tag{6.A.2}$$

where $(q_{\text{ext}})_i$ represents the external inflow or demand at the junction node. There are N of these junction equations. For each primary loop, the energy equation can be written for the pipe sections in the loop as follows:

$$\left(\sum h_L\right)_1 = \left(\sum E_p\right)_1, \quad l = 1, 2, \ldots, \text{LP} \tag{6.A.3}$$

where h_L is the energy loss in a pipe in the loop (minor losses included) and E_p is the energy put into the liquid by a pump in the loop. For the case wherein no pumps exist within the loop, the sum of the energy losses within the loop becomes zero. $F_{\max} - 1$ independent energy equations can be written for paths between any two fixed grade nodes as follows:

$$(\Delta E)_f = \left(\sum h_L\right)_f - \left(\sum E_p\right)_f, \quad f = 1, 2, \ldots, F_{\max} - 1 \tag{6.A.4}$$

where ΔE is the difference in total grade between the two fixed grade nodes. Note that Eq. (6.A.3) can be considered as a special case of Eq. (6.A.4) where the difference in total grade, ΔE, is zero for a path which forms a closed loop. Jointly, there are LP + $F_{\max} - 1$ of these path equations; Eqs. (6.A.1)–(6.A.3), otherwise known as loop equations, constitute a set of p simultaneous nonlinear algebraic equations which describe steady-state flow analysis for the solution of the flow rate in each pipe. Path equations can be further modified so as to express them in terms of the flow rates.

The energy loss in a pipe, h_L, is the sum of the line loss h_{LP} and the minor loss h_{LM}. The line loss expressed in terms of the flow rate is given by

$$h_{\text{LP}} = K_p q^n \tag{6.A.5}$$

where K_p is a constant which is a function of line length (L), diameter (D), and roughness (C), or friction factor (f), and n is an exponent. The values of K_p and n depend on the energy loss expression used in the analysis. Using the Hazen–Williams equation,

$$K_p = \frac{XL}{C^{1.852}D^{4.87}} \tag{6.A.6}$$

and $n = 1.852$. In this equation, $X = 4.73$ for English units or $X = 10.69$ for SI units. Using the Darcy–Weisbach equation,

$$K_p = \frac{8fL}{gD^5\pi^2} \tag{6.A.7}$$

and $n = 2$. The minor loss in a pipe section expressed in terms of flow rate is given by

$$h_{\text{LM}} = K_M q^2 \tag{6.A.8}$$

where K_M is a constant which is the sum of the minor loss coefficients which, in turn, are functions of the number and type of fittings used.

The energy put into the liquid by a pump can be described by operating data. The within-range operation can be mathematically represented by a polynomial as follows:

$$E_p = A + Bq + Cq^2 \tag{6.A.9}$$

where A, B, and C are coefficients describing the characteristics of the pumps. Combining Eqs. (6.A.5), (6.A.8), and (6.A.9), and making the appropriate substitutions in Eq. (6.A.4), with Eq. (6.A.3) taken as a special case of Eq. (6.A.4), we have

$$(\Delta E)_f = \left(\sum (K_p q^n + K_M q^2)\right)_f - \left(\sum (A + Bq + Cq^2)\right)_f,$$

$$f = 1, \ldots, \text{LP} + F_{\text{max}} - 1 \tag{6.A.10}$$

A set of M simultaneous equations in terms of the unknown flow rates is formed by the N continuity equations (6.A.2) and the LP + F_{max} − 1 energy equations (6.A.10). The solution of the above equations, jointly called loop equations, involves the use of numerical or iterative methods, as the unknown flowrates could not be explicitly expressed in terms of the other variables in the system of equations. Wood and Charles (1972) suggested that the linearization scheme is the most reliable and efficient algorithm in solving the loop equations. A discussion of the algorithm follows.

6.A.2 Algorithm for the Solution of the Loop Equations: The Linear Method

A simple gradient method that handles the nonlinear flow rate in Eq. (6.A.10) is used in KYPIPE (Wood, 1980). Consider a single pipe section within a given path. A single term in Eq. (6.A.10) would then represent the grade difference across a pipe section carrying a flow rate q such that

$$f(q) = K_p q^n + K_M q^2 - (A + Bq + Cq^2) \tag{6.A.11}$$

Equation (6.A.11) is linearized using a first-order Taylor series approximation about the point $\mathbf{q} = \mathbf{q}_i$, where subscript i refers to the previous iteration or an initial guess. If we let $\Delta E = f(\mathbf{q})$, where $\mathbf{q} = \{q_1, q_2, \ldots, q_j \ldots\}$ = set of flow rates in the pipes contained in path f, then

$$f(\mathbf{q}) = f(\mathbf{q}_i) + \sum_j \frac{\partial f}{\partial \mathbf{q}}\bigg|_{\mathbf{q}_j=\mathbf{q}_{j,i}} (\mathbf{q}_j - \mathbf{q}_{j,i}) \tag{6.A.12}$$

where

$$f(\mathbf{q}_i) = \sum_j K_p \mathbf{q}_{j,i}^n + \sum_j K_M \mathbf{q}_{j,i}^2 - \sum_j (A + B\mathbf{q}_{j,i} + C\mathbf{q}_{j,i}^2) \tag{6.A.13}$$

and the gradient of the general form of Eq. (6.A.11) is

$$\frac{\partial f}{\partial \mathbf{q}} = \sum_j nK_p \mathbf{q}_j^{n-1} + \sum_j 2K_M \mathbf{q}_j - \sum_j (B + C\mathbf{q}_{j,i}) \tag{6.A.14}$$

Equation (6.A.12) is employed to formulate LP + $F_{\max}$ − 1 energy equations, which when combined with the already linear N continuity equations (6.A.2) form a set of M simultaneous linear equations in terms of flow rate in each pipe.

A unit flow in all pipes is assumed initially and the system of equations is solved using a modified sparse matrix solver. The computed flow rates become the new "assumed" flow rates and are used to evaluate the linearized equations and obtain a second solution. This procedure or trial is repeated until no significant change in the "assumed," and computed flow rates is observed or a maximum number of trials is reached. The first criterion is met if the relative accuracy, defined as the sum of the changes in flow rate between the last two trials divided by the sum of the flow rates, becomes less than a specified value (default value = 0.005). Given the final values of pipe flows, starting from a fixed grade node, Eqs. (6.A.5) and/or (6.A.8) can be written for each pipe to compute pressure heads.

The solution tends to oscillate about the actual solution such that the average of the previous two iterations tend to be very close to the true solution (Wood and Charles, 1972). Thus, q_i is redefined as

$$q_i = \left(\frac{q_{i-1} - q_{i-2}}{2}\right) \tag{6.A.15}$$

and this new value is used in Eq. (6.A.12). Because all flows are computed simultaneously, convergence occurs faster than other procedures (Wood, 1980). A high degree of accuracy is achieved using only four to eight trials even for a very large system.

APPENDIX B: COMPUTATION OF BASIS ELEMENTS

The following discussion deals with gradient computations. First, the Jacobian [$\partial \mathbf{g}/\partial \mathbf{H}$, $\partial \mathbf{g}/\partial \mathbf{D}$] will be derived taking into consideration the pipe, tank, and pump components of the system. Second, the derivative of the implicit derivatives of the reduced objective with respect to **H** and **D** will be shown.

The structure of the left partition of the Jacobian is called the basis matrix **B**. The node system of equations and tank continuity equations are solved for the unknown nodal total heads and tank water levels at all time steps during an extended period simulation.

Water reaches a node by way of conveyance through pipes converging at the node (Figure 6.6). The node equation for a node i with only pipe connections for any time step is

$$G_i\text{:}\quad \sum_j q_{ij} + Q_{ext} = \sum_j \frac{K_{ij} C_{ij} D_{ij}^{2.63}}{L_{ij}^{0.54}} [\text{sign}(H_i - H_j)][|H_i - H_j|]^{0.54} = 0 \tag{6.B.1}$$

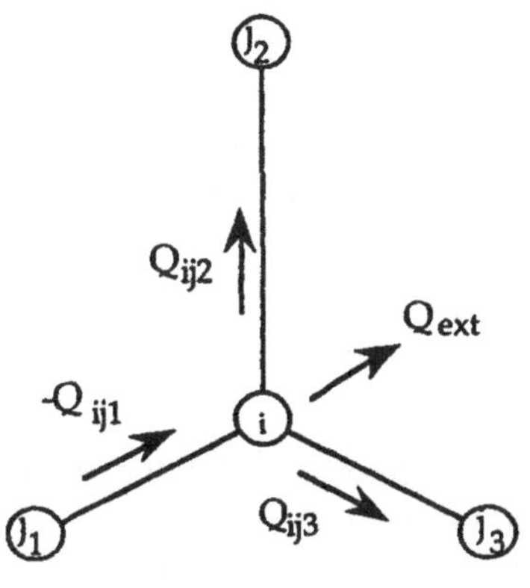

Figure 6.6 Node connected with three pipes (Brion, 1990).

where K_{ij} is a pipe parameter equal to $(4.737L)/(C^{1.852}D^{4.87})$ for English units and $(10.675L)/(C^{1.852}D^{4.87})$ for SI units; C_{ij} is the Hazen–Williams roughness coefficient; D_{ij} is the length of pipe ij; L_{ij} is the length of pipe ij; H_i is the total head at node i, and H_j is the total head at node j (or tank level if pipe ij connects node i to a storage tank j). Double subscript ij denotes a pipe connecting nodes i and j. The order in which these subscripts are written determine flow direction. Flow is positive if there is outflow from node i, and negative otherwise. External source Q_{ext} is positive for demand and negative for supply.

The derivative of Eq. (6.B.1) with respect to any H_j (including tank levels) is

$$\frac{\partial g_i}{\partial h_j} = \frac{0.54(K_{ij}C_{ij}D_{ij}^{2.63})}{(L_{ij}^{0.54})[\text{sign}(H_i - H_j)][|H_i - H_j|]^{-0.46}(-1)}$$
$$= \frac{-0.54(K_{ij}C_{ij}D_{ij}^{2.63})}{(L_{ij}^{0.54})[\text{sign}(H_i - H_j)][|H_i - H_j|]^{0.54}}$$
$$= \frac{-0.54\ \text{abs}(q_{ij})}{H_i - H_j} \tag{6.B.2}$$

Similarly, the relation exists for $\partial G_i/\partial H_i$ and can be shown to be

$$\frac{\partial g_i}{\partial h_i} = \sum_i \frac{0.54\ \text{abs}(q_{ij})}{H_i - H_j} \tag{6.B.3}$$

A variation of Eq. (6.B.3) exists for a node i connected to a fixed grade node f. Two cases will be discussed: a node connected to a fixed grade node without a pump in between, and a node connected to a fixed grade node with a pump in between. The no pump case merely requires the addition of the single term $0.54\ \text{abs}(q_{if})/(H_i - H_f)$ in Eq. (6.B.3), where H_f is the water surface elevation at the fixed grade node.

If pumps are online and if the KYPIPE pump representation is used, the derivation is as follows. Pumps operating within its three-point operating data has an exponential Q–H curve of the form

$$Q_p = -\left(\frac{H1 - H_p}{C}\right)^{1/B} \tag{6.B.4}$$

where H1 is the pump cutoff head, or the head above which the pump can no longer sustain the system pressure requirement, and B and C are pump curve parameters expressed as functions of the three-point ($H1 - Q1$, $H2 - Q2$, $H3 - Q3$) pump data. They are computed by

$$B = \frac{\log [(H1 - H3)/(H1 - H2)]}{\log(Q3)]} \tag{6.B.5}$$

$$C = \frac{H1 - H2}{Q_2^B} \tag{6.B.6}$$

The pump head H_p is the energy imparted by the pump on the fluid and is a linear function of the head at the downstream node i,

$$H_p = H_i + h_{L_{fi}} - H_f \tag{6.B.7}$$

where $h_{L_{fi}}$ is the head loss due to friction along pipe fi. Taking the derivative of Eq. (6.B.4) with respect to H_i yields

$$\frac{\partial Q_p}{\partial H_i} = -\left(\frac{1}{BC^{(1/B)}}\right)(H1 - H_p)^{(1/B)-1}\frac{\partial(H1 - H_p)}{\partial H_i}$$
$$= \left(\frac{1}{BC^{(1/B)}}\right)(H1 - H_p)^{(1/B)-1}\frac{\partial(H_p)}{\partial H_i} \tag{6.B.8}$$

Differentiating Eq. (6.B.7) gives

$$\frac{\partial H_p}{\partial H_i} = \frac{\partial H_i}{\partial H_i} + \frac{\partial h_{L_{fi}}}{\partial H_i} - \frac{\partial H_f}{\partial H_i} \tag{6.B.9}$$

The first term in Eq. (6.B.9) is identity, whereas the last term is zero. For suction pipes, the second term is negligible but whose contribution can be derived as follows. In English units, if we let $K_p = (4.737L)(C^{1.852}D^{4.87})$, then

$$\frac{\partial h_{L_{fi}}}{\partial H_i} = \frac{\partial(K_p Q_p^{1.852})}{\partial H_i}$$
$$= 1.852 K_p Q_p^{0.852}\frac{\partial Q_p}{\partial H_i} \tag{6.B.10}$$

Substituting Eq. (6.B.10) into Eq. (6.B.11) and then into Eq. (6.B.8), we obtain

$$\frac{\partial Q_p}{\partial H_i} = \left(\frac{1}{BC^{1/B}}\right)(H1 - H_p)^{(1/B)-1}\left(1 + 1.852\, K_p Q_p{}^{0.852}\frac{\partial Q_p}{\partial H_i}\right) \tag{6.B.11}$$

If we let $X1 = 1.852\, K_p\, Q_p^{0.852}$and $X2 = [(H1 - H_p)^{(1/B)-1}]/[BC^{1/B}]$, we get

$$\frac{\partial Q_p}{\partial H_i} = X2\left(1 + X1\,\frac{\partial Q_p}{\partial H_i}\right)$$

$$= X2 + X1X2\,\frac{\partial Q_p}{\partial H_i} \tag{6.B.12}$$

Finally, we have

$$\frac{\partial Q_p}{\partial H_i} = \frac{X2}{1 - X1X2} \tag{6.B.13}$$

If the pump operates above its third operating point, KYPIPE represents the pump head discharge relationship by a straight-line formula

$$Q_p = -\left(\frac{(H_p - A)}{S}\right) \tag{6.B.14}$$

where

$$S = -BC(Q3)^{B-1} \tag{6.B.15}$$

$$A = H3 - SQ3 \tag{6.B.16}$$

Proceeding as before,

$$\frac{\partial Q_p}{\partial H_i} = \left(-\frac{1}{S}\right)\frac{\partial(H_p - A)}{\partial H_i}$$

$$= \left(-\frac{1}{S}\right)\frac{\partial H_p}{\partial H_i} \tag{6.B.17}$$

By substituting Eq. (6.B.10) into Eq. (6.B.9) and then into Eq. (6.B.17), we obtain

$$\frac{\partial Q_p}{\partial H_i} = \left(-\frac{1}{S}\right)\left(1 + 1.852K_pQ_p^{0.852}\,\frac{\partial Q_p}{\partial H_i}\right) \tag{6.B.18}$$

$$\frac{\partial Q_p}{\partial H_i} = \left(-\frac{1}{S}\right)\left(1 + X1\,\frac{\partial Q_p}{\partial H_i}\right)$$

$$= -\left(\frac{1}{S} + \left(\frac{X1}{S}\right)\frac{\partial Q_p}{\partial H_i}\right) \tag{6.B.19}$$

Finally,

$$\frac{\partial Q_p}{\partial H_i} = -\frac{1}{S(1 + X1/S)}$$
$$= -\frac{1}{S + X1} \quad (6.B.20)$$

Equations (6.B.13) or (6.B.20) is added to Eq. (6.B.3) in cases wherein a source pump delivers flow into node i.

The analysis of tanks considers a sequence of demands with which tank levels vary with time. During each time step, the tank level is considered fixed and flow distribution is computed. A storage tank may be considered as a node with its water level as a state variable (Fig. 6.7). Continuity at the storage tank is a mass balance equation from one time step to the next and can be defined as

$$E_{st} = E_{st-1} - q_{sjt-1}\frac{D_t}{A_s} \quad (6.B.21)$$

where

E_{st} = water level in storage tank s at time step t
E_{st-1} = water level in storage tank s at time step $t - 1$
q_{sj-1} = flow rate in pipe connecting storage tank s to the rest of the system via node j and is considered positive if tank is emptying
D_t = duration of time between time step t and time step $t - 1$
A_s = horizontal cross-sectional area of tank s

Using the Hazen–Williams equation and transposing all terms on one side of the equality sign, we have

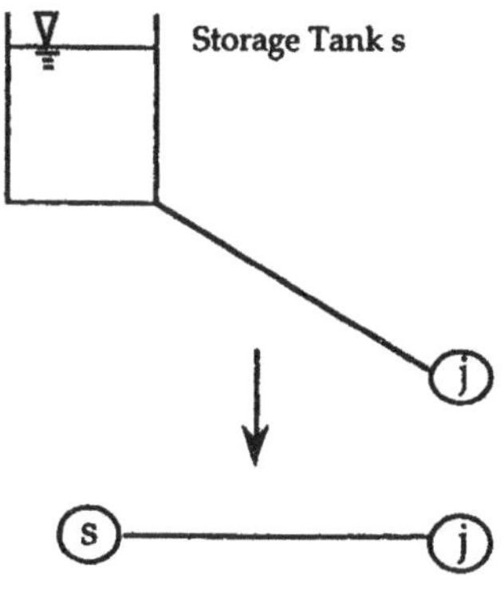

Figure 6.7 Tank representation (Lansey, 1987).

$$G_{st}: \quad -E_{st} + E_{st-1} \frac{K_{sj}C_{sj}D_{sj}^{2.63}}{L_{sj}^{0.54}} \cdot [\text{sign}(E_{st-1} - H_{jt-1})][|E_{st-1} - H_{jt-1}|]^{0.54} = 0 \quad (6.B.22)$$

This equation links the time sequence of system equations corresponding to the different demand patterns. The required derivatives used in computing the Lagrange multipliers are

$$\frac{\partial g_{st}}{\partial E_{st}} = -1 \quad (6.B.23)$$

$$\begin{aligned} \frac{\partial g_{st}}{\partial E_{st-1}} &= 1 - 0.54[\text{sign}(E_{st-1} - H_{jt-1})]K_p \left(\frac{D_t}{A_s}\right) \\ &\quad \times [|E_{st-1} - H_{jt-1}|]^{-0.46} \\ &= 1 - \left(\frac{0.54|q_{sjt-1}|}{(E_{st-1} - H_{jt-1})}\right)\left(\frac{D_t}{A_s}\right) \end{aligned} \quad (6.B.24)$$

$$\begin{aligned} \frac{\partial g_{st}}{\partial H_{jt-1}} &= -0.54[\text{sign}(E_{st-1} - H_{jt-1})]K_p \left(\frac{D_t}{A_s}\right) \\ &\quad \times [|E_{st-1} - H_{jt-1}|]^{-0.46}(-1) \\ &= \left(\frac{0.54|q_{sjt-1}|}{E_{st-1} - H_{jt-1}}\right)\left(\frac{D_t}{A_s}\right) \end{aligned} \quad (6.B.25)$$

$$\begin{aligned} \frac{\partial g_{st}}{\partial D_t} &= -[\text{sign}(E_{st-1} - H_{jt-1})] \left(\frac{K_p}{A_s}\right) [|E_{st-1} - H_{jt-1}|]^{-0.54} \\ &= -\frac{q_{sjt-1}}{A_s} \end{aligned} \quad (6.B.26)$$

where K_p is as previously defined. The procedure applied to compute the Lagrange multipliers for multiple time steps is discussed in Sect. 6.3. Essentially, computation of the element of the state space is done forward in time, whereas the multipliers are solved backward in time. Equations (6.B.24) and (6.B.25) may be converted to apply from time steps $T - 1$ to 1. The more useful forms are

$$\frac{\partial g_{st+1}}{\partial E_{st}} = 1 - \left(\frac{0.54|q_{sjt}|}{E_{st} - H_{jt}}\right)\left(\frac{D_{t+1}}{A_s}\right) \quad (6.B.27)$$

$$\frac{\partial g_{st+1}}{\partial H_{jt}} = \left(\frac{0.54|q_{sjt}|}{E_{st} - H_{jt}}\right)\left(\frac{D_{t+1}}{A_s}\right) \quad (6.B.28)$$

The partial derivative of the original objective function with respect to the control variable is the gradient of the pumping cost. It is always positive because the pumping cost increases as pumps are used for longer durations. Mathematically, it is given by

$$\frac{\partial f}{\partial D_{pt}} = \frac{1}{\text{EFF}_{pt}} \text{UC}_t \frac{(0.746\ \gamma\ Q_{pt} H_{pt})}{550} \tag{6.B.29}$$

where all terms are defined previously.

The partial derivative of the original objective function with respect to state variable H can be derived by examining the effect of pumps on the rest of the network. First, the objective function is expressed in terms of the state variable H. The pump may be considered as a special type of pipe with a head–flow relationship mathematically defined by the pump characteristic curve. If placed online, additional energy is brought into the flow. A pipe with a pump will still experience energy loss by friction in the direction of flow. Thus, the total head at the downstream end of a pipe with a pump is equal to the total head at the upstream end minus the frictional head loss plus the pump head or

$$H_{\text{ds}} = H_{\text{us}} - h_L + H_p \tag{6.B.30}$$

The downstream (ds) end would be a node in the system, whereas the upstream (us) end would be a fixed grade node. By using the previous notation, Eq. (6.B.30) is transformed into

$$H_p = H_i - H_f + h_{Lfi} \tag{6.B.31}$$

H_i is a state variable and h_{Lfi} is a function of the pipe flow which is also the pump discharge. Because pump discharge is a function of pump head, we can deduce that it is also a function of the state of the system by way of Eq. (6.B.31). By the same reasoning, pump efficiency is also a function of the state of the system. For illustration purposes, the partial derivative of the cost function with respect to the state variables will be derived for the type of pump head–discharge-efficiency relationship used in the first and second example applications.

KYPIPE fits an exponential curve to define the pump characteristics (Fig. 6.8). Based on a user specified set of three operating points, KYPIPE determines the coefficients of the exponential function defined in Eq. (6.B.4). Above the third operating point, KYPIPE uses an alternate curve by extending the characteristic beyond the third point using the linear form Eq. (6.B.14). Assuming a maximum efficiency, $\text{EFF}_{\max}$, at the second operating point that linearly decreases to a minimum, EFF_{min}, at the other

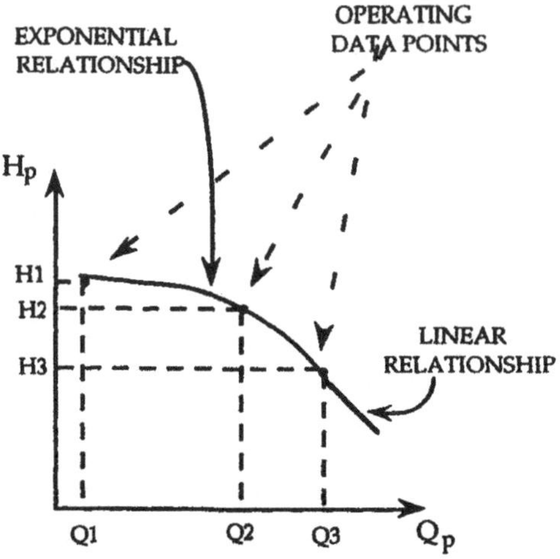

Figure 6.8 Pump *H–Q* curve representation in KYPIPE (Lansey, 1987).

operating points; a triangular efficiency "curve" is defined (Fig. 6.9) and has the form

$$\begin{aligned} \text{EFF} &= \text{EFF}_{\max} && \text{if } Q_p = Q2 \\ &= aQ_p + b && \text{if } Q1 < Q_p < Q3 \\ &= \text{EFF}_{\min} && \text{otherwise} \end{aligned} \tag{6.B.32}$$

where a and b are the slope and intercept, respectively. The three regions marked in Figures 6.8 and 6.9 require three different types of gradients for the cost function. They are

Case I

$$\frac{\partial f}{\partial H_i} = K4C4 \left\{ \frac{(k6H_{ip}^2) + k4(H_{1p} - H_{1pd})}{(k6H_{1p} + k4)^2} \right\} \tag{6.B.33}$$

Case II

$$\frac{\partial f}{\partial H_i} = K4C4 \left\{ \frac{(k5H_{ip}^2) + k2(H_{1p} - H_{1pd})}{(k5H_{1p} + k2)^2} \right\} \tag{6.B.34}$$

Case III

$$\frac{\partial f}{\partial H_i} = K2C4\{2H_p - A\} \tag{6.B.35}$$

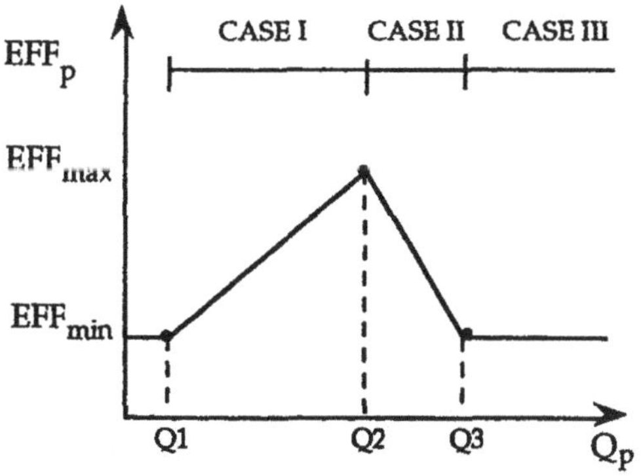

Figure 6.9 Triangular pump efficiency curve (Lansey, 1987).

where

$$H_{1p} = (H1 - H_p)^{1/B}$$

$$H_{1pd} = \left(-\frac{1}{B}\right)(H1 - H_p)^{(1/B)-1}$$

$$k1 = \frac{\mathrm{EFF}_{\max} - \mathrm{EFF}_{\min}}{Q2 - Q1}$$

$$k2 = \mathrm{EFF}_{\max} - k1Q2$$

$$k3 = \frac{\mathrm{EFF}_{\max} - \mathrm{EFF}_{\min}}{Q2 - Q3}$$

$$k4 = \mathrm{EFF}_{\max} - k3Q2$$

$$k5 = \frac{k1}{C^{1/B}}$$

$$k6 = \frac{k3}{C^{1/B}}$$

$$K1 = \mathrm{UC}_t \frac{0.746\ \gamma D_{pt}}{550}$$

$$K2 = \frac{K1}{\mathrm{EFF}_{\min} S}$$

$$K3 = \frac{K1}{\mathrm{EFF}_{\min} C^{1/B}}$$

$$K4 = \frac{K1}{C^{1/B}}$$

$$C1 = 1.852K_pQ_p^{0.852}$$

$$C2 = \frac{H_{1pd}}{C^{1/B}}$$

$$C3 = C1C2$$

$$C4 = 1 + \frac{C3}{1 - C3}$$

Other forms of functions relating H_p, Q_p, EFF_p, and even the unit pumping cost, UC_p (if a per usage rate structure is used), require different forms of the gradient of the cost function with respect to the state variable.

REFERENCES

Bazaraa, M.S. and Shetty, C.M., *Nonlinear Programming: Theory and Algorithms*, John Wiley & Sons, New York, 1979.

Brion, L.M., Methodology for Optimal Operation of Pumping Stations in Water Distribution Systems, Ph.D. dissertation, The University of Texas at Austin, Austin, 1990.

Brion, L.M. and Mays, L.W., Methodology for Optimal Operation of Pumping Station in Water Distribution Systems, *Journal of Hydraulic Engineering*, Vol. 117, No. 11, pp. 1551–1571, 1991.

Carpentier, P. and Cohen, G., Decomposition, Coordination and Aggregation in the Optimal Control of a Large Water Supply Network, *Proceedings of the Ninth Triennial World Congress of IFAC*, J. Gertler and L. Keviczky (eds.), Pergamon Press, Elmford, N.Y, 1985, Vol. 6, pp. 3207–3212.

Chase, D.V. and Ormsbee, L.E., Optimal Pump Operation of Water Distribution System with Multiple Storage Tanks. Proceedings, ASCE Conference on Water Resources Planning and Management, Sacramento, California, 1989, pp. 733–736.

Cohen, G., Optimal Control of Water Supply Networks, in *Optimization and Control of Dynamic Operational Research Models*, S. G. Tzafestas (ed.), North-Holland, Amsterdam, 1982, Vol. 4, pp. 251–276.

Coulbeck, B. and Orr, C.H., Optimized Pumping in Water Supply Systems, *Proceedings of the Ninth Triennial World Congress of IFAC*, J. Gertler and L. Keviczky (eds.), Pergamon Press, Elmsford, NY, 1985, Vol. 6, pp. 3175–3180.

Coulbeck, B. and Sterling, M.J.H., Optimised Control of Water Distribution Systems, *IEE Proceedings*, Vol. 125, No. 9, pp. 1039–1044, 1978.

Duan, N., Mays, L.W., and Lansey, K.E., Optimal Reliability-Based Design and

Analysis of Pumping Systems for Water Distribution Systems, *Journal of Hydraulic Engineering*, Vol. 116, No. 2, pp. 249–268. 1990.

Fallside, F. and Perry, P.F., Hierarchical Optimization of Water-Supply Network, *IEE Proceedings*, Vol. 122, No. 2, pp. 202–208, 1975.

Fletcher, R., An Ideal Penalty Function for Constrained Optimization, *Journal of the Institute of Mathematics and its Applications*, Vol. 15, pp. 319–342, 1975.

Hsin, J.K., The Optimal Control of Deterministic Econometric Planning Models, Ph.D. dissertation, The University of Texas at Austin, Austin, 1980.

Joalland, G. and Cohen, G., Optimal Control of a Water Distribution Network by Two Multilevel Methods, *Automatica*, Vol. 16, pp. 83–88, 1980.

Lansey, K., Optimal Design of Large-Scale Water Distribution Systems, Ph.D. Dissertation, Dept. of Civil Engineering, University of Texas at Austin, 1987.

Lansey, K.E. and L.W. Mays, Optimization Model for Water Distribution System Design, *Journal of Hydraulic Engineering*, Vol. 115, No. 10, pp. 1401–1418, Oct. 1989.

Lasdon, L.S. and Waren, A.D., *GRG2 User's Guide*, The University of Texas at Austin, Austin, 1986.

Lasdon, L.S., Waren, A.D., Jain, A., and Ratner, M.S., Design and Testing of a Generalized Reduced Gradient Code for Nonlinear Programming, *ACM Transactions on Mathematical Software*, Vol. 4, No. 1, pp. 34–50, 1978.

Luenberger, D.G., *Linear and Nonlinear Programming*, 2nd ed., Addison-Wesley, Reading, MA, 1984.

Mantell, J., and Lasdon, L.S., AGRG Algorithm for Econometric Central Problems, *Annals of Economic and Social Management*, Vol. 6, pp. 581–597, 1978.

Ormsbee, L.E. (ed.), Energy Efficient Operation of Water Distribution Systems, Report by the ASCE Task Committee on the Optimal Operation of Water Distribution System, Research Report No. UKCE 9104, Dept. of Civil Engineering, University of Kentucky, Lexington, 1991.

Ormsbee, L.E., Walski, T.M., Chase, D.V., and Sharp, W.W., Techniques for Improving Energy Efficiency at Water Supply Pumping Stations, Technical Report EL-87-16, Environmental Laboratory, U.S. Army Engineer Waterways Experiment Station, Vicksburg, MS, November, 1987.

Powell, M.J.D, Algorithms for Nonlinear Constraints That Use Lagrangian Functions, *Mathematical Programming*, Vol. 14, No. 2, pp. 224–248, 1978.

Sabet, M.H., and Helweg, O.J., Cost Effective Operation of Urban Water Supply System Using Dynamic Programming, *Water Resources Bulletin*, Vol. 21, No. 1, pp. 75–81, 1985.

Solanas, J.L., and Montolio, J.M., The Optimum Operation of Water Systems, *International Conference, Computer Applications for Water Supply and Distribution*, Leicester Polytechnic, Leicester, England, 1987.

Solanas, J.L., and M. Vergés, "Approximations Procedure and its Application to Automatic Operational Control of Water Distribution Systems," *IFAC-IFORS Symposium*, Varan, Bulgaria, October 1974.

Wood, D.J., *Computer Analysis of Flow in Pipe Networks Including Extended Period Simulation—User's Manual*, University of Kentucky, Lexington, 1980.

Wood, D.J., and Charles, C., Hydraulic Network Analysis Using Linear Theory, *Journal of the Hydraulics Division*, Vol. 98, No. 7, pp. 1157–1170, 1972.

Zessler, U. and Shamir, U., Optimal Operation of Water Distribution Systems, Unpublished report, Technion—Israel Institute of Technology, Haifa, 1985.

Chapter 7
Freshwater Inflows to Estuaries

7.1 PROBLEM IDENTIFICATION

In many areas of the country, particularly the Gulf Coast states, California, and elsewhere in the world, the freshwater discharge of rivers has become a limited commodity, for which the need for freshwater inflow to maintain the productivity of coastal estuaries must compete with the demands of upstream users (viz. municipal and industrial uses, and agriculture). The desired approach to water resources management is to optimize flow into the estuary (by minimizing the total volume of flow, or by maximizing the diversions and storage within limits of water rights and capacity, or both) while preserving an acceptable habitat in specific regions of the estuary to accommodate the requirements of key organisms. Salinity has been long established as an index to ecological habitat in an estuary because it measures the relative proportion of fresh water to sea water. Even for those organisms which are euryhaline (i.e., whose physiology can accommodate wide excursions of salt concentration), salinity still provides a useful habitat index because of other "information" contained in the freshwater ratio, such as nutrient supply, sediment and detritus, or stenohaline components of the food web.

A key element in this optimization problem is the mathematical relation between salinity in the estuary and flow, $S = \Phi(Q)$. Usually, the relation is based on statistical association (i.e., a regression form established from field data). The Texas Water Development Board (TWBD) has made particularly extensive application of this approach in establishing freshwater inflow requirements, as a part of its Bays and Estuaries Program. The work of the TWDB probably represents the most extensive incorporation of water requirement for estuaries within a larger water resources management context, and the Texas bays are an excellent model for similar problems elsewhere.

The statistical regression $S = \Phi(Q)$ proves to be extremely noisy because of the variability in salinity. In the case of the Texas bays, nearly the entire possible range of salinity values can be found in the historical field data for any given value of concurrent inflow. The reasons for this are twofold. First, the value of salinity in a given region of the bay is dependent on several other factors in addition to freshwater inflow, notably the various hydrodynamic circulation processes, including tides, responses of the bay to meteorological forcing, and the effect of density currents particularly operating in conjunction with deep-draft ship channels. Second, the time scale of response of salinity is typically much longer than the variability of freshwater inflow. The value of salinity is the integrated response to perhaps several months of the freshwater inflow "signal."

It should also be noted that the optimization problem as summarized above is, in fact, time varying, primarily because the salinity requirements of key organisms in the estuary will vary with season through the year, depending on the life stage of the organism and its presence or absence within the estuary. (Many of the important commercial species are anadromous, migrating into or out of the estuary.) The salinity limits for a specific organism are based on the statistical association between the presence of that organism in the estuary (as reflected in catch data or harvest data) and salinity, or on the physiological dependence on salinity as revealed in laboratory studies. Thus far, the optimization problem has only been treated on a steady-state basis. Accommodation of the seasonal variation in salinity requirement was made by the TWDB by subdividing the year into several seasons and solving the steady-state problem separately for each season. The most general formulation of the problem, however, should accommodate not only seasonal variation in salinity limits of the organisms but also seasonal variation in upstream water demands and the specific time response of salinity to freshwater inflow.

The essential weakness in the above formulation is the mathematical expression of salinity dependence on freshwater inflow to the bay. This chapter reformulates the problem, replacing the statistical regression $S = \Phi(Q)$ with a mathematical model of hydrodynamic transport, relating salinity at a given point in the estuary to a time-varying boundary condition of riverine inflow. Such an approach has the following advantages:

1. More accurate and self-consistent definition of salinity as a function of flow, enabling greater precision in the optimization results
2. Explicit incorporation of physical processes other than freshwater inflow affecting salinity in the real system, including tides, meteorology, and internal circulations
3. The ability to accommodate time variation in the response of salinity to freshwater inflow, so as to readily generalize to the full time-varying problem (although the optimization problem can also be solved in a steady-state framework with steady inflows)
4. The ability to accommodate generalization to full time variation in upstream water demands, including seasonality of irrigation and long-term demographic changes
5. The ability to consider either averaged inflow, prespecified scenarios of inflow, or long-term simulations using real hydrological data

In some estuaries, a direct measure of organism abundance is available in the data on commercial fishery landings taken from the estuary. This "harvest" data can be employed as an index of populations of key organisms and analyzed statistically to establish its dependence on freshwater inflow,

$$H_k = f(Q) \tag{7.1.1}$$

Although this might appear superior to the indirect salinity-index approach, the causal connection between flow and harvest may be obscured by unmeasureable parameters of the fishing process such as effort, selectivity, and skill, and may be corrupted by poor reporting or the difference between locality of landing (i.e., port) and locality of catch, to say nothing of other environmental variables unrelated to inflow. This regression therefore tends to be noisy and statistically uncertain. On the other hand, it is directly pertinent to the problem, and when the data are available, it should be accommodated within the optimization problem, either as an objective function or as a constraint.

7.2 PROBLEM FORMULATION

7.2.1 Hydrodynamic Transport Simulator for Estuaries

The essence of this chapter is to develop a general methodology for the estuarine freshwater resources management, so that for discussion purposes, the hydrodynamic transport model needed for simulation of temporal and spatial variation of salinity is not restricted to a particular model. The selection of an appropriate model depends on a number of factors such as efficiency, accuracy, complexity, and availability of the model. Even if a desired hydrodynamic transport model has been chosen and applied in the simulation, a better model can always be used to replace it in the future as more efficient models are developed. The formulation of hydrodynamic and transport governing equations varies slightly for each model, depending on the various assumptions and approximations introduced. The model used for discussion purposes is a two-dimensional, finite-difference model. Such a model as HYD-SAL (see Appendix 7.A) is used as an example for the formulation of governing equations and their finite-differencing approximations.

The governing equations for the two-dimensional horizontal model are the vertical-averaged equations of momentum, continuity and salinity mass budget: the momentum equation in the x-direction,

$$\frac{\partial q_x}{\partial t} - \Omega q_y = -gd\frac{\partial h}{\partial y} - fqq_x + X_w \tag{7.2.1}$$

the momentum equation in the y-direction,

$$\frac{\partial q_y}{\partial t} - \Omega q_x = -gd\frac{\partial h}{\partial y} - fqq_y + Y_w \tag{7.2.2}$$

the continuity equation,

$$\frac{\partial q_x}{\partial x} + \frac{\partial q_y}{\partial y} + \frac{\partial h}{\partial t} = r - e \tag{7.2.3}$$

and the conservation (transport) equation,

$$\frac{\partial s}{\partial t} + \frac{\partial (Us)}{\partial x} + \frac{\partial (Vs)}{\partial y} = \frac{\partial}{\partial x} E_x \frac{\partial s}{\partial x} - \frac{\partial}{\partial y} E_y \frac{\partial s}{\partial y} \tag{7.2.4}$$

where

t = time
x and y = horizontal Cartesian coordinates
q_x and q_y = depth-averaged flow components in x- and y-directions, respectively, per unit width
Ω = Coriolis parameter equal to $2\omega \sin \varphi$
ω = angular rotation of the earth
ϕ = latitude
g = gravitational acceleration
h = water surface elevation

d = water depth equal to $h - z$
z = bottom elevation
f = bottom friction term from the Manning equation
q = flow per unit width equal to $\sqrt{q_x^2 + q_y^2}$
X_w = wind stress per unit density of water in the x-direction equal to $KV_w^2 \cos \theta$
Y_w = wind stress per unit density of water in the y-direction equal to $KV_w^2 \sin \theta$
K = a wind-stress coefficient
V_w = wind velocity at 10 m above the water surface
q = wind direction with respect to the x-axis
r = rainfall intensity
e = evaporation rate
U and V = net velocities over a tidal cycle
s = vertical-averaged salinity
E_x and E_y = horizontal dispersion coefficients in the x- and y-directions, respectively

In the momentum equations, the advective terms are neglected and the water density is treated as a constant. The assumption of constant density considerably simplifies the governing equations by decoupling salinity from the momentum equations, but at the expense of neglecting salinity-induced accelerations. The remaining terms in the momentum equations are the inertia, the Coriolis acceleration, gravity, friction, and wind stress. The precipitation and evaporation terms are also added in the continuity equation for the mass conservation. The transport equation is a linear second-order partial differential equation (PDE) of the convective-dispersion equation. The dispersion coefficients are introduced to absorb the density–current fluxes.

Boundary conditions are imposed around the periphery of the estuary, including water–land boundaries, partial internal boundaries (e.g., submerged reefs for hydrodynamic equations only), freshwater flows (e.g., river flows, diversions, and return flows), and open saltwater ocean boundaries (tidal excitation). For salinity, $s = s_o$ is imposed at the ocean boundaries, a von Neumann condition (zero flux) at land boundaries, and an open-boundary condition at the inflow points. These boundary conditions can all be functions of time.

The hydrodynamic equations are nonlinear first-order partial differential equations to solve for three unknowns of flow flux in the x- and y-directions and water surface elevation (q_x, q_y, and h). A fully explicit method used for solving the hydrodynamic equations is a time-centered difference scheme involving time stepping of the "leap frog" type for computations of flows and water surface elevations. Knowing the values at time t, the unknowns q_x, q_y, and h can be solved at time $t + 1$ [derived from Eqs. (7.2.1)–(7.2.3):

$$q_x^{t+1}(i, j) = \frac{1}{C_x^{t-1}} \left[q_x^{t-1}(i, j) + g\Delta t \left\{ \frac{d^t(i, j) + d^t(i + 1, j)}{2} \right\} \cdot \left\{ \frac{h^t(i, j) - h^t(i + 1, j)}{\Delta x} \right\} \right] + \frac{1}{C_x^{t-1}} \left[X_w^t(i, j)\Delta t + \Omega \bar{q}_y^{t-1}(i, j)\Delta t \right] \quad (7.2.5)$$

$$q_y^{t+1}(i, j) = \frac{1}{C_y^{t-1}} \left[q_y^{t-1}(i, j) + g\Delta t \left\{ \frac{d^t(i, j) + d^t(i, j + 1)}{2} \right\} \cdot \left\{ \frac{h^t(i, j) - h^t(i, j + 1)}{\Delta y} \right\} \right] + \frac{1}{C_y^{t-1}} \left[Y_w^t(i, j)\Delta t - \Omega \bar{q}_x^{t-1}(i, j)\Delta t \right] \quad (7.2.6)$$

$$h^{t+2}(i, j) = h^t(i, j) + \Delta t \left[\frac{q_x^{t+1}(i - 1, j) - q_x^{t+1}(i, j)}{\Delta x} + \frac{q_x^{t+1}(i, j - 1) - q_x^{t-1}(i, j)}{\Delta y} \right] + \Delta t \left[r^{t+1}(i, j) - e^{t+1}(i, j) \right] \quad (7.2.7)$$

where

$$d(i, j) = h(i, j) - z(i, j)$$

$$\bar{q}_x(i, j) = \frac{q_x(i, j) + q_x(i, j + 1) + q_x(i - 1, j + 1) + q_x(i - 1, j)}{4}$$

$$\bar{q}_y(i, j) = \frac{q_y(i, j) + q_y(i + 1, j) + q_y(i, j - 1) + q_y(i + 1, j - 1)}{4}$$

$$C_x = 1 + \frac{gn^2(i, j)}{\left\{ 2.21 \left[\frac{d^t(i, j) + d^t(i + 1, j)}{2} \right]^{1/3} \right\}} \cdot \Delta t \frac{[\{q_x^{t-1}(i, j)\}^2 + \{\bar{q}_y^{t-1}(i, j)\}^2]^{1/2}}{\left[\frac{d^t(i, j) + d^t(i + 1, j)}{2} \right]^2}$$

$$C_y = 1 + \frac{gn^2(i, j)}{\left\{2.21\left[\dfrac{d^t(i, j) + d^t(i, j+1)}{2}\right]^{1/3}\right\}} \cdot \Delta t \frac{[\{q_y^{t-1}(i, j)\}^2 + \{\bar{q}_x^{t-1}(i, j)\}^2]^{1/2}}{\left[\dfrac{d^t(i, j) + d^t(i, j+1)}{2}\right]^2}$$

The alternating direction implicit (ADI) method is used to solve the transport equation. Thus, theoretically, it is unconditionally stable for any size of time or spatial step. The linear system of equations result in a tridiagonal matrix which is efficiently solved using the Thomas algorithm. The ADI method is carried out in two steps. At time step $t + 1$, the x-derivatives are written in implicit form and y-derivatives in explicit form. At time step $t + 2$, the direction is switched so that the y-derivatives are written in implicit form and x-derivatives in explicit form. The resultant two sets of simultaneous equations are solved directly without iteration.

At time step $t + 1$, the conservation equation (7.2.4) can be approximated in the x-direction as

$$\begin{aligned}
&s^{t+1}(i-1, j)\left\{-E_x^{t+1}(i-1, j)\left[\frac{\Delta t}{\Delta x^2}\right] - U^{t+1}(i-1, j)\left[\frac{\Delta t}{2\Delta x}\right]\right\} \\
&\quad + s^{t+1}(i, j)\left\{1 + E_x^{t+1}(i, j)\left[\frac{\Delta t}{\Delta x^2}\right] + E_x^{t+1}(i-1, j)\left[\frac{\Delta t}{\Delta x^2}\right]\right. \\
&\quad \left. - U^{t+1}(i-1, j)\left[\frac{\Delta t}{2\Delta x}\right] + U^{t+1}(i, j)\left[\frac{\Delta t}{2\Delta x}\right]\right\} + s^{t+1}(i+1, j) \\
&\quad \cdot \left\{-E_x^{t+1}(i, j)\left[\frac{\Delta t}{\Delta x^2}\right] + U^{t+1}(i, j)\left[\frac{\Delta t}{2\Delta x}\right]\right\} \\
&= s^{t+1}(i, j-1)\left\{E_y^t(i, j-1)\left[\frac{\Delta t}{\Delta x^2}\right] + V^t(i, j-1)\left[\frac{\Delta t}{2\Delta x}\right]\right\} \\
&\quad + s^t(i, j)\left\{1 - E_y^t(i, j)\left[\frac{\Delta t}{\Delta x^2}\right] - E_y^t(i, j-1)\left[\frac{\Delta t}{\Delta x^2}\right]\right. \\
&\quad \left. + V^t(i, j-1)\left[\frac{\Delta t}{2\Delta x}\right] - V^t(i, j)\left[\frac{\Delta t}{2\Delta x}\right] + K\Delta t\right\} \\
&\quad + s^t(i, j+1)\left\{-E_y^t(i, j)\left[\frac{\Delta t}{\Delta x^2}\right] - V^t(i, j)\left[\frac{\Delta t}{2\Delta x}\right]\right\}
\end{aligned}$$

Similarly, the implicit approximation can be written in the y-direction at time step $t + 2$. The resultant linear algebraic equations for the solution of s^{t+1} (or s^{t+2}) can be solved by inversion of a tridiagonal matrix.

7.2.2 Constraints

The mathematical programming model can have the objective of minimizing the sum of freshwater inflows, Q_{tj}, for month t and river j

$$\text{Min} \sum_j \sum_t Q_{tj} \tag{7.2.8}$$

subject to the following constraints:

1. The nonlinear relationship of estuary salinity and freshwater inflow, Eqs. (7.2.1)–(7.2.4):

$$\mathbf{G}(\mathbf{Q}, \mathbf{s}) = \mathbf{0} \tag{7.2.9}$$

2. Upper ($\bar{s}$) and lower ($\underline{s}$) bounds on the monthly average salinity at a specified location in the estuary, for each river j:

$$\underline{s}_{tj} \leq s_{tj} \leq \bar{s}_{tj} \tag{7.2.10}$$

3. Lower limits on the tth monthly inflows for the jth river, QI_{tj}, to express seasonal biological requirements (e.g., of the estuarine marsh inundation):

$$Q_{tj} \geq QI_{tj} \tag{7.2.11}$$

4. The sum of monthly flows must be less than or equal to the upper limit of the total annual inflow, QT_j, from each river j:

$$\sum_t Q_{tj} \leq QT_j \tag{7.2.12}$$

5. Upper and lower limits on mean monthly flows in seasons for each river j:

$$\underline{QS}_{jm} \leq QS_{jm} \leq \overline{QS}_{jm} \tag{7.2.13}$$

where $QS_{jm} \equiv (1/N_m) \Sigma_{t \in M_m} Q_{tj}$; M_m is the set of months in season m and N_m is the number of months in season m.

6. The nonlinear regression relationship between the harvest of organism k and the seasonal inflow in river j:

$$H_k = \Psi_k(QS_{jm}) \tag{7.2.14}$$

7. Lower limits on annual fish harvest, $\underline{H}_k$, by species k:

$$H_k \geq \underline{H}_k \tag{7.2.15}$$

8. Upper and lower limits on monthly inflows ($\overline{Q}_{tj}$ and $\underline{Q}_{tj}$) from each river:

$$\underline{Q}_{tj} \le Q_{tj} \le \overline{Q}_{tj} \tag{7.2.16}$$

Monthly mean salinity bounds are specified for selected locations. There are two types of upper and lower limits on monthly salinity selected to provide a salinity range. The first type is based on the bounds for viable metabolic and reproductive activity, and the second salinity upper bound selected is the median monthly historical salinity level, or equal to the first type salinity upper bound if it is lower than the median monthly historical salinity level.

7.2.3 Alternative Management Model Strategies

Four alternative formulations of the optimization model can be applied to achieve different management objectives, as summarized below. Other management objectives are possible, and can be similarly formulated within the general framework of (7.2.8)–(7.2.16).

Alternative I The basic formulation of the problem for estuarine management is to minimize the total annual freshwater inflow subject to salinity level control, which will accomplish the requirements of nutrient transport, habitat maintenance, and marsh inundation requirement. The corresponding mathematical model can be formulated as

$$\text{Min} \sum_{j} \sum_{t} Q_{tj} \tag{7.2.17}$$

subject to constraints (7.2.9)–(7.2.11) and (7.2.15)–(7.2.16).

Alternative II Maintenance of the fishery harvest. The objective is to minimize the total annual freshwater inflow while satisfying minimum seasonal flow needs to maintain the annual commercial harvest of key species at desired levels, and meeting viability limits for salinity. The constraints for Alternative II are (7.2.9)–(7.2.11) and (7.2.13)–(7.2.16).

Alternative III Enhancement of the fishery harvests. It is to maximize the total annual commercial harvest of a selected organism k while meeting viability limits for salinity, satisfying minimum seasonal flow needs, and limiting an annual combined inflow no greater than its historical mean value. The objective is to

$$\text{Max } \mathbf{QS}^T\hat{\boldsymbol{\beta}}_{HK} \tag{7.2.18}$$

subject to (7.2.9)–(7.2.11) and (7.2.15)–(7.2.16), where $\mathbf{QS}^T$ is the transpose of vector of the seasonal freshwater inflow and $\hat{\boldsymbol{\beta}}_{HK}$ is the vector of estimated coefficients of the harvest regression equation for species k.

The periodic inundation of deltaic marshes serves to maintain shallow protected habitats for postlarval and juvenile stages of several important estuarine species, provides a suitable fluid medium for nutrient exchange processes, and acts as a transport mechanism to move detrital materials from the deltaic marsh into the open estuary (Texas Department of Water Resources, 1980; Valiela and Teal, 1974).

Alternative IV Minimize the total annual freshwater inflow subject to the salinity restriction. This is similar to Alternative I except the minimum seasonal flow (marsh inundation) requirement [constraint (7.2.11)] is removed.

7.2.4 Chance-Constraint Formulation for Harvest Equation

The regression equations in the optimization model for salinity and harvest are subject to uncertainty due to the variance in the basic data. This uncertainty arises because for the population of observations associated with the sampling process, there is a probability distribution of salinity of commercial harvest for each level of freshwater inflow. The basic application of chance constraints in stochastic programming is to account for the uncertainty of the regression due to random variation in the regression variables by formulating the corresponding constraints into probabilistic form and then transforming them into their deterministic equivalents (Charnes and Cooper, 1959, 1962, 1963; Charnes and Sterdy, 1966; Jagannathan, 1974; Miller and Wagner, 1965; Sengupta, 1972). In the environmental and water resources area, there are a number of papers on water quality models and reservoir design and operation models using chance constraints (Bao and Mays, 1994a,b; Bao et al., 1989; Tung et al., 1990; Fujiwara et al., 1986; Houck, 1979; Ellis, 1987, Ellis et al., 1985, 1986; Lohani and Thanh, 1978, 1979).

In the problem formulation, these stochastic constraints are transformed into probabilistic statements so that each chance constraint states the probability that the constraint will be satisfied with a specified reliability level. The harvest constraint (7.2.15) can be rewritten in chance-constraint form as

$$P_r\{H_k \geq \underline{H_k}\} \geq p_k \qquad (7.2.19)$$

where the harvest H_k is a random variable due to the uncertainty induced by the regression equation (7.2.14); p_k is the desired or required reliability. The chance constraint (7.2.19) must be transformed into an equivalent deterministic form in order to implement the optimization algorithm. The harvest regression equations are either multiple linear models or transformed linear models after logarithmic transformation of H_k and QS_{jm}, depending on the species of fish. The commercial fish harvest can be written in a linear or nonlinear form, depending on the species [again, using the regressions of the Texas Department of Water Resources (1980)], see Table 7.1:

$$H_k = (\mathbf{QS})_j^T \cdot \boldsymbol{\beta}_{H_{kj}} \tag{7.2.20}$$

or

$$\ln (H_k) = [\ln(\mathbf{QS}_j]^T \cdot \boldsymbol{\beta}_{H_{k\phi}} \tag{7.2.21}$$

The harvest chance constraint (7.2.19) is determined using Eqs. (7.2.20) and (7.2.21):

$$P_r\{(\mathbf{QS})_j^T \cdot \boldsymbol{\beta}_{H_{kj}} \geq \underline{H_k}\} \geq p_k \tag{7.2.22}$$

or

$$P_r\{[\ln(\mathbf{QS})_j]^T \cdot \boldsymbol{\beta}_{H_{kj}} \geq \ln(\underline{H_k})\} \geq p_k \tag{7.2.23}$$

The deterministic form of Inequalities (7.2.22) and (7.2.23) are, respectively,

$$t_{n-v,1-p_k}\hat{\sigma}_s\sqrt{(\mathbf{QS}_j)^T[(\mathbf{QSD}_j)^T \cdot (\mathbf{QSD}_j)]^{-1}(\mathbf{QS}_j) + 1} + (\mathbf{QS}_j)^T \cdot \hat{\boldsymbol{\beta}}_{H_{kj}} \geq \underline{H_k} \tag{7.2.24}$$

and

$$t_{n-v,1-p_k}\hat{\sigma}_{s_{ij}}\sqrt{[\ln(\mathbf{QS}_j)]^T\{[\ln(\mathbf{QSD}_j)]^T[\ln(\mathbf{QSD}_j)]\}^{-1}[\ln(\mathbf{QS}_j)] + 1} + \ln(\mathbf{QS}_j)^T \cdot \hat{\boldsymbol{\beta}}_{H_{kj}} \geq \ln(\underline{H_k}) \tag{7.2.25}$$

where $t_{n-v,1-p_k}$ is the quantile of t − random variable with $n - v$ degrees of freedom and probability of $1 - p_k$, $\hat{\sigma}_{H_k}$ is the estimated standard error associated with the harvest regression equations, $\mathbf{QSD}_j$ is a matrix of the observed data of seasonal freshwater inflow used for the harvest regression equations, and $\ln(\mathbf{QSD}_j)$ is a matrix in which each element is the logarithmic transform of the corresponding one in $\mathbf{QSD}_j$.

The chance-constrained model for various alternatives is obtained by using the associated objective along with constraints (7.2.24) and (7.2.25), replacing the respective regression relationships. Derivation of the deterministic equivalent of chance constraints based on regression equations is shown in Appendix 7.B and Tung et al. (1990).

Table 7.1 Modified Regression Equations of Fishery Harvest and Freshwater Inflow Relations (derived from historical gaged flow and commercial harvest records)

Index k for Fish Species	Equations	$\hat{\sigma}_k$	Inflow used in equations
1. All shellfish	$H_1 = 3109.5 - 3.782QS_1 + 2.553QS_2 - 12.14QS_3$	±482.6	a*
2. Spotted seatrout	$\ln(H_2) = 7.21 - 1.247\ln(QS_1) + 1.153\ln(QS_2) - 0.404\ln(QS_4)$	±0.290	b**
3. Red drum	$\ln(H_3) = 4.134 + 0.697\ln(QS_2) - 0.869\ln(QS_3)$	±0.287	b
4. All penaeid shrimp	$\ln(H_4) = 1888.6 - 1.061\,QS_1 + 1.088QS_2 - 1.071QS_5$	±463.3	c***
5. Blue crab	$\ln(H_5) = 289.5 + 1.725QS_3 + 0.429QS_4 + 0.202QS_5$	±298.3	c

Where H_k = commercial harvest of species k in thousands of pounds; QS = mean monthly freshwater inflow during the season (acre/ft): QS_1 = January–March, QS_2 = April–June, QS_3 = July–August, QS_4 = September–October, QS_5 = November–December; and $\hat{\sigma}_k$ = standard error.

*Using freshwater inflow at the Lavaca Delta.
**Using freshwater inflow at the Colorado Delta.
***Using combined freshwater inflows from both Lavaca Delta and Colorado Delta.
Source: Texas Department of Water Resources (1980).

7.3 PROBLEM SOLUTION

7.3.1 Overview

The overall optimization model can be stated in the following general nonlinear programming format using an objective to minimize freshwater inflows or to maximize fishery harvest:

$$\text{Optimize } f(\mathbf{Q}, \mathbf{s}, \mathbf{H}) \tag{7.3.1}$$

subject to the following constraints:

(a) Hydrodynamic transport equations (7.2.1)–(7.2.4) that relate salinity, **s** (vector in spatial and temporal domains) to the freshwater inflow, **Q**:

$$\mathbf{G}(\mathbf{Q}, \mathbf{s}) = \mathbf{0} \tag{7.3.2}$$

where **Q** is a vector of the independent variable (control variable) as a function of time and **s** is a vector of the dependent variable (state variable) as a function of time and location.

(b) Regression equations that relate inflow to fishery harvest:

$$\mathbf{h}(\mathbf{Q}, \mathbf{H}) = \mathbf{0} \tag{7.3.3}$$

where **H** is a vector of the fishery harvest for different species.

(c) Constraints that define limitations on freshwater inflows due to upstream demands and water uses, and historical ranges:

$$\underline{\mathbf{Q}} \leq \mathbf{Q} \leq \overline{\mathbf{Q}} \tag{7.3.4}$$

where **Q** and the limitations are defined as the general terms that they can be interpreted as monthly, seasonal, and annual flows. The marsh inundation requirements are also included in this expression, which are basically lower bounds of flows during certain time periods.

(d) Constraints that define limitations on salinity:

$$\underline{\mathbf{s}} \leq \mathbf{s} \leq \overline{\mathbf{s}} \tag{7.3.5}$$

The problem posed is a discrete-time optimal control problem in which the constraints that relate the state variables (salinities) to the control variables (freshwater inflows) are grouped as a simulator; this is separated from the original constraint set and is solved implicitly. For each iteration in the process of optimization, the optimizer computes the new values of control variables and passes that information to the simulator to update the corresponding state variables. A reduced optimization problem is then formed with a smaller number of decision variables and constraints. The control variables are the freshwater inflows as a function of time. The state variables are the salinities as a function of time and location in the bay and estuary. During each iteration of the optimizer, a set of control or decision variables, the freshwater inflows for each time period, are sent to the simulator, as shown in Figure 7.1. The purpose of the estuarine hydrodynamic transport model is to simulate the flow circulation in the bay system and to compute the salinity spatial distribution in the bay for the time period of interest for given freshwater inflow and other boundary conditions. The hydrodynamic transport model then solves for the salinities for each location in the bay and estuary at each time period. Solution of the simulator is performed to evaluate the embedded hydrodynamic transport in the optimization problem. Basically, the state variables (salinities) and the control variables (freshwater inflows) are related through the hydrodynamic transport model. In essence, the simulator equations are used to express the states in terms of the controls, yielding a much smaller nonlinear optimization problem.

One of the key elements of the above problem formulation is the relation **G**(**Q**, **s**) whereby salinity levels in the estuary are defined in terms

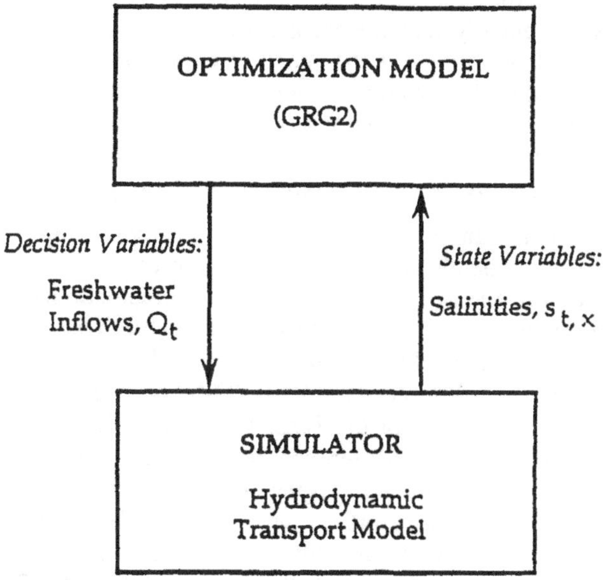

Figure 7.1 Optimizer-simulator interface (Bao, 1992).

of a particular sequence of inflows. The hydrodynamic model embedded within this procedure should satisfy the following desired criteria:

1. The model should be capable of representing an estuarine system with complex circulation, to offer a fair level of complexity in the salinity–inflow relation and therefore in the optimization methodology.
2. The model should be capable of exhibiting a significantly filtered response to time variations in freshwater inflow, including time lags and inertia, in order to differentiate the salinity–inflow association from the simple regression forms used in past studies.
3. The model should be representative of a real estuarine system, so as to allow demonstration of the methodology in a case-study format.
4. The model should facilitate generalization to a more sophisticated high-resolution estuarine model for detailed applications of the optimization methodology.

In addition to the general requirements for an estuarine hydrodynamic model, the following criteria are considered, in order of priority, when selecting a simulation model:

1. The hydrodynamic transport model needs to be called by the optimizer so frequently that the most restrictive requirement for a suitable simulation model is the speed of execution of the code.
2. The model should be capable of representing an estuarine system with complex circulation, temporal, and spatial salinity variability.
3. The model should be capable of simulating long-term salinity values such as monthly averaged salinity in the bay system.

The above requirements can be met for most applications using a two-dimensional, horizontal, depth-averaged, tidal, hydrodynamic, transport model, implemented for one of the Texas bays. The computational model to be employed is one of several models currently available. These include the finite-difference models developed in the Galveston Bay Project (Ward and Espey, 1971), the finite-difference model HYD-SAL developed by the Texas Water Development Board for the Texas bays (Texas Department of Water Resources, 1980), the finite-difference model developed by RAND (Leendertse, 1967a, 1967b; Ward and Espey, 1971), and the quasi-two-dimensional, finite-difference Dynamic Estuary Model (DEM) developed for Sabine Lake estuarine system (Brandes et al., 1975). Available two-dimensional, finite-element models tested for estuaries are FESWMS-2DH (Froelich, 1989), GEVIS (developed by the Notre Dame University in

1990), TXBLEND (Matsumoto, 1992a), and the simplified finite element model, FETEX (Matsumoto, 1991), among others.

7.3.2 Reduced Problem

For illustration purposes, Alternative II, for minimizing the total annual freshwater inflow, is selected to demonstrate the formulation of the optimization problem and solution procedure. The independent (decision) variables are the monthly averaged freshwater inflows from each river connected to the bay system. Thus, even in the original general format [Eq. (7.3.1)], the objective function is a function of flow vector, **Q** only. The problem formulated below, however, is still defined as the "reduced" problem for the reasons that (1) it can be viewed as the coefficients associated with **s** (salinity vector) terms in the objective function are set to zero, (2) the size of the optimization problem is dramatically reduced because the G constraints in Eq. (7.3.2) are solved implicitly by a separate hydrodynamic transport simulator, and (3) this notation make it more convenient for the description of model formulation and structure hereafter. The reduced problem consists of the "reduced" objective function

$$\text{Minimize}\, f(\mathbf{Q}, \mathbf{s}(\mathbf{Q})) = \text{Min}\, F(\mathbf{Q}) \tag{7.3.6}$$

subject to constraints of harvest (7.3.3), bounds of inflows (7.3.4), and salinity limits (7.3.5).

7.3.3 Solution Procedure

In order to force satisfaction of the salinity bound constraints in the optimizer, these bounds on the state variables (salinities) are incorporated into the objective function using the augmented Lagrangian algorithm. Such an approach not only forces the state bounds to be satisfied but also reduces the number of constraints. Because only inequality bound-type salinity constraints need to be incorporated, the objective function with the augmented Lagrangian function (see Chapter 3) is

$$\text{Min}\, L(\mathbf{s}(\mathbf{Q}), \mathbf{Q}, \boldsymbol{\mu}, \boldsymbol{\sigma}) = \mathrm{F}(\mathbf{Q}) + \frac{1}{2}\sum \sigma_i \left\{\min\left[0, c_i - \frac{\mu_i}{\sigma_i}\right]\right\}^2 \left(\frac{1}{2}\sum_i \frac{\mu_i^2}{\sigma_i}\right) \tag{7.3.7}$$

where i is the index for each bound constraint; s_i and m_i are the penalty weights and Lagrangian multipliers for the ith bound, respectively; and c_i is the violation of the bounds either above or below the minimum, defined as

$$c_i = \min[s_i - \underline{s}_i, \bar{s}_i - s_i] \tag{7.3.8}$$

The reduced optimization problem with augmented Lagrangian terms for minimizing freshwater inflows solved by GRG2 is the objective equation (7.3.7)

$$\text{Minimize } L(\mathbf{s}(\mathbf{Q}), \mathbf{Q}, \boldsymbol{\mu}, \boldsymbol{\sigma}) \tag{7.3.9}$$

subject to

$$\mathbf{h}(\mathbf{Q}, \mathbf{s}(\mathbf{Q}), \mathbf{H}) = \mathbf{0} \tag{7.3.10}$$

$$\underline{\mathbf{Q}} \leq \mathbf{Q} \leq \overline{\mathbf{Q}} \tag{7.3.11}$$

which are the constraints on harvest and the bounds on the freshwater inflows, respectively. The solution to this reduced problem is a two-step procedure. The overall problem is

$$z = \min_{\sigma,\mu} \left[\min_{\mathbf{Q} \in \mathbf{S}_Q} L(s(\mathbf{Q}), \mathbf{Q}, \boldsymbol{\mu}, \boldsymbol{\sigma})\right] \tag{7.3.12}$$

where $\mathbf{S}_Q$ is the set of feasible fresh water inflows as given by (7.3.11). For given values of vectors $\boldsymbol{\sigma}$ and $\boldsymbol{\mu}$, the reduced problem, Eqs. (7.3.9), (7.3.10) and (7.3.11), is then solved using a nonlinear optimizer, which is based on the reduced gradient method. The outer problem is iterated by updating the values of $\boldsymbol{\sigma}$ and $\boldsymbol{\mu}$ for the next solution run of the inner problem. The overall optimization is attained when $\boldsymbol{\sigma}$ and $\boldsymbol{\mu}$ both converge.

The updating formula used for $\boldsymbol{\mu}$ is

$$\mu_i^{k+1} = \begin{cases} \mu_i^k - \sigma_i c_i & \text{if } c_i < \dfrac{\mu_i}{\sigma_i} \\ 0 & \text{otherwise} \end{cases} \tag{7.3.13}$$

where k is the number of the current iteration. The value of σ_i is normally adjusted once during early iterations and then kept constant.

The overall solution procedure is further illustrated through the flowchart in Figure 7.2. There are two loops in this procedure, with the outer loop determining the Lagrangian multipliers (dual variables) and penalty weights. The inner loop solves the reduced augmented Lagrangian problem using the nonlinear programming (NLP) optimizer GRG2, whose dual variables and penalty weights are fixed at the values determined by the outer loop. Once an inner loop is finished, the convergence criterion is checked by looking at the size of the salinity bound infeasibility. If it is small enough, the procedure terminates; otherwise, the procedure returns to the outer loop and updates the dual variables and penalty weights and then goes to the inner loop and solves the new reduced augmented Lagrangian

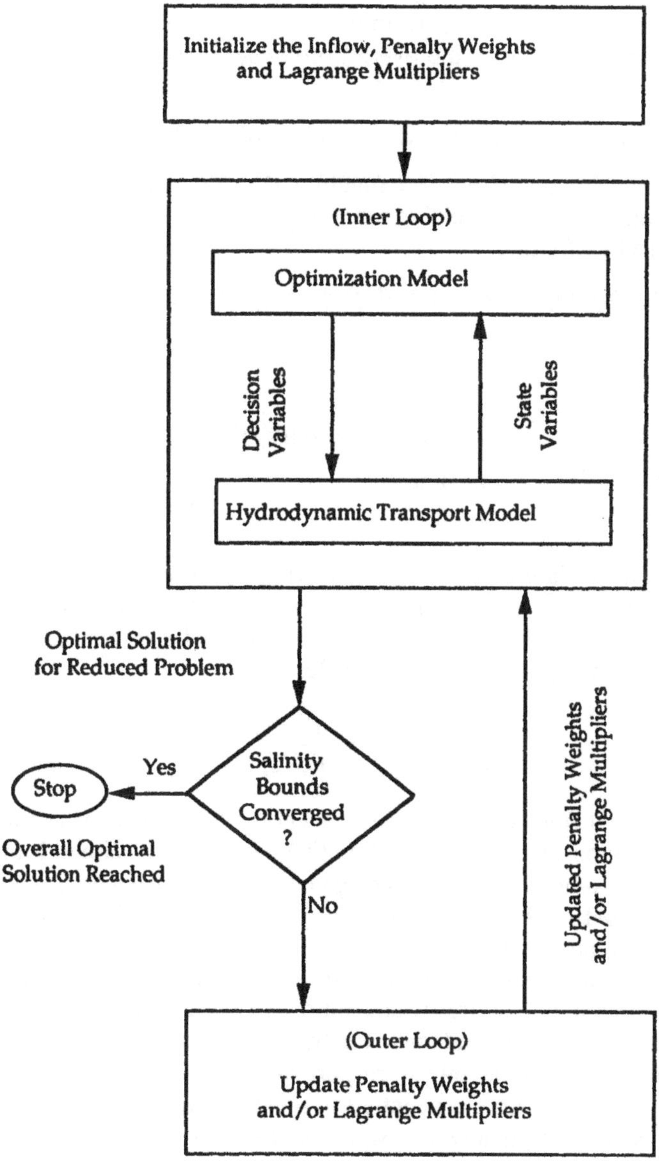

Figure 7.2 Overall solution procedure (Bao, 1992).

again with the updated **σ** and **μ** from the outer loop. This process continues until an optimal solution of the overall problem is found.

7.3.4 Computation of Reduced Gradients of AL Problem

The augmented Lagrangian (AL) function [Eq. (7.3.7)] is a function of flow (**Q**), salinity [**s**(**Q**)], and Lagrangian parameters **μ** and **σ**, which is also expressed as follows:

$$\text{Min } L(\mathbf{s}(\mathbf{Q}), \mathbf{Q}, \boldsymbol{\mu}, \boldsymbol{\sigma}) = f(\mathbf{Q}) + \sum_i l_i(c_i\{s_i(\mathbf{Q})\}, \mu_i, \sigma_i) \qquad (7.3.14)$$

where

$$\sum_i l_i(\mathbf{s}, \boldsymbol{\mu}, \boldsymbol{\sigma}) = \sum_i \begin{cases} -\mu_i c_i(s) + \dfrac{1}{2}\sigma_i[c_i(s)]^2 & \text{if } c_i(s) < \dfrac{\mu_i}{\sigma_i} \\ -\dfrac{1}{2}\dfrac{\mu_i}{\sigma_i} & \text{if } c_i(s) \geq \dfrac{\mu_i}{\sigma_i} \end{cases} \qquad (7.3.15)$$

and the salinity violation vector **c** is a function of salinity **s** and the salinity bounds (7.3.8). From Eq. (7.3.8), the salinity violation term $c_i(\mathbf{s}) = c_i(s_i)$ or $c_i(\mathbf{s}) = c_i(s_i(\mathbf{Q}))$. The gradients of the augmented Lagrangian function can be derived by applying the chain rule:

$$\frac{\partial L}{\partial \mathbf{Q}} = \frac{\partial f}{\partial \mathbf{Q}} + S_i \frac{\partial l}{\partial c_i}\frac{\partial c_i}{\partial s_i}\frac{\partial s_i}{\partial \mathbf{Q}} \qquad (7.3.16)$$

where $\partial l/\partial c_i$ is a function of m_i and s_i. Hence, $(\partial l/\partial c_i)$ is constant for the inner optimization problem. From Eq. (7.3.8), $\partial c_i/\partial s_i$ is either 1 or -1. Thus, the key component for the computation of the (reduced) gradients of the augmented Lagrangian is the partial derivatives of the salinity with respect to the monthly flow, $\partial \mathbf{s}/\partial \mathbf{Q}$.

The spatial and temporal salinities in the bay system are computed by solving the simulator. The freshwater inflows, **Q**, are part of the boundary conditions (water–land boundaries) for the hydrodynamic model, and the salinity values in the river inlets are part of the boundary conditions for the transport model (source concentration boundaries). In order to compute the matrix $\partial \mathbf{s}/\partial \mathbf{Q}$ analytically, a new set of simulator equations need to be derived and the analytical solution may be very difficult, if not impossible. The computation of $\partial \mathbf{s}/\partial \mathbf{Q}$ is carried out by finite-difference methods, either forward differencing or central differencing. More specifically, the $\partial \mathbf{s}/\partial \mathbf{Q}$ are computed by perturbation of **Q** and running the hydrodynamic transport simulator repeatedly.

7.3.5 CPU Concern

The computation of the reduced gradient is done by the forward difference or the central difference method through calling of the hydrodynamic transport model to simulate the temporal and spatial salinity variability in the nonlinear optimizer. In order to update the objective function, 12 calls to the hydrodynamic transport simulator are required with each simulating for a period of 1 month.

Theoretically, if the central difference is used, it requires $24 \times 12 \times 2 = 576$ calls of the hydrodynamic transport simulator in order to update the AL reduced gradients, where 24 is the number of decision variables (monthly river flows, assuming two rivers); 12 is the number of months to be simulated for each variable (Q) to be perturbed to obtain $\partial L/\partial \mathbf{Q}$, which is on an annual basis, Eq. (7.3.16); and 2 results from the fact that the central difference requires monthly flows to be perturbed on both sides for computation of the AL reduced gradients. Although the number estimated above for the simulation requirement can be reduced by 50% by running the simulation only for the remaining months, 288 calls of the simulations are still extremely expensive in CPU time for only updating the AL gradients once.

In work done by Bao (1992) the simulation results using HYD-SAL (Appendix 7.A) indicate that the impact of a monthly flow perturbation in month t on river j of the salinities in the bay system for the remaining months ($t = t + 1, t + 2, \ldots, 12$) is so small that it might be mainly affected due to the numerical computation errors (less than 1.0E-8). Therefore the effect of the flow perturbation from previous months is considered as negligible. Hence, the number of hydrodynamic transport simulation calls for updating the AL reduced gradient matrix can be reduced from 576 to 48 for the central difference method for not simulating the salinity in the bay for the remaining months.

Other test run results by Bao (1992) indicate that the difference of the computed AL reduced gradients between the forward and the central difference methods is insignificant. The forward difference method is sufficient for the purposes of the AL reduced gradient computations. Thus, the number of simulation calls to the hydrodynamic transport model can be further reduced to 24.

The test results indicate that over 95% of the CPU time for the model run is required in the hydrodynamic transport model runs for flow and salinity simulations. This is confirmed based on a comparison of CPU time requirements of LAV2106 and HYD-SAL (Appendix 7.A) and the estimation of the number of calls of the hydrodynamic transport model. Although this dramatical reduction in the number of hydrodynamic transport simulations (from 576 to 24) will save the CPU time significantly, it is still an extremely intensive computation effort for the whole model. The inner optimization model of GRG2 requires 7–60 iterations before the optimal solution is found for the given augmented Lagrangian parameters (initial multiplier, initial penalty, and penalty multiplier). Each iteration may require one or more updates of the reduced gradient and many updates for computing the objective functions. The number of the simulation calls is then multiplied by the number of outerloop iterations for updating the augmented Lagrangian parameters and rerun to the inner optimizer.

7.3.6 Gradient Approximation Scheme

The frequent number of simulations require such high CPU time that it is too expensive to run the model. Some innovative modification is needed to reduce the CPU time in simulation. Many considerations and attempts are made to approaches for solving this problem, which are briefly described below as examples. (1) Increase the grid size from one to two nautical miles to reduce the number of grids. The problem with this approach is that the grid will be too coarse to have reasonable resolution of the simulation results. The water–land boundaries are also very difficult to fit with this grid network, not to mention the ship channel. (2) Reduce modeling area from the whole estuary to part of the bay system, such as the upper Lavaca bay and part of the Matagorda bay. This is quite reasonable for solving the problem for this application but does not solve the real problem which would limit the model from application in the future. (3) Run the simulations separately and build a database to establish the relationship between the freshwater inflow and salinities in the bay system. Intuitively, this approach is pragmatic; however, the use of the actual salinity versus flow as the entities for the database might also cause as high an uncertainty as in the cases of salinity regression equations.

The approximation scheme for computing the AL gradients presented here is based on the premise that the change of the salinity derivatives with respect to flow is relatively small compared with the flow changes within a certain flow range. In another words, for a set of given flows, the higher order of salinity derivatives (second partial derivatives) are negligible. This assumption is not proven in theory, but the fact that the linearity in the formed transport PDE (second order though) and and the fully explicit time-centered differencing for the nonlinear hydrodynamic PDEs might suggest that the assumption be a close guess.

Figure 7.3 is a flowchart of the procedure for the approximation scheme for computing the AL gradients and the objective functions. By the finite-difference method, the gradients of the AL objective function

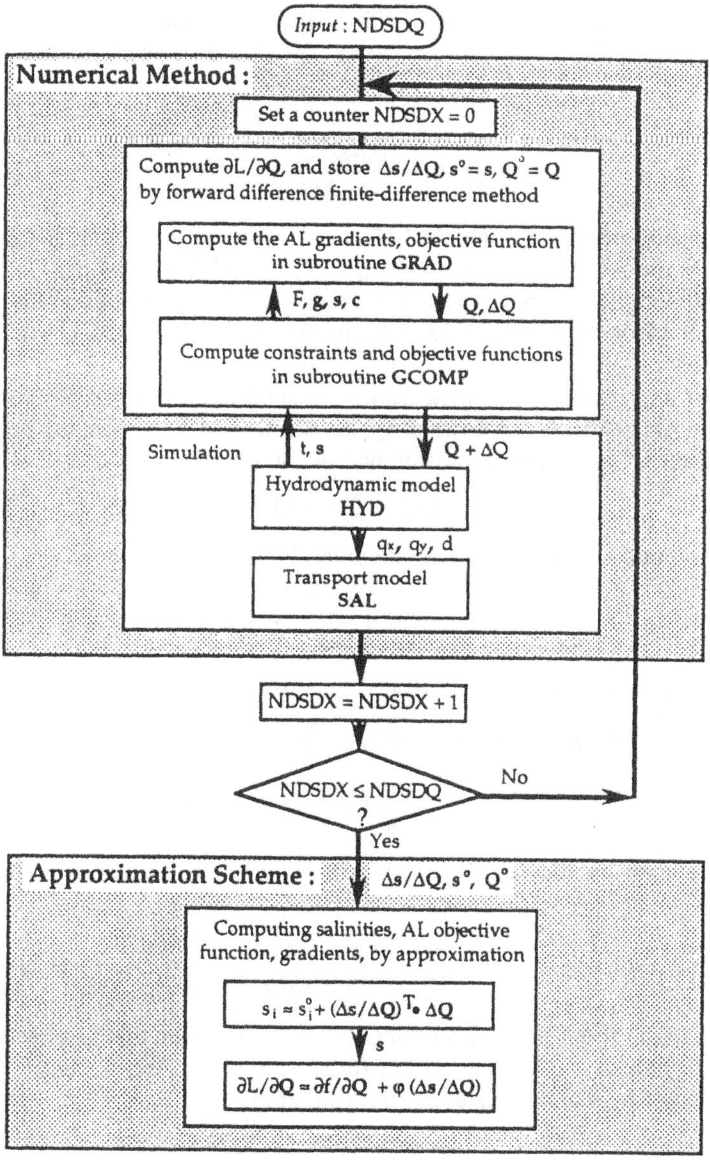

Figure 7.3 Flowchart of approximation scheme for computing the gradients and objective function (Bao, 1992).

$L(\mathbf{s}(\mathbf{Q}), \mathbf{Q}, \boldsymbol{\mu}, \boldsymbol{\sigma})$ [Eq. (7.3.14)], with respect to monthly inflows, $\mathbf{Q}$ ($\mathbf{Q} = \{Q_1, Q_2, \ldots, Q_n\}$) is computed by the forward difference method,

$$\frac{\partial L}{\partial Q_m} = \frac{L\Big|_{Q_m+\Delta Q_{mm}} - L\Big|_{Q_m}}{\Delta Q_m} \tag{7.3.17}$$

for variable element Q_m ($m = 1, 2, \ldots, n$). The simulator is called to compute the AL terms in the objective, $\Sigma_i\, l_i(c_i\{s_i(\mathbf{Q})\}\mu_i, \sigma_i)$ [Eq. (7.3.14)] by simulating the salinities and computing the salinity violation terms. The resultant Jacobian matrix of $\partial\mathbf{s}/\partial\mathbf{Q}$, and the vectors of $\mathbf{s}$ and $\mathbf{Q}$ are stored as $\Delta\mathbf{s}/\Delta\mathbf{Q}$, $\mathbf{s}^0$, and $\mathbf{Q}^0$.

The approximation scheme can be described as follows. To compute the new AL gradients for flows of $\mathbf{Q}$, the changes of flow from the previous evaluation $\mathbf{Q}^0$ is simply the difference of the two, as

$$\Delta\mathbf{Q} = \mathbf{Q}^0 - \mathbf{Q} \tag{7.3.18}$$

The associated salinities, $\mathbf{s}$, are computed by

$$s_i \approx s_i^c + \left(\frac{\Delta\mathbf{s}}{\Delta\mathbf{Q}}\right)^T \cdot \Delta\mathbf{Q} \tag{7.3.19}$$

and the updated objective function is

$$\begin{aligned} L(\mathbf{s}(\mathbf{Q}), \mathbf{Q}, \boldsymbol{\mu}, \boldsymbol{\sigma}) &= f(\mathbf{Q}) + \sum_i l_i(c_i\{s_i(\mathbf{Q})\}, \mu_i, \sigma_i) \\ &= \sum_m Q_m + \sum_i l_i\{c_i(s_i), \mu_i, \sigma_i\} \end{aligned} \tag{7.3.20}$$

The gradients of the AL objective with respect to the monthly flow are approximated by

$$\frac{\partial L}{\partial\mathbf{Q}} = \frac{\partial f}{\partial\mathbf{Q}} + \frac{\Delta\mathbf{s}}{\Delta\mathbf{Q}} \tag{7.3.21}$$

where $\partial L/\partial\mathbf{Q}$ [Eq. (7.3.16)] is a function of the Jacobian matrix of salinities $\Delta\mathbf{s}/\Delta\mathbf{Q}$. Once the computation of $\partial L/\partial\mathbf{Q}$ is completed, the $\mathbf{s}^0$ and $\mathbf{Q}^0$ values are updated by $\mathbf{s}$ and $\mathbf{Q}$ in the current iteration.

7.4 APPLICATION

Bao (1992) developed a compter code, OPTFLOW, that interfaces GRG2 and HYD-SAL (Appendix 7.A) for determining the optimal freshwater inflows to bays and estuaries. Bao (1992) and Bao and Mays (1994b) presented application of the model to the Lavaca-Tres Palacios estuary in Texas (i.e., Matagorda Bay) and its secondary (e.g., Lavaca Bay), and tertiary (e.g., Cox Bay) systems, shown in Figure 7.4. The major freshwater inflow sources are the Colorado River, which principally affects the eastern segment of Matagorda Bay, and the Lavaca River, which principally influences Lavaca Bay.

The regression equations for fishery harvest (see Table 7.1) and the monthly mean salinity bounds are specified for selected locations. For the Matagorda Bay system, these are two types of upper and lower limits on monthly salinity which determine a salinity range. The first type is based on the bounds for viable metabolic and reproductive activity. The second type of upper bound selected is the lesser of the historical median monthly

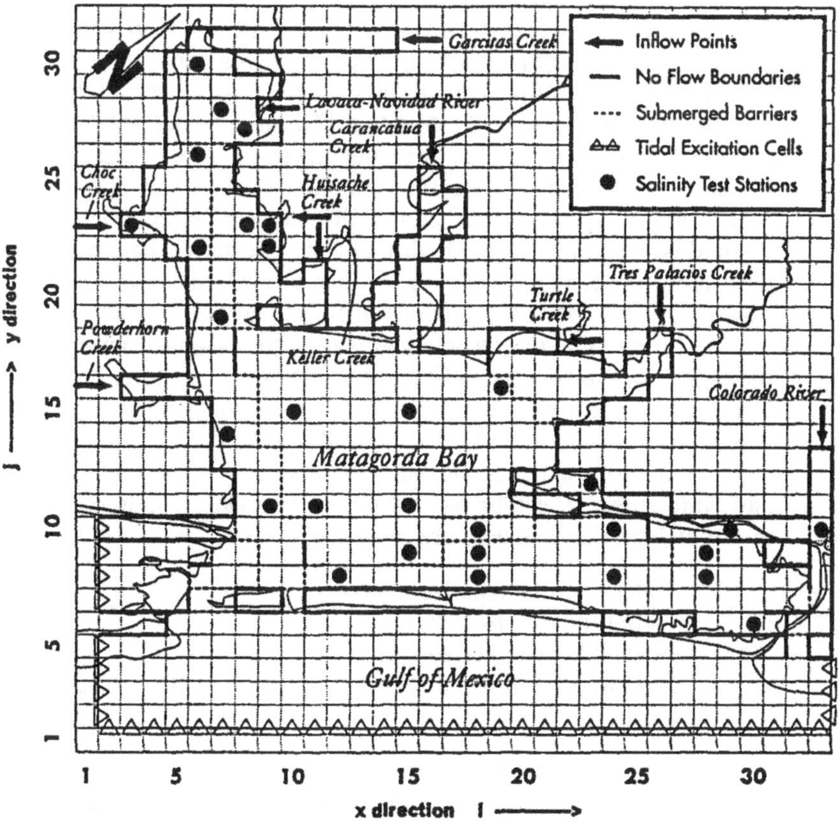

Figure 7.4 Computation grid for Lavaca–Tres Palacios Estuary (Bao, 1992).

salinity level or the first-type salinity upper bound (i.e., viability limits) (Texas Department of Water Resources, 1980).

APPENDIX A: HYD-SAL SIMULATION MODELS

The finite-difference model HYD-SAL consists of two separate but linked models: a tidal hydrodynamic model (HYD) and a salinity transport model (SAL). The input and output of the models and their linkage are shown in Figure 7.5. Major efforts have been devoted to the development and applications of these models (Masch and Associates, 1971; Masch and Brandes, 1971; Texas Department of Water Resources, 1980). HYD and SAL have been applied to four bay systems in Texas, including San Antonio, Matagorda (Lavaca–Tres Palacios), Corpus Christi–Aransas–Copano, and Galveston (Texas Department of Water Resources, 1979, 1980).

The hydrodynamic model (HYD) is developed for vertically well-mixed estuaries to solve the two-dimensional dynamic equations of motion and the unsteady continuity equation. These are nonlinear partial differential equations to solve for three unknowns of flow flux in the x- and y-directions and depth (or the tidal amplitude). The transport equation for SAL is a linear, second-order partial differential equation. The fully explicit method is used to solve the hydrodynamic equations. The explicit method used is a time-centered difference scheme involving time stepping of the "leap frog" type for computations of flows and water levels. The alternating direction implicit (ADI) method is used to solve the transport equation; therefore, it is unconditionally stable for any size of time or spatial step. The linear system equations result in a tridiagonal matrix which is efficiently solved using the Thomas algorithm (Masch and Associates, 1971).

The hydrodynamic model also incorporates the Coriolis acceleration and wind stress. The four basic types of boundary conditions considered in the hydrodynamic model are as follows:

1. Water–land boundaries
2. Partial internal boundaries
3. Artificial ocean boundaries
4. Freshwater inflow, diversion, and return flow magnitudes and location.

The salinity model is simplified as the convective-dispersion equation based on the principle of mass conservation. The effect of evaporation and precipitation on salinity is considered in SAL. Similarly, the boundary conditions for the salinity model are listed as follows:

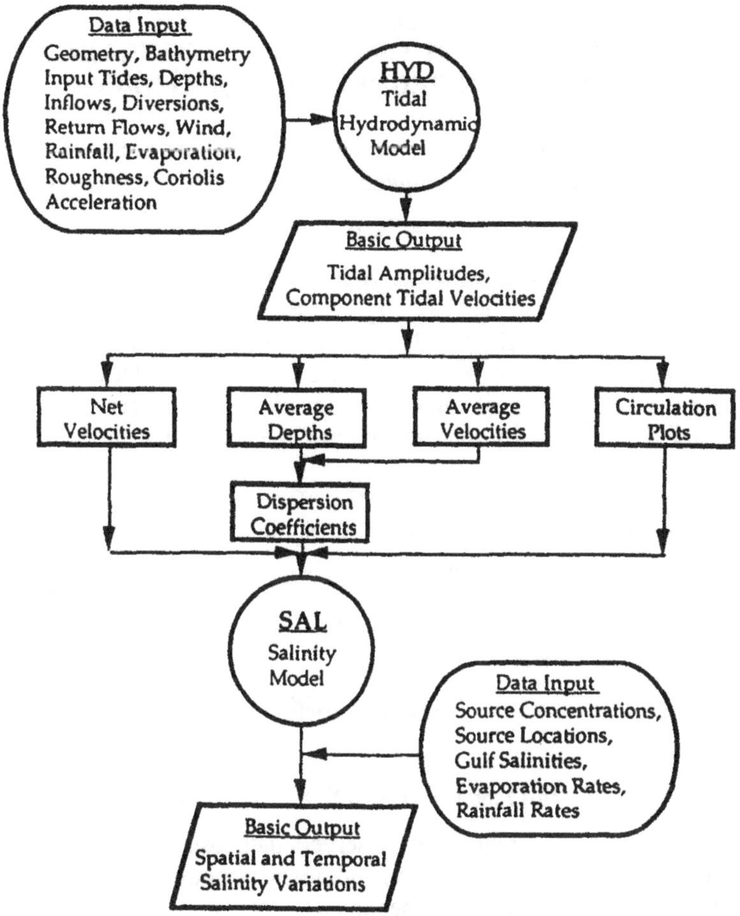

Figure 7.5 Input and output of the HYD and SAL models and their linkage (Texas Department of Water Resources, 1980).

1. Water–land boundaries
2. Impermeable internal boundaries
3. Source concentration boundaries

The HYD and SAL models were modified and run as separate models, so that the output of HYD is used as input to SAL and as a combined model (HYD-SAL) to simulate long-term salinity pattern (monthly and annual). For monthly simulation, the CPU time for execution is about 2.5 min on a Sun4/390 workstation and 12 sec on a Cray Y-MP8/864.

APPENDIX B: DERIVATION OF DETERMINISTIC EQUIVALENT OF CHANCE CONSTRAINTS BASED ON REGRESSION EQUATIONS

In order to transform the chance constraint (7.2.19) into their deterministic equivalent forms, first consider a general multiple linear regression model,

$$Y = \mathbf{X}^T\boldsymbol{\beta} + \varepsilon \tag{7.B.1}$$

where Y is the dependent variable, $\mathbf{X}$ is a $\nu \times 1$ column vector of independent variables, $\{1, x_1, x_2, \ldots, x_{\nu-1}\}^T$; $\boldsymbol{\beta}$ is a $\nu \times 1$ column vector of regression parameters, $\{\beta_0, \beta_1, \beta_2, \ldots, \beta_{\nu-1}\}^T$, ε is the model error with $E(\varepsilon) = 0$, and $\text{Var}(\varepsilon) = \sigma^2$. Because ε is a random variable, the true value of Y and the coefficients of regression equation, $\boldsymbol{\beta}$ are never known. Replacing the Y, $\boldsymbol{\beta}$, and ε by their estimators, the regression model becomes

$$\hat{Y} = \mathbf{X}^T\hat{\boldsymbol{\beta}} + \hat{\varepsilon} \tag{7.B.2}$$

For a given set of independent variables, $\mathbf{x}_0$, the corresponding dependent variable Y_0 can be estimated as

$$\hat{Y}_0 = \mathbf{x}_0^T\hat{\boldsymbol{\beta}} \tag{7.B.3}$$

with the associated mean

$$\text{E}(\hat{Y}_0|\mathbf{x}_0) = \mathbf{x}_0^T\hat{\boldsymbol{\beta}}$$

and variance

$$\text{Var}(\hat{Y}_0|\mathbf{x}_0) = \sigma^2[\mathbf{x}_0^T(\mathbf{X}^T\mathbf{X})^{-1}\mathbf{x}_0 + 1]$$

where $\mathbf{X}$ is an $n \times \nu$ matrix of observed data used in developing the regression equations. Replacing the unknown population variance by its estimator, the predicted variance becomes

$$\text{Var}(\hat{Y}_0|\mathbf{x}_0) = \hat{\sigma}^2[\mathbf{x}_0^T(\mathbf{X}^T\mathbf{X})^{-1}\mathbf{x}_0 + 1]$$

Consider a chance constraint

$$P_r\{\underline{Y} \leq Y_0\} \geq p \tag{7.B.4}$$

by standardizing

$$P_r\left\{\frac{Y_0 - E(\hat{Y}_0|\mathbf{x}_0)}{\sqrt{\text{Var}(\hat{Y}_0|\mathbf{x}_0)}} \geq \frac{\mathbf{Y} - \mathbf{x}_0^T\hat{\boldsymbol{\beta}}}{\sqrt{\hat{\sigma}^2\{\mathbf{x}_0^T(\mathbf{X}^T\mathbf{X})^{-1}\mathbf{x}_0 + 1\}}}\right\} \geq p$$

which can be rearranged as

$$P_r\left\{T_{n-\nu} \leq \frac{\mathbf{Y} - \mathbf{x}_0^T\hat{\boldsymbol{\beta}}}{\sqrt{\hat{\sigma}^2\{\mathbf{x}_0^T(\mathbf{X}^T\mathbf{X})^{-1}\mathbf{x}_0 + 1\}}}\right\} \leq 1 - p \tag{7.B.5}$$

Knowing the reliability p, the standard student distribution deviate can be easily computed. Hence, the deterministic equivalent of the chance constraint is

$$\frac{\mathbf{Y} - \mathbf{x}_0^T\hat{\boldsymbol{\beta}}}{\sqrt{\hat{\sigma}^2\{\mathbf{x}_0^T(\mathbf{X}^T\mathbf{X})^{-1}\mathbf{x}_0 + 1\}}} \leq t_{n-v,1-p}$$

or

$$t_{n-v,1-p}\hat{\sigma}\sqrt{\{\mathbf{x}_0^T(\mathbf{X}^T\mathbf{X})^{-1}\mathbf{x}_0 + 1\}} + \mathbf{x}_0^T\hat{\boldsymbol{\beta}} \geq \mathbf{Y} \qquad (7.B.6)$$

with $n - v$ degrees of freedom, and probability of $1 - p$.

Consider the case that the constraint is bounded on both sides:

$$P_r\{\underline{Y} \leq Y_0 \leq \overline{Y}\} \geq p$$

then

$$F_{T,n-v}\left[\frac{\overline{Y} - \mathbf{x}_0^T\hat{\boldsymbol{\beta}}}{\hat{\sigma}\sqrt{\{\mathbf{x}_0^T(\mathbf{X}^T\mathbf{X})^{-1}\mathbf{x}_0 + 1\}}}\right] - F_{T,n-v}\left[\frac{\overline{Y} - \mathbf{x}_0^T\hat{\boldsymbol{\beta}}}{\hat{\sigma}\sqrt{\{\mathbf{x}_0^T(\mathbf{X}^T\mathbf{X})^{-1}\mathbf{x}_0 + 1\}}}\right] \geq p \qquad (7.B.7)$$

However, the explicit expression of the deterministic equivalent of this type of chance constraint cannot be derived. The deterministic equivalents of the commercial harvest constraints can be obtained by substitution of the corresponding variables and parameters into Eq. (7.B.6). The salinity constraints can be written in the form of Eq. (7.B.7). The fact that this salinity constraint has only an implicit form must be considered when selecting a programming algorithm.

REFERENCES

Bao, Y. X. and Mays, L. W., New Methodology for Optimization of Freshwater Inflows to Estuaries, *Journal of Water Resources Planning and Management*, Vol. 120, No. 2, pp. 199–217, 1994a.

Bao, Y. X. and Mays, L. W., Optimization of Freshwater Inflows to the Lavaca-Tres Palacios, Texas, Estuary, *Journal of Water Resources Planning and Management*, Vol. 120, No. 2, pp. 218–236, 1994b.

Bao, Y. X., Methodology for Determining the Optimal Freshwater Inflows into Bays and Estuaries, Ph.D. Dissertation, Department of Civil Engineering, The University of Texas at Austin, 1992.

Bao, Y. X., Tung, Y. K., Mays, L. W., and Ward, W. H. Jr., Analysis of the Effect of Freshwater Inflows on Estuary Fishery Resources, Technical Memorandum 89-2, Report to Texas Water Development Board, by the Center for Research in Water Resources, The University of Texas at Austin, Austin, 1989.

Brandes, R. J., Johnson, A. E., Icemena, K. R., and F. D. Masch, Computer Program Documentation for the Dynamic Estuary Model with Application to Sabine Lake Estuarine System, Final Report to Texas Water Development Board, by Water Resources Engineers, Inc., Austin, TX, April 1975.

Charnes, A. and Cooper, W. W., Chance-Constrained Programming, *Management Science*, Vol. 6, pp. 73–79, 1959.

Charnes, A. and Cooper, W. W., Chance Constraints and Normal Deviates, *Journal of American Statistics Association*, Vol. 57, pp. 143–148, 1962.

Charnes, A. and Cooper, W. W., Deterministic Equivalents for Optimizing and Satisficing Under Chance Constraints, *Operations Research*, Vol. 11, No. 1, pp. 18–39, 1963.

Charnes, A. and Sterdy, A. C., A Chance-Constrained Model for Real-Time Control in Research and Development, *Management Science*, Vol. 12, No. 8, pp. B-353–B-362, 1966.

Ellis, J. H., Stochastic Water Quality Optimization Using Imbedded Chance Constraints, *Water Resources Research*, Vol. 23, No. 12, pp. 2227–2238, 1987.

Ellis, J. H., McBean, E. A., and Farquhar, G. J., Chance-Constrained/Stochastic Linear Programming Model for Acid Rain Abatement, 1, Complete Conlinearity and Nonconlinearity, *Atmosphere Environment*, Vol. 19, pp. 925–937, 1985.

Ellis, J. H., McBean, E. A., and Farquhar, G. J., Chance-Constrained/Stochastic Linear Programming Model for Acid Rain Abatement, 2, Limited Conlinearity, *Atmosphere Environment*, Vol. 20, pp. 501–511, 1986.

Froehich, D. C., Finite Element Surface-Water Modeling System: Two-Dimensional Flow in a Horizontal Plane Users Manual, Technical Report, FHWA-RD-88-177, Federal Highway Administration Office of Research, Development, and Technology, McLean, VA, 1989.

Fujiwara, O., Gnanendran, S. K., and Ohgaki, S., River Quality Management Under Stochastic Streamflow, *Journal of Environmental Engineering*, Vol. 112, pp. 185–198, 1986.

Houck, M. H., A Chance Constrained Optimization Model for Reservoir Design and Operation, *Water Resources Research*, Vol. 15, No. 5, pp. 1011–1016, 1979.

Jagannathan, R., Chance-Constrained Programming with Joint Constraints, *Operation Research*, Vol. 22, pp. 358–372, 1974.

Leendertse, J. J., Aspects of a Computational Model for Long-Period Water-Wave Propagation, The RAND Corp., Santa Monica, CA, 1967a.

Leendertse, J. J., Aspects of a Computational Model for Well-Mixed Estuaries and Coastal Seas, Report No. RM 5294 - PR, The Rand Corporation, Santa Monica, CA, 1967b.

Lohani, B. N. and Thanh, N. C., Stochastic Programming Model for Water Quality Management in a River, *Journal of the Water Pollution Control Federation*, Vol. 50, pp. 2175–2182, 1978.

Lohani, B. N. and Thanh, N. C., Probabilistic Water Quality Control Policies, *Journal of Environmental Engineering*, Vol. 5, pp. 713–725, 1979.

Masch, F. D. and Associates, Tidal Hydrodynamic and Salinity Models for San Antonio and Matagorda Bays, Texas, A Report to Texas Water Development Board, Austin, 1971.

Masch, F. D. and Brandes, R. J., Tidal Hydrodynamic Simulation in Shallow Estuaries, Report to Office of Water Resources Research, U.S. Department of the Interior, Technical Report, HYD 12-7102, Hydraulic Engineering Laboratory, University of Texas, Austin, 1971.

Matsumoto, J., Mathematical Description of the FETEX Model Based on a New Computational Method: a Simplified Finite Element Method or a Generalized Finite Difference Method, Technical Report of the Texas Water Development Board, Austin, 1991.

Matsumoto, J., *User's Manual for the Texas Water Development Board's Circulation and Salinity Model: TXBLEND*, Texas Water Development Board, Austin, TX, 1992a.

Matsumoto, J., Guadalupe Estuary Example Analysis, in *Freshwater Inflows to Texas Bays and Estuaries: Ecological Relationships and Methods for Determination of Needs*, W. L. Longley (ed.), Texas Water Development Board and Texas Parks and Wildlife Department, Austin, TX, 1992b.

Miller, B. L. and Wagner, H. M., Chance-Constrained Programming with Joint Constraints, *Operation Research*, Vol. 13, pp. 930–945, 1965.

Sengupta, J. K., Chance-Constrained Linear Programming with Chi-Square Type Deviates, *Management Science*, Vol. 19, pp. 337–349, 1972.

Texas Department of Water Resources, Mathematical Simulation Capabilities in Water Resource Systems Analysis, Report LP-16, 1979.

Texas Department of Water Resources, Lavaca-Tres Palacios Estuary: A Study of the Influence of Freshwater Flows, Report LP-106, 1980.

Tung, Y. K, Bao, Y. X., Mays, L. W., and Ward, W. H., Optimization of Freshwater Inflow to Estuaries, *Journal of Water Resources Planning and Management*, Vol. 116, No. 4, pp. 567–584, 1990.

Valiela, I. and Teal, J. M., Nutrient Limitation in Salt Marsh Vegetation, in *Ecology of Halophytes*, R. J. Reimold and W. H. Queen (eds.), Academic Press, New York, 1974, pp. 547–563.

Ward, G. H. and Espey, W. H., *Estuarine Modeling: An Assessment*, EPA 16070 DZV, U.S. Government Printing Office, Washington, DC, 1971.

Chapter 8
Optimal Control by Feedback Control Methods

8.1 DYNAMIC PROGRAMMING

Dynamic programming (DP) transforms a sequential or multistage decision problem that may contain many interrelated decision variables into a series of single-stage problems, each containing only one or a few variables. In other words, the dynamic programming technique decomposes an N decision problem into a sequence of N separate, but interrelated, single-decision subproblems. Decomposition is very useful in solving large, complex problems by decomposing a problem into a series of smaller subproblems and then combining the solutions of the smaller problems to obtain the solution of the entire model composition. The reason for using decomposition is to solve a problem more efficiently, which can lead to significant computational savings. As a rule of thumb, computations increase exponentially with the number of variables, but only linearly with the number of subproblems.

Dynamic programming can overcome the shortcomings of an exhaustive enumeration procedure using the following concepts (Bellman and Dreyfus, 1962).

1. The problem is decomposed into subproblems and the optimal alternative is selected for each subproblem so that it is never necessary to enumerate all combinations of the problem in advance.
2. Because optimization is applied to each subproblem, nonoptimal combinations are automatically eliminated.

3. The subproblems should be linked together in a special way so that it is never possible to optimize over infeasible combinations.

Referring to Figure 8.1, the basic elements and terminologies in a dynamic programming formulation are introduced as follows (Mays and Tung, 1992):

1. Stages (n) are the points of the problem where decisions are to be made. If a decision-making problem can be decomposed into N subproblems, there will be N stages in the dynamic programming formulation.
2. Decision variables (d_n) are courses of action to be taken for each stage. The number of decision variables, d_n, in each stage is not necessarily equal to 1.
3. State variables (S_n) are variables describing the state of a system at any stage n. A state variable can be discrete or continuous, finite or infinite. Referring to Figure 8.1, at any stage n, there are input states, S_n, and output states, S_{n+1}. The state variables of the system in a dynamic programming model have the function of linking succeeding stages so that when each stage is optimized separately, the resulting decision is automatically feasible for the entire problem. Furthermore, it allows one to make optimal decisions for the remaining stages without having to check the effect of future decisions for decisions previously made.
4. State return (r_n) is a scalar measure of the effectiveness of decision making in each stage. It is a function of the input state, the output state, and the decision variables of a particular stage; that is, $r_n = r(S_n, S_{n+1}, d_n)$.
5. Stage transformation or state transition (t_n) is a single-valued transformation which expresses the relationships among the input state, the output state, and the decision. In general, through the stage

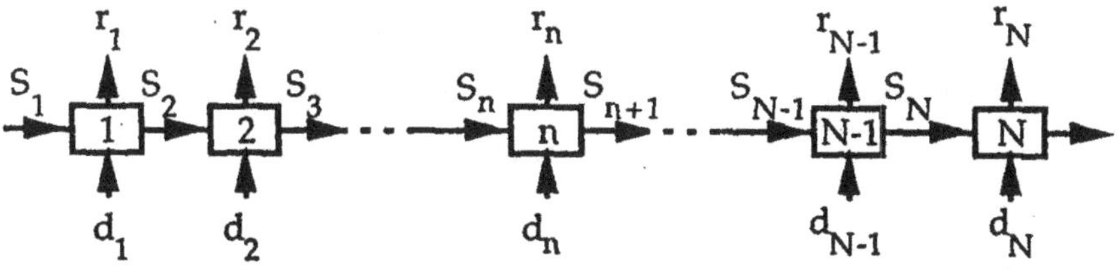

Figure 8.1 Sequential representation of serial dynamic programming problems.

transformation, the output state at any stage n can be expressed as the function of the input state and the decision as

$$S_{n+1} = t_n(S_n, d_n) \tag{8.1.1}$$

The basic features that characterize all dynamic programming problems are as follows (Mays and Tung, 1992):

1. The problem is divided into stages, with decision variables at each stage.
2. Each stage has a number of states associated with it.
3. The effect of the decision at each stage is to produce return, based on the stage return function, and to transform the current state variable into the state variable for the next stage, through the state transform function.
4. Given the current state, an optimal policy for the remaining stages is independent of the policy adopted in previous stages. This is called Bellman's principle of optimality, which serves as the backbone of dynamic programming.
5. The solution begins by finding the optimal decision for each possible state in the last stage (called the backward recursive) or in the first stage (called the forward recursive). A forward algorithm computationally advances from the first to the last stage, whereas a backward algorithm advances from the last stage to the first.
6. A recursive relationship that identifies the optimal policy for each state at any stage n can be developed, given the optimal policy for each state at the next stage, $n + 1$. This backward recursive equation, referring to Figure 8.1, can be written as

$$\begin{aligned} f_n^*(S_n) &= \underset{d_n}{\text{opt.}} \{r_n(S_n, d_n) \circ f_{n+1}^*(S_{n+1})\} \\ &= \underset{d_n}{\text{opt.}} \{r_n(S_n, d_n) \circ f_{n+1}^*[t_n(S_n, d_n)]\} \end{aligned} \tag{8.1.2}$$

   where $\circ$ represents an algebraic operator which can be $+$, $-$, $\times$, or $\div$, whichever is appropriate to the problem.

The recursive equation for a forward algorithm is stated as

$$f_n^*(S_n) = \underset{d_n}{\text{opt.}} \{r_n(S_n, d_n) \circ f_{n-1}^*(S_{n-1})\} \tag{8.1.3}$$

The recursive equation for the backward dynamic programming technique can be written as

$$f_n^*(S_n) = \begin{cases} \underset{d_n}{\text{opt.}} \, [r_n(S_n, d_n)], & \text{for } n = N \quad (8.1.4a) \\ \underset{d_n}{\text{opt.}} \, [r_n(S_n, d_n) \circ f_{n+1}^*(S_{n+1})] & \text{for } n = 1 \text{ to } N-1 \quad (8.1.4b) \end{cases}$$

Although dynamic programming possesses several advantages in solving water resources problems, especially for those involving the analysis of multistage processes, it has two disadvantages: the computer memory and time requirements. These disadvantages could become especially severe under two situations: (1) when the number of state variables is large and (2) when the dynamic programming is applied in a discrete fashion to a continuous state space. The problem associated with the latter case is that there exist difficulties in obtaining the true optimal solution without a considerable increase in discretization of the state space. With the advancement in computer technology, those disadvantages are becoming less and less severe.

An increase in the number of discretizations and/or state variables would geometrically increase the number of evaluations of the recursive formula and core-memory requirement per stage. This problem of rapid growth of computer time and core-memory requirement associated with multiple-state variable dynamic programming problems is referred to as the curse of dimensionality. From the problem-solving viewpoint, the problem of increased computer time is of much less concern than that of the increased computer storage requirement. Therefore, the rapid growth in memory requirements associated with multiple-state-variable problems can make the difference between solvable and unsolvable problems.

8.2 FEEDBACK METHOD OF OPTIMAL CONTROL FOR LINEAR SYSTEMS

The general optimal control problem for hydrosystems is stated as

$$\text{Optimize} \quad f(\mathbf{x}_t, \mathbf{u}_t, \mathbf{t}) \tag{8.2.1}$$

subject to the state equation

$$\mathbf{x}_t = \mathbf{g}_t(\mathbf{x}_t, \mathbf{x}_{t-1}, \mathbf{u}_t) \qquad t = 1, \ldots, T \tag{8.2.2}$$

Many hydrosystems optimal control problems can be formulated to minimize the sum of the squared deviations of a state variable from a specified target of the state variable, subject to the state equation. These problems constitute a linear deterministic control problem which consist of minimizing a quadratic loss function measuring the preference subject to the state equation that defines the dynamics of the system given as

$$\underset{\mathbf{u}_{(t)}}{\text{Minimize}}\ Z = \text{Min} \sum_{t=1}^{T} (\mathbf{x}_t - \mathbf{r}_t)^T \mathbf{P}_t (\mathbf{x}_t - \mathbf{r}_t) \tag{8.2.3}$$

subject to the state equation

$$\mathbf{x}_t = \mathbf{A}_t \mathbf{x}_{t-1} + \mathbf{C}_t \mathbf{u}_t + \mathbf{b}_t \tag{8.2.4}$$

in which $\mathbf{x}_t$ is the n-dimensional state variable vector, where $\mathbf{x}_0$ is the vector of initial (known) state variables; $\mathbf{r}_t$ is an n-dimensional vector of target values for the state variable at time t; $\mathbf{P}_t$ is an $n \times n$ positive semidefinite penalty matrix for deviating from target $\mathbf{r}_t$ at time t; $\mathbf{A}_t$ is an $n \times n$ matrix of known elements; $\mathbf{C}_t$ is an $n \times n$ matrix of known elements; $\mathbf{m}$ is the number of control variables; $\mathbf{b}_t$ is an $n \times 1$ vector of known constants; $\mathbf{u}_t$ is an m-dimensional vector of control variables; and superscript T refers to a transpose of the matrix. The above problem defined by Eqs. (8.2.3) and (8.2.4) is a linear-quadratic optimal control problem.

The optimal solution to the above optimal control problem is the time sequence of the control variables, $\mathbf{u}_t$, $t = 1, \ldots, T$, which are the decision variables. The feedback method to solve these optimal control problems is a dynamic programming approach consisting of a stage-by-stage optimization of the objective function subject to the system state equation. The control solves the above optimal control problem by deriving a set of feedback rules from a set of recursive equations (Chow, 1981). Dynamic programming requires that the objective function be separable in order to perform the stage-by-stage optimization. The quadratic objective function satisfies the requirement of separability.

The loss for period T is conditioned upon information up to $T - 1$, which is a function of $\mathbf{u}_T$,

$$\Psi_T = (\mathbf{x}_T - \mathbf{r}_T)^T \mathbf{P}_T (\mathbf{x}_T - \mathbf{r}_T) \tag{8.2.5a}$$

$$= \mathbf{x}_T^T \mathbf{H}_T \mathbf{x}_T - 2\mathbf{x}_T^T \mathbf{a}_T + \mathbf{c}_T \tag{8.2.5b}$$

where

$$\mathbf{P}_T = \mathbf{H}_T \tag{8.2.6}$$

$$\mathbf{P}_T \mathbf{r}_T = \mathbf{a}_T \tag{8.2.7}$$

$$\mathbf{c}_T = \mathbf{r}_T^T \mathbf{P}_T \mathbf{r}_T \tag{8.2.8}$$

in which $\mathbf{H}_T$ is an $n \times n$ matrix at time T and $\mathbf{a}_T$ is an n-dimensional vector at time T. Substituting $\mathbf{A}_T\ \mathbf{x}_{T-1} + \mathbf{C}_T\ \mathbf{u}_T + \mathbf{b}_T$ for $\mathbf{x}_T$ in Eq. (8.2.5) and minimizing with respect to $\mathbf{u}_T$ by differentiation results in

$$\hat{\mathbf{u}}_T = \mathbf{V}_T\mathbf{x}_{T-1} + \mathbf{w}_T \tag{8.2.9}$$

where

$$\mathbf{V}_T = -(\mathbf{C}_T^T\mathbf{H}_T\mathbf{C}_T)^{-1}(\mathbf{C}_T^T\mathbf{H}_T\mathbf{A}_T) \tag{8.2.10}$$

$$\mathbf{w}_T = -(\mathbf{C}_T^T\mathbf{H}_T\mathbf{C}_T)^{-1}\mathbf{C}_T^T(\mathbf{H}_T\mathbf{b}_T - \mathbf{a}_T) \tag{8.2.11}$$

The minimum expected loss for the last period is obtained by substituting for $\mathbf{u}_T$ in $\mathbf{\Psi}_T$:

$$\begin{aligned}\hat{\mathbf{\Psi}}_T = {} & \mathbf{x}_{T-1}^T(\mathbf{A}_T + \mathbf{C}_T\mathbf{V}_T)^T\mathbf{H}_T(\mathbf{A}_T + \mathbf{C}_T\mathbf{V}_T)\mathbf{x}_{T-1} \\ & + 2\mathbf{x}_{T-1}^T(\mathbf{A}_T + \mathbf{C}_T\mathbf{V}_T)^T(\mathbf{H}_T\mathbf{b}_T - \mathbf{a}_T) \\ & + (\mathbf{b}_T + \mathbf{C}_T\mathbf{w}_T)^T\mathbf{H}_T(\mathbf{b}_T + \mathbf{C}_T\mathbf{w}_T) \\ & - 2(\mathbf{b}_T + \mathbf{C}_T\mathbf{w}_T)^T\mathbf{a}_T + \mathbf{c}_T\end{aligned} \tag{8.2.12}$$

To obtain the optimal for the last two periods, consider that $\hat{\mathbf{u}}_T$ has been computed that would yield the minimum $\hat{\mathbf{\Psi}}_T$ and that by the principle of optimality of dynamic programming, $\mathbf{u}_{T-1}$ is needed to minimize

$$\mathbf{\Psi}_{T-1} = [(\mathbf{x}_{T-1} - \mathbf{r}_{T-1})^T\mathbf{P}_{T-1}(\mathbf{x}_{T-1} - \mathbf{r}_{T-1}) + \hat{\mathbf{\Psi}}_T] \tag{8.2.13a}$$

$$= [\mathbf{x}_{T-1}\mathbf{H}_{T-1}\mathbf{x}_{T-1}^T - 2\mathbf{x}_{T-1}^T\mathbf{a}_{T-1} + \mathbf{c}_{T-1}] \tag{8.2.13b}$$

where the equation (8.2.12) for $\hat{\mathbf{\Psi}}_T$ has been defined and

$$\mathbf{H}_{T-1} = \mathbf{P}_{T-1} + (\mathbf{A}_T + \mathbf{C}_T\mathbf{V}_T)^T\mathbf{H}_T(\mathbf{A}_T + \mathbf{C}_T\mathbf{V}_T) \tag{8.2.14}$$

$$\mathbf{h}_{T-1} = \mathbf{P}_{T-1}\mathbf{r}_{T-1} - (\mathbf{A}_T + \mathbf{C}_T\mathbf{V}_T)^T(\mathbf{H}_T\mathbf{b}_T - \mathbf{a}_T) \tag{8.2.15}$$

$$\begin{aligned}\mathbf{c}_{T-1} = {} & \mathbf{r}_{T-1}^T\mathbf{P}_{T-1}\mathbf{r}_{T-1} + (\mathbf{b}_T + \mathbf{C}_T\mathbf{w}_T)^T\mathbf{H}_T(\mathbf{b}_T + \mathbf{C}_T\mathbf{w}_T) \\ & - 2(\mathbf{b}_T + \mathbf{C}_T\mathbf{w}_T)^T\mathbf{a}_T + \mathbf{c}_T\end{aligned} \tag{8.2.16}$$

Because Eq. (8.2.13b) is identical to Eq. (8.2.5b) with the subscript T replaced by $T - 1$, the solution for $\hat{\mathbf{u}}_{T-1}$ is identical with Eq. (8.2.9) with the subscript T replaced by $T - 1$, where $\mathbf{V}_{T-1}$ and $\mathbf{w}_{T-1}$ are defined by Eqs. (8.2.10) and (8.2.11), respectively, with a similar change in time subscripts. Accordingly, $\hat{\mathbf{\Psi}}_{T-1}$ is given by Eq. (8.2.12) with the subscripts T replaced by $T - 1$. When solving the problem for the last three periods, $\hat{\mathbf{u}}_T$ and $\hat{\mathbf{u}}_{T-1}$ have been found that would yield the minimum expected loss $\hat{\mathbf{\Psi}}_{T-1}$ for the last two periods.

By the principle of optimality, we only need to minimize

$$\mathbf{\Psi}_{T-2} = [(\mathbf{x}_{T-2} - \mathbf{r}_{T-2})^T\mathbf{P}_{T-2}(\mathbf{x}_{T-2} - \mathbf{r}_{T-2}) + \hat{\mathbf{\Psi}}_{T-1}] \tag{8.2.17}$$

with respect to $\mathbf{u}_{T-2}$ and so forth. At the end of this process, $\hat{\mathbf{u}}$ is determined from $\hat{\mathbf{u}} = \mathbf{V}_1\mathbf{x}_0 + \mathbf{w}_1$, as the optimal policy for the first period and the associated minimum loss $\hat{\Psi}_1$ for all periods (or from period 1 forward). Computationally, solve Eqs. (8.2.10) and (8.2.14) with t replacing T for $\mathbf{V}_t$ and $\mathbf{H}_t$ backward in time, for $t = T, T-1, \ldots, 1$. Then solve Eqs. (8.2.11) and (8.2.15) with t replacing T for $\mathbf{w}_t$ and $\mathbf{a}_t$ backward in time, for $t = T, T-1, \ldots, 1$. Finally, the solution of Eq. (8.2.16) with t replacing T backward in time yields $\mathbf{c}$, which is used to evaluate $\hat{\Psi}_1$ given by Eq. (8.2.12) with 1 replacing T.

The expression $\hat{\Psi}_T$ given by Eq. (8.2.12) can be used to obtain the values (shadow prices or dual variables) of the initial resources $\mathbf{x}_{t-1}$. The vector of dual variables (shadow prices) is the derivation of $-\hat{\Psi}_t$ (negative loss or benefits) with respect to $\mathbf{x}_{t-1}$, namely

$$-\frac{\partial \hat{\Psi}_t}{\partial \mathbf{x}_{t-1}} = -2(\mathbf{A}_t\mathbf{C}_t\mathbf{V}_t)^T[\mathbf{H}_t(\mathbf{A}_t + \mathbf{C}_t\mathbf{V}_t)\mathbf{x}_{t-1} + \mathbf{H}_t\mathbf{b}_t - \mathbf{a}_t] \quad (8.2.18)$$

The algorithm is summarized below.

1. Initialization of the recursive equations for $t = T$ are

$$\mathbf{H}_T = \mathbf{P}_T \quad (8.2.19)$$

$$\mathbf{a}_T = \mathbf{P}_T\mathbf{r}_T \quad (8.2.20)$$

$$\mathbf{c}_T = \mathbf{h}_T^T\mathbf{P}_T\mathbf{r}_T \quad (8.2.21)$$

   where $\mathbf{H}_T$ is an $n \times n$ matrix at time T and $\mathbf{a}_T$ is an n-dimensional vector at time T.

2. The following recursive equations are solved backward in time from the terminal period $t = T, \ldots, 1$ to derive the feedback coefficients $\mathbf{V}_t$ and

$$\mathbf{V}_t = -(\mathbf{C}_t^T\mathbf{H}_t\mathbf{C}_t)^{-1}(\mathbf{C}_t\mathbf{H}_t\mathbf{A}_t) \quad (8.2.22)$$

$$\mathbf{w}_t = -(\mathbf{C}_t^T\mathbf{H}_t\mathbf{C}_t)^{-1}\mathbf{C}^T(\mathbf{H}_t\mathbf{b}_t - \mathbf{a}_t) \quad (8.2.23)$$

$$\mathbf{H}_{t-1} = \mathbf{P}_{t-1} + (\mathbf{A}_t + \mathbf{C}_t\mathbf{V}_t)^T\mathbf{H}_t(\mathbf{A}_t + \mathbf{C}_t\mathbf{V}_t) \quad (8.2.24)$$

$$\mathbf{a}_{t-1} = \mathbf{P}_{t-1}\mathbf{r}_{t-1} - (\mathbf{A}_t + \mathbf{C}_t\mathbf{V}_t)^T(\mathbf{H}_t\mathbf{b}_t - \mathbf{a}_t) \quad (8.2.25)$$

$$\mathbf{c}_{t-1} = \mathbf{r}_{t-1}^T\mathbf{P}_{t-1}\mathbf{r}_{t-1} + (\mathbf{b}_t + \mathbf{c}_t\mathbf{w}_t)^T\mathbf{H}_t(\mathbf{b}_t + \mathbf{C}_t\mathbf{w}_t) - 2(\mathbf{b}_t + \mathbf{C}_t\mathbf{w}_t)^T\mathbf{a}_t + \mathbf{c}_t \quad (8.2.26)$$

   where $\mathbf{V}_t$ is an $m \times n$ feedback coefficient matrix at time t and $\mathbf{w}_t$ is an m-dimensional feedback vector at time t.

3. After the feedback coefficients $\mathbf{V}_t$ and $\mathbf{w}_t$ have been computed, the optimal control variable, $\mathbf{u}_t$ can be computed using the following

feedback rule:

$$\hat{\mathbf{u}}_t = \mathbf{V}_t\mathbf{x}_{t-1} + \mathbf{w}_t \tag{8.2.27}$$

for $t = 1, \ldots, T$ and the state equation can be solved for $t = 1, \ldots, T$

$$\mathbf{x}_t = \mathbf{A}_t\mathbf{x}_{t-1} + \mathbf{C}_t\mathbf{w}_t + \mathbf{b}_t \tag{8.2.28}$$

8.3 GROUNDWATER MANAGEMENT PROBLEMS

Makinde-Odusola and Marino (1989) employed the feedback method of optimal control (discussed in Sec. 8.2) to model groundwater hydraulic management problems. The generalized mathematical formulation of the two-dimensional groundwater hydraulic management problem defines the state variable vector $\mathbf{x}_t$ as the piezometric head vector $\mathbf{h}_t$ and the control vector $\mathbf{u}_t$ as the vector of pumping rates $\mathbf{q}_t$ so that the problem is

$$\text{Minimize(or maximize) } f(\mathbf{h}_t, \mathbf{q}_t, t) \tag{8.3.1}$$

subject to

$$\mathbf{h}_t = \mathbf{g}_t(\mathbf{h}_{t-1}, \mathbf{q}_t), \quad t = 1, \ldots, T \tag{8.3.2}$$

The feedback method of control described in Sec. 8.2 applied to the groundwater management problem would constitute a linear deterministic control problem of the form

$$\underset{\mathbf{q}_t}{\text{Min}}\ Z = \sum_{t=1}^{T} (\mathbf{h}_t - \mathbf{r}_t)^T\mathbf{P}_t(\mathbf{h}_t - \mathbf{r}_t) \tag{8.3.3}$$

subject to

$$\mathbf{h}_t = \mathbf{A}_t\mathbf{h}_{t-1} + \mathbf{C}_t\mathbf{q}_t + \mathbf{b}_t \tag{8.3.4}$$

where $\mathbf{r}_t$ is an n-dimensional vector of targets for the state variable; $\mathbf{P}_t$ is an $n \times n$ positive semidefinite penalty matrix for deviating from target $\mathbf{r}_t$ at time t; $\mathbf{A}_t$ and $\mathbf{C}_t$ are matrices with known elements; m is the number of control variables; and n is the number of state variables.

The recursive equations that lead to the feedback rule are initialized as follows:

$$\mathbf{H}_T = \mathbf{P}_T \tag{8.3.5}$$

$$\mathbf{a}_T = \mathbf{P}_T\mathbf{r}_T \tag{8.3.6}$$

where $\mathbf{H}_T$ and $\mathbf{a}_T$ have been defined in Sec. 8.2.

Feedback coefficients $\mathbf{V}_t$ and $\mathbf{w}_t$ are derived by solving the recursive Eqs. (8.2.22)–(8.2.25) backward in time for $\mathbf{a}_t = \mathbf{x}_t$. Once the feedback rule coefficients have been computed for all time periods, the optimal pumping strategy, $\mathbf{q}_t$, is determined using the feedback rule (8.2.9) and the initial piezometric head vector, $\mathbf{h}_0$:

$$\mathbf{q}_t = \mathbf{V}_t\mathbf{h}_{t-1} + \mathbf{w}_t \tag{8.3.7}$$

Many groundwater management problems can be formulated as Eqs. (8.3.3) and (8.3.4), so that the feedback method of control can be used for the solution procedure. The state variable is the vector of piezometric heads $\mathbf{h}_t$ at all simulation nodes; the control variable is the vector of pumping (or recharge rates) $\mathbf{q}_t$; and the target state vector, $\mathbf{r}_t$, is the optimized piezometric head. If the hydraulic management problem is to estimate the rate of recharge to the aquifer, the target state vector, $\mathbf{r}_t$, should be set equal to the historical (or estimated) piezometric head (Makinde-Odusola and Marino, 1989). Yazdanian and Peralta (1986) discussed other methods for obtaining the vector of target piezometric heads, $\mathbf{r}_t$.

Makinde-Odusola and Marino (1989) solved the groundwater flow equation (state equation) numerically using SUTRA (Voss, 1984), which is a model for simulating two-dimensional, saturated–unsaturated, fluid, density-dependent, groundwater flow. This model also solves the energy or reactive adsorptive single-species solute transport. A two-dimensional, finite element scheme is used for the spatial discretization of the governing groundwater flow equation, and a finite-difference scheme is used for the temporal discretization of the groundwater flow equation.

Some advantages of the feedback control method include (Makinde-Odusola and Marino, 1989) the following: (1) the ease in specification and interpretation of the objective function parameters; (2) stochasticity in either the parameters or in the state variable can be handled; (3) the objective function form exploits the duality between control and parameter estimation; (4) the objective function can be used to impose physical hydraulic constraints; and (5) the incorporation of operational experience in specification of the vector of target piezometric heads.

8.4 FEEDBACK METHOD OF OPTIMAL CONTROL FOR NONLINEAR SYSTEMS

Consider an optimal control problem with a quadratic objective function of the form

$$\text{Minimize } Z = \text{Min} \sum_{t=1}^{T} (\mathbf{x}_t - \mathbf{r}_t)^T \mathbf{P}_t (\mathbf{x}_t - \mathbf{r}_t) \tag{8.4.1}$$

subject to the $\mathbf{u}_t$ nonlinear state equation of the form

$$\mathbf{x}_t = \mathbf{g}(\mathbf{x}_t, \mathbf{x}_{t-1}, \mathbf{u}_t, \boldsymbol{\eta}_t) + \boldsymbol{\varepsilon}_t \tag{8.4.2}$$

where $\mathbf{x}_t$ is the vector of state variables; $\mathbf{r}_t$ is a vector of specified target values; $\boldsymbol{\eta}_t$ is the vector of parameters that are not subject to control; and $\boldsymbol{\varepsilon}_t$ is a vector of random disturbances with mean zero and variance σ and is distributed independently through time. For purposes of the following discussion, the elements of $\boldsymbol{\eta}_t$ are given, leaving $\boldsymbol{\varepsilon}_t$ as the only random variable.

The quadratic objective function can be expressed as

$$Z = \sum_{t=1}^{T} (\mathbf{x}_t^T \mathbf{P}_t \mathbf{x}_t - 2\mathbf{x}_t^T \mathbf{P}_t \mathbf{r}_t + \mathbf{r}_t^T \mathbf{P}_t \mathbf{r}_t) \tag{8.4.3}$$

The problem is to *minimize* the expectation of Z so that, using DP, the optimal control problem for the last period T is to minimize

$$\begin{aligned} \boldsymbol{\Psi}_T &= \mathbf{E}_{T-1}(\mathbf{x}_T^T \mathbf{P}_T \mathbf{x}_T - 2\mathbf{x}_T^T \mathbf{P}_T \mathbf{r}_T + \mathbf{r}_T^T \mathbf{P}_T \mathbf{r}_T) \\ &= \mathbf{E}_{T-1}(\mathbf{x}_T^T \mathbf{H}_T \mathbf{x}_T - 2\mathbf{x}_T^T \mathbf{a}_T + \mathbf{c}_T) \end{aligned} \tag{8.4.4}$$

with respect to $\mathbf{x}_T$, where

$$\mathbf{H}_T = \mathbf{P}_T \tag{8.4.5}$$

$$\mathbf{a}_T = \mathbf{P}_T \mathbf{r}_T \tag{8.4.6}$$

$$\mathbf{c}_T = \mathbf{r}_T^T \mathbf{P}_T \mathbf{r}_T \tag{8.4.7}$$

The following steps are required to solve the optimal control problem for period T:

Step 1: Start with a nominal (trial) policy (control) $\tilde{\mathbf{u}}_T$ and set $\boldsymbol{\varepsilon}_T$ equal to zero; then linearize Eq. (8.4.2) about $\mathbf{x}_{T-1} = \mathbf{x}_{T-1}^{\circ}$ (given); $\mathbf{x}_T = \mathbf{x}_T^*$, and $\mathbf{u}_T = \tilde{\mathbf{u}}_T$, so that the solution of the system is

$$\mathbf{x}_T^* = \mathbf{g}(\mathbf{x}_T^*, \mathbf{x}_{T-1}^{\circ}, \tilde{\mathbf{u}}_T, \boldsymbol{\eta}_T) \tag{8.4.8}$$

where $\mathbf{x}_T^*$ is solved by an iterative method such as Gauss–Siedel. A linearized version of Eq. (8.4.2) is

$$\mathbf{x}_T = \mathbf{x}_T^* + \mathbf{B}_{1T}(\mathbf{x}_T - \mathbf{x}_T^* + \mathbf{B}_{2T}(\mathbf{x}_{T-1} - \mathbf{x}_{T-1}^{\circ}) + \mathbf{B}_{3T}(\mathbf{u}_T - \tilde{\mathbf{u}}_T) + \boldsymbol{\varepsilon}_T \tag{8.4.9}$$

The jth column of $\mathbf{B}_{1T}$ consists of the partial derivatives of the vector function $\mathbf{g}$ with respect to the jth element of $\mathbf{x}_T$ evaluated at the given values $\mathbf{x}_T^*$, $\mathbf{x}_{T-1}^\circ$, $\tilde{\mathbf{u}}_T$, and $\boldsymbol{\eta}_T$. The same is true for the jth columns of $\mathbf{B}_{2T}$ and $\mathbf{B}_{3T}$.

Step 2: Equation (8.4.9) can be rearranged (solved) to obtain the linearized approximation

$$\mathbf{x}_T = \mathbf{A}_T\mathbf{x}_{T-1} + \mathbf{C}_T\mathbf{u}_T + \mathbf{b}_T + \mathbf{y}_T \tag{8.4.10}$$

where

$$(\mathbf{A}_T\mathbf{C}_T\mathbf{y}_T) = (\mathbf{I} - \mathbf{B}_{1T})^{-1}(\mathbf{B}_{2T}\mathbf{B}_{3T}\boldsymbol{\varepsilon}_T) \tag{8.4.11a}$$

$$\mathbf{b}_T = \mathbf{x}_T^* - \mathbf{A}_T\mathbf{x}_{T-1}^\circ - \mathbf{C}_T\tilde{\mathbf{u}}_T \tag{8.4.11b}$$

and $\mathbf{y}_T$ is a random vector that is serially independent and identically distributed.

The matrix $\mathbf{I} - \mathbf{B}_{IT}$ is

$$\mathbf{I} = \mathbf{B}_{1T} = \begin{bmatrix} \mathbf{I} - \mathbf{B}_{1T}^* & \mathbf{0} \\ \mathbf{0} & \mathbf{I} \end{bmatrix} \tag{8.4.12}$$

in which the order of $\mathbf{B}_{1T}^*$ is the number of simultaneous state equations excluding the identities.

Step 3: Equation (8.4.4) is minimized with respect to $\mathbf{u}_T$ subject to Eq. (8.4.10) by differentiating Eq. (8.4.4) with respect to $\mathbf{u}_T$ and interchanging the order of taking the expectation and differentiation:

$$\begin{aligned}\frac{\partial \boldsymbol{\Psi}_T}{\partial \mathbf{u}_T} &= 2\mathbf{E}_{T-1}\left[\left(\frac{\partial \mathbf{x}_T^T}{\partial \mathbf{u}_T}\right)\mathbf{H}_T\mathbf{x}_T - \left(\frac{\partial \mathbf{x}_T^T}{\partial \mathbf{u}_T}\right)\mathbf{a}_T\right] \\ &= 2\mathbf{E}_{T-1}[\mathbf{C}_T^T\mathbf{H}_T(\mathbf{A}_T\mathbf{x}_{T-1} + \mathbf{C}_T\mathbf{u}_T + \mathbf{b}_T + \mathbf{y}_T) \\ &\quad - \mathbf{C}_T^T\mathbf{a}_T] = \mathbf{0}\end{aligned} \tag{8.4.13}$$

where Eq. (8.4.10) has been substituted for $\mathbf{x}_T$ and used to compute $\partial\mathbf{x}_T^T/\partial\mathbf{u}_T = \mathbf{c}_T^T$. Equation (8.4.13) is solved for $\mathbf{u}_T$ and now is referred to as $\hat{\mathbf{u}}_T$,

$$\hat{\mathbf{u}}_T = \mathbf{V}_T\mathbf{x}_{T-1} + \mathbf{w}_T \tag{8.4.14}$$

where

$$\mathbf{V}_T = -(\mathbf{E}_{T-1}\mathbf{C}_T^T\mathbf{H}_T\mathbf{C}_T)^{-1}(\mathbf{E}_{T-1}\mathbf{C}_T^T\mathbf{H}_T\mathbf{A}_T) \tag{8.4.15}$$

$$\mathbf{w}_T = -(\mathbf{E}_{T-1}\mathbf{C}_T^T\mathbf{H}_T\mathbf{C}_T)^{-1}(\mathbf{E}_{T-1}\mathbf{C}_T^T\mathbf{H}_T\mathbf{b}_T - \mathbf{E}_{T-1}\mathbf{C}_T^T\mathbf{a}_T) \tag{8.4.16}$$

In the linear approximation (8.4.10), $\mathbf{A}_T$, $\mathbf{C}_T$, and $\mathbf{b}_T$ are not functions of $\boldsymbol{\varepsilon}_T$ and, as a result, are not random. The expectation signs in Eqs. (8.4.15) and (8.4.16) can be dropped.

Step 4: The solution from Eq. (8.4.14), $\hat{\mathbf{u}}_T$ replaces the initial guess $\tilde{\mathbf{u}}_T$ in step 1; then repeat steps 1 through 4 until there is convergence in $\tilde{\mathbf{u}}_T$. Even when convergence occurs, the solution is not truly optimal because the approximate form (8.4.10) is used with constant coefficients $\mathbf{A}_T$, $\mathbf{C}_T$, and $\mathbf{b}_T$. Replace Eq. (8.4.9) by

$$\mathbf{x}_T = \tilde{\mathbf{x}}_T + \mathbf{B}_{1T}(\mathbf{x}_T - \tilde{\mathbf{x}}_T) + \mathbf{B}_{2T}(\mathbf{x}_{T-1} - \mathbf{x}^{\circ}_{T-1}) + \mathbf{B}_{3T}(\mathbf{u}_T - \hat{\mathbf{u}}_T) \tag{8.4.17}$$

Step 5: Using Eq. (8.4.10) for $\mathbf{x}_T$ and Eq. (8.4.14) for $\mathbf{u}_T$, the minimum objective for period T is, from Eq. (8.4.4),

$$\begin{aligned}\hat{\boldsymbol{\Psi}}_T = {} & \mathbf{x}^T_{T-1}\mathbf{E}_{T-1}(\mathbf{A}_T + \mathbf{C}_T\mathbf{V}_T)^T\mathbf{H}_T(\mathbf{A}_T + \mathbf{C}_T\mathbf{V}_T)\mathbf{x}_{T-1} \\ & + 2\mathbf{x}^T_{T-1}\mathbf{E}_{T-1}(\mathbf{A}_T + \mathbf{C}_T\mathbf{V}_T)^T(\mathbf{H}_T\mathbf{b}_T - \mathbf{a}_T) \\ & + \mathbf{E}_{T-1}(\mathbf{b}_T + \mathbf{C}_T\mathbf{w}_T)^T\mathbf{H}_T(\mathbf{b}_T + \mathbf{C}_T\mathbf{w}_T) \\ & + \mathbf{E}_{T-1}\mathbf{y}^T\mathbf{H}_T\mathbf{y}_T - 2\mathbf{E}_{T-1}(\mathbf{b}_T + \mathbf{C}_T\mathbf{w}_T)^T\mathbf{a}_T \\ & + \mathbf{E}_{T-1}\mathbf{C}_T\end{aligned} \tag{8.4.18}$$

Applying the principle of optimality in dynamic programming, minimizing with respect to $\mathbf{u}_{T-1}$.

$$\begin{aligned}\boldsymbol{\Psi}_{T-1} &= \mathbf{E}_{T-2}(\mathbf{x}^T_{T-1}\mathbf{P}_{T-1}\mathbf{x}_{T-1} - 2\mathbf{x}^T_{T-1}\mathbf{P}_{T-1}\mathbf{r}_{T-1} + \mathbf{r}^T_{T-1}\mathbf{P}_{T-1}\mathbf{r}^T_{T-1} + \hat{\boldsymbol{\Psi}}_T) \\ &= \mathbf{E}_{T-2}(\mathbf{x}^T_{T-1}\mathbf{H}_{T-1}\mathbf{x}_{T-1} - 2\mathbf{x}^T_{T-1}\mathbf{a}_{T-1} + \mathbf{c}_{T-1})\end{aligned} \tag{8.4.19}$$

Substituting Eq. (8.4.18) for $\hat{\boldsymbol{\Psi}}_T$,

$$\begin{aligned}\mathbf{H}_{T-1} &= \mathbf{P}_{T-1} + \mathbf{E}_{T-1}(\mathbf{A}_T + \mathbf{C}_T\mathbf{V}_T)^T\mathbf{H}_T(\mathbf{A}_T + \mathbf{C}_T\mathbf{V}_T) \\ &= \mathbf{P}_{T-1} + \mathbf{E}_{T-1}(\mathbf{A}^T_T\mathbf{H}_T\mathbf{A}_T) + \mathbf{V}^T_T(\mathbf{E}_{T-1}\mathbf{C}^T_T\mathbf{H}_T\mathbf{A}_T)\end{aligned} \tag{8.4.20}$$

$$\begin{aligned}\mathbf{a}_{T-1} &= \mathbf{P}_{T-1}\mathbf{r}_{T-1} + \mathbf{E}_{T-1}(\mathbf{A}_T + \mathbf{C}_T\mathbf{V}_T)^T(\mathbf{a}_T - \mathbf{H}_T\mathbf{b}_T) \\ &= \mathbf{P}_{T-1}\mathbf{r}_{T-1} + \mathbf{E}_{T-1}(\mathbf{A}_T + \mathbf{C}_T\mathbf{V}_T)^T\mathbf{a}_T - \mathbf{E}_{T-1}(\mathbf{A}^T_T\mathbf{H}_T\mathbf{b}_T) \\ &\quad - \mathbf{V}^T_T(\mathbf{E}_{T-1}\mathbf{C}^T_T\mathbf{H}_T\mathbf{b}_T)\end{aligned} \tag{8.4.21}$$

$$\begin{aligned}\mathbf{c}_{T-1} = {} & \mathbf{E}_{T-1}(\mathbf{b}_T + \mathbf{C}_T\mathbf{w}_T)^T\mathbf{H}_T(\mathbf{b}_T + \mathbf{C}_T\mathbf{w}_T) \\ & - 2\mathbf{E}_{T-1}(\mathbf{b}_T + \mathbf{C}_T\mathbf{w}_T)^T\mathbf{a}_T \\ & + \mathbf{r}_{T-1}^T\mathbf{P}_{T-1}\mathbf{r}_{T-1} + \mathbf{E}_{T-1}\mathbf{y}_T^T\mathbf{H}_T\mathbf{y}_T + \mathbf{E}_{T-1}\mathbf{c}_T\end{aligned} \tag{8.4.22}$$

The second line of Eq. (8.4.18) has the same form as Eq. (8.4.4); therefore, the steps in the solution for $\mathbf{u}_T$ with $T - 1$ replacing T yield an optimal solution $\hat{\mathbf{u}}_{T-1}$ in the form (8.4.14), and for the corresponding minimum two-period loss, $\hat{\mathbf{\Psi}}_{T-1}$ from Eq. (8.4.14) the process continues backward in time until $\hat{\mathbf{u}}_1$ and $\hat{\mathbf{\Psi}}_1$ are obtained.

Summarizing the procedure:

Step 1: Use initial guesses $\tilde{\mathbf{u}}_1, \tilde{\mathbf{u}}_2, \ldots, \tilde{\mathbf{u}}_T$ to solve the system (8.4.2) with $\boldsymbol{\varepsilon}_t = 0$ to obtain $\mathbf{x}_1^\circ, \mathbf{x}_2^\circ, \ldots, \mathbf{x}_{T-1}^\circ$.

Step 2: Linearize the system (8.4.8) as shown in Eq. (8.4.4) for $t = T, T - 1, \ldots, 1$, noting that $\mathbf{x}_t^* = \mathbf{x}_t^\circ$ has been computed in step 1. Compute $\mathbf{A}_t$, $\mathbf{C}_t$, and $\mathbf{b}_t$ by Eqs. (8.4.11a) and (8.4.11b).

Step 3: Compute $\mathbf{V}_t$ and $\mathbf{H}_{t-1}$ for $t = T, T - 1, \ldots, 1$ using Eqs. (8.4.15) and (8.4.20), respectively. Use Eq. (8.4.21) to compute $\mathbf{a}_{t-1}$ and Eq. (8.4.16) to compute $\mathbf{w}_t$ backward in time.

Step 4: Use the feedback control equations, $\hat{\mathbf{u}}_t = \mathbf{V}_t\mathbf{x}_{t-1} + \mathbf{w}_t$ and the system Eq. (8.4.2) to compute successively $\hat{\mathbf{u}}_1, \mathbf{x}_1^\circ, \hat{\mathbf{u}}_2, \mathbf{x}_2^\circ, \ldots$. Return to step 1 using the $\hat{\mathbf{u}}_t$ as the initial guesses. Repeat steps 1–4 until $\hat{\mathbf{u}}_t$ converges.

Step 5: Use Eq. (8.4.22) to compute $\mathbf{c}_{t-1}$ backward in time. Solve for $\hat{\mathbf{\Psi}}_t$.

REFERENCES

Bellman, R. and Dreyfus, S., *Applied Dynamic Programming*, Princeton University Press, Princeton, NJ, 1962.

Chow, G. C., *Econometric Analysis by Control Methods*, John Wiley and Sons, New York, 1981.

Dyer P. and McReynolds, S., *The Computational Theory of Optimal Control*, Academic Press, New York, 1970.

Makinde-Odusola, B. A. and Marino, M. A., Optimal Control of Groundwater by the Feedback, *Water Resources Research*, Vol. 25., No. 6, pp. 1341–1352, 1989.

Mays, L.W., and Tung, Y.K., *Hydrosystems Engineering and Management*, McGraw-Hill, New York, 1992.

Voss, C., A Finite Element Simulation Model for Saturated–Unsaturated, Fluid-Density-Dependent Groundwater Flow with Energy Transport or Chemically Reactive Single Species Solute Transport, U.S. Geological Survey Resource Investigation Report 84-4369, 1984.

Yazdanian, A. and Peralta, R. C., Sustained-Yield Groundwater Planning by Goal Programming, *Ground Water*, Vol. 24, No. 2, pp. 157–165, 1986.

Chapter 9
Differential Dynamic Programming

9.1 DIFFERENTIAL DYNAMIC PROGRAMMING ALGORITHM

9.1.1 Introduction

This section presents the differential dynamic programming method for discrete-time optimal control problems. The term differential dynamic programming (DDP) used by Jacobson and Mayne (1970) broadly refers to stagewise nonlinear programming procedures. Earlier works that basically developed DDP procedures for unconstrained discrete-time control problems, including those by Bellman and Dreyfus (1962), Mayne (1966), Gershwin and Jacobson (1970), Dyer and McReynolds (1970), Jacobson and Mayne (1970), Yakowitz and Rutherford (1984), and Yakowitz (1989), Ohno (1978), and Murray and Yakowitz (1981), have contributed to DDP techniques for constrained optimal control problems.

Yakowitz and Rutherford (1984) summarized the following;

> Our opinion is that a little-known technique called "differential dynamic programming" offers the potential of enormously expanding the scale of discrete-time optimal control problems which are subject to numerical solution. Among the attractive features of this method are that no discretization of control or state space is used; the memory requirements grow as m^2 and the computational requirements as m^3, with m being the dimension of the control variable; the successive approximation converges globally under lenient smoothness assumption and the convergence is quadratic if certain convexity assumptions hold.

The objective of DDP is to minimize a quadratic approximation instead of solving the actual control problem. Yakowitz and Rutherford (1984) pointed out the following properties of DDP for unconstrained problems:

(a) DDP overcomes the curse of dimensionality (computational burden and memory requirements grow exponentially with state and control dimensions) in that the computational requirements grow as m^3N and memory requirements as mnN, where n and m are the state and control variable dimensions, respectively, and N is the number of decision times.
(b) Under lenient conditions, DDP is globally convergent.
(c) No discretization of state or control spaces is required.
(d) The convergent rule of the DDP algorithm is quadratic for control problems in which the Hessian matrix of the objective function is convex in a neighborhood of the solution.

The basic optimal control problem considered here is stated as follows:

$$\underset{\mathbf{u}}{\text{Min}}\ Z = \sum_{t=1}^{T} f_t(\mathbf{x}_t, \mathbf{u}_t, t) \tag{9.1.1}$$

subject to

$$\mathbf{x}_{t+1} = \mathbf{g}_t(\mathbf{x}_t, \mathbf{u}_t, t), \quad t = 1, \ldots, T \tag{9.1.2}$$

9.1.2 Algorithm Definition

Define the current or known control policy as $\bar{\mathbf{u}}_t$ for $t = 1, \ldots, T$ and the current state trajectory or $\bar{\mathbf{x}}_t$ for $t = 1, \ldots, T + 1$ is obtained by Eq. (9.1.2) with known $\mathbf{x}_1$. The initial state is $\mathbf{x}_1$. For any function $W(\mathbf{x}, \mathbf{u})$ defined by the control and state variables, let QP($W(\mathbf{x}, \mathbf{u})$) denote the linear and quadratic path of the Taylor's series expansion of $W(\ )$ about $(\bar{\mathbf{u}}, \bar{\mathbf{x}})$. The quadratic for time T where the DDP backward recursion begins is

$$\begin{aligned} L(\mathbf{x}, \mathbf{u}, T) &= \text{QP}(f(\mathbf{x}, \mathbf{u}, T)) \\ &= \frac{1}{2}\,\delta\mathbf{x}^T\left(\frac{\partial^2 f}{\partial \mathbf{x}^2}\right)\delta\mathbf{x} + \delta\mathbf{x}^T\left(\frac{\partial^2 f}{\partial \mathbf{x}\,\partial \mathbf{u}}\right)\delta\mathbf{u} + \frac{1}{2}\,\delta\mathbf{u}^T\left(\frac{\partial^2 f}{\partial \mathbf{u}^2}\right)\delta\mathbf{u} \\ &\quad + \left(\frac{\partial f}{\partial \mathbf{u}}\right)\delta\mathbf{u} + \frac{\partial f}{\partial \mathbf{x}}\,\delta\mathbf{x} \end{aligned} \tag{9.1.3}$$

where $\delta\mathbf{x} = (\mathbf{x} - \bar{\mathbf{x}}_T)$ and $\delta\mathbf{u} = (\mathbf{u} - \bar{\mathbf{u}}_T)$ are state and input perturbations and the gradients and the Hessian of $f(\mathbf{x}, \mathbf{u}, T)$ are evaluated at $\bar{\mathbf{x}}_T$ and $\bar{\mathbf{u}}_T$. Equation (9.1.3) can be presented in a more compact form as

$$L(\mathbf{x}, \mathbf{u}, T) = \delta\mathbf{x}^T\mathbf{A}_T\delta\mathbf{x} + \delta\mathbf{x}^T\mathbf{B}_T\delta\mathbf{u} + \delta\mathbf{u}^T\mathbf{C}_T\delta\mathbf{u} + \mathbf{D}_T^T\delta\mathbf{u} + \mathbf{E}_T^T\delta\mathbf{x} \quad (9.1.4)$$

The idea of DDP is to minimize the quadratic approximation instead of the actual control problem value function, thereby obtaining a computer amenable function which is at the expense of involving truncation error. A necessary condition that a control $\mathbf{u}^*$ minimizes $\mathbf{L}(\mathbf{x}, \mathbf{u}, T)$ is

$$\nabla_{\mathbf{u}}L(\mathbf{x}, \mathbf{u}, T)^T = 2\mathbf{C}_T\delta\mathbf{u} + \mathbf{B}_T\delta\mathbf{x} + \mathbf{D}_T = 0 \quad (9.1.5)$$

Then the optimal control $\mathbf{u}^*$ can be found from Eq. (9.1.5), under the assumption that $\mathbf{C}_T$ is nonsingular:

$$\delta\mathbf{u}(\mathbf{x}, T) = (\mathbf{u}^* - \bar{\mathbf{u}}_T) \quad (9.1.6a)$$

$$= -\tfrac{1}{2}\mathbf{C}_T^{-1}(\mathbf{D}_T + \mathbf{B}_T\delta\mathbf{x}) \quad (9.1.6b)$$

$$= \boldsymbol{\alpha}_T + \boldsymbol{\beta}_T\delta\mathbf{x} \quad (9.1.6c)$$

where $\boldsymbol{\alpha}_T = (-\tfrac{1}{2})\mathbf{C}_T^{-1}\mathbf{D}_T$ and $\boldsymbol{\beta}_T = (-\tfrac{1}{2})\mathbf{C}_T^{-1}\mathbf{B}_T$

The optimal value function is

$$F(\mathbf{x}, T) = \min_{\mathbf{u}} Z = \min f(\mathbf{x}_T, \mathbf{u}_T, T) \quad (9.1.7)$$

which is approximated by the quadratic as

$$V(\mathbf{x}; T) = L(\mathbf{x}, \mathbf{u}(\mathbf{x}, T), T)$$
$$= L(\mathbf{x}, \bar{\mathbf{u}}_T + (\boldsymbol{\alpha}_T + \boldsymbol{\beta}_T\delta\mathbf{x}), T) \quad (9.1.8)$$

The following quadratic $V(\mathbf{x}; T)$ is determined by substituting Eq. (9.1.6) into Eq. (9.1.4),

$$V(\mathbf{x}; T) = \delta\mathbf{x}^T\mathbf{P}_T\delta\mathbf{x} + \mathbf{Q}_T\delta\mathbf{x} \quad (9.1.9)$$

where

$$\mathbf{P}_T = \mathbf{A}_T - \tfrac{1}{4}\mathbf{B}_T^T\mathbf{C}_T^{-1}\mathbf{B}_T \quad (9.1.10)$$

$$\mathbf{Q}_T^T = -\tfrac{1}{2}\mathbf{D}_T^T\mathbf{C}_T^{-1}\mathbf{B}_T + \mathbf{E}_T \quad (9.1.11)$$

as long as $\mathbf{C}_T$ is nonsingular

The DDP backward recursion procedure is performed for $t = T, T - 1, \ldots, 1$ using the quadratic

$$L(\mathbf{x}, \mathbf{u}, t) = \mathrm{QP}[f(\mathbf{x}, \mathbf{u}, t) + V(\mathbf{g}(\mathbf{x}, \mathbf{u}, t); t + 1)] \quad (9.1.12)$$

The quadratic approximate optimal return function $V(\mathbf{x}, t + 1)$ is defined as

$$V(\mathbf{x}, t + 1) = (\delta\mathbf{x})^T\mathbf{P}_{t+1}\delta\mathbf{x} + \mathbf{Q}_{t+1}\delta\mathbf{x} \quad (9.1.13)$$

Similar to Eq. (9.1.4), the quadratic can be expressed as

$$L(\mathbf{x}, \mathbf{u}, t) = \delta\mathbf{x}^T\mathbf{A}_t\delta\mathbf{x} + \delta\mathbf{u}^T\mathbf{B}_t\delta\mathbf{x} + \delta\mathbf{u}^T\mathbf{C}_t\delta\mathbf{u} + \mathbf{D}_t^T\delta\mathbf{u} + \mathbf{E}_t^T\delta\mathbf{x} \quad (9.1.14)$$

The following expressions of the coefficients $\mathbf{A}_t$, $\mathbf{B}_t^T$, $\mathbf{C}_t$, $\mathbf{D}_t^T$ and $\mathbf{E}_t^T$ are derived in Appendix 9.A:

$$\mathbf{A}_t = \frac{1}{2}\left(\frac{\partial^2 f}{\partial \mathbf{x}^2}\right)_t + \left(\frac{\partial \mathbf{g}}{\partial \mathbf{x}}\right)_t^T \mathbf{P}_{t+1}\left(\frac{\partial \mathbf{g}}{\partial \mathbf{x}}\right)_t + \frac{1}{2}\sum_{i=1}^{n}(Q_{t+1})_i\left(\frac{\partial^2 \mathbf{g}}{\partial x_i^2}\right) \quad (9.1.15)$$

$$\mathbf{B}_t^T = \left(\frac{\partial^2 f}{\partial \mathbf{x}\,\partial \mathbf{u}}\right)_t + 2\left(\frac{\partial \mathbf{g}}{\partial \mathbf{x}}\right)_t^T \mathbf{P}_{t+1}\left(\frac{\partial \mathbf{g}}{\partial \mathbf{x}}\right)_t + \frac{1}{2}\sum_{i=1}^{n}(Q_{t+1})_i\left(\frac{\partial^2 g}{\partial x_i\,\partial u_i}\right) \quad (9.1.16)$$

$$\mathbf{C}_t = \frac{1}{2}\left(\frac{\partial^2 f}{\partial \mathbf{u}^2}\right)_t + \left(\frac{\partial \mathbf{g}}{\partial \mathbf{x}}\right)_t^T \mathbf{P}_{t+1}\left(\frac{\partial \mathbf{g}}{\partial \mathbf{x}}\right)_t + \frac{1}{2}\sum_{i=1}^{n}(Q_{t+1})_i\left(\frac{\partial^2 \mathbf{g}}{\partial u_i^2}\right) \quad (9.1.17)$$

$$\mathbf{D}_t^T = \left(\frac{\partial f}{\partial \mathbf{u}}\right)_t + \mathbf{Q}_{t+1}\left(\frac{\partial \mathbf{g}}{\partial \mathbf{u}}\right)_t \quad (9.1.18)$$

$$\mathbf{E}_t^T = \left(\frac{\partial f}{\partial \mathbf{x}}\right)_t + \mathbf{Q}_{t+1}\left(\frac{\partial \mathbf{g}}{\partial \mathbf{x}}\right)_t \quad (9.1.19)$$

The first-order derivatives of $f(\mathbf{x}, \mathbf{u}, t)$ in the above equations are components of the gradient of $f(\mathbf{x}, \mathbf{u}, t)$; the second-order derivatives are the components of the Hessian of $f(\mathbf{x}, \mathbf{u}, t)$; the first-order derivatives of $\mathbf{g}(\mathbf{x}, \mathbf{u}, t)$ are components of the Jacobian of $\mathbf{g}(\mathbf{x}, \mathbf{u}, t)$; and the second-order derivatives of $\mathbf{g}(\mathbf{x}, \mathbf{u}, t)_i$ for $1 \le i \le n$ are the blocks of the Hessian matrices of the coordinates of $\mathbf{g}(\mathbf{x}, \mathbf{u}, t)$. All derivatives are evaluated about the current states and controls.

The first-order necessary condition for optimality is

$$\nabla_u \mathbf{L}(\mathbf{x}, \mathbf{u}, t) = \mathbf{0} \quad (9.1.20)$$

so that the minimizing strategy for the quadratic $\mathbf{L}(\mathbf{x}, \mathbf{u}, t)$ is

$$\partial\mathbf{u}(\mathbf{x}; t) = \boldsymbol{\alpha}_t + \boldsymbol{\beta}_t(\mathbf{x} - \bar{\mathbf{x}}_t) \quad (9.1.21)$$

where

$$\boldsymbol{\alpha}_t = -\tfrac{1}{2}\mathbf{C}_t^{-1}\mathbf{D}_t^T \quad (9.1.22)$$

and

$$\boldsymbol{\beta}_t = -\tfrac{1}{2}\mathbf{C}_t^{-1}\mathbf{B}_t \quad (9.1.23)$$

The approximating polynomial for the optimal return function is

$$V(\mathbf{x}; t) = L(\mathbf{x}, \mathbf{u}(\mathbf{x}, t), t)$$
$$= (\delta\mathbf{x})^T\mathbf{P}_t(\mathbf{x} - \bar{\mathbf{x}}_t) + \mathbf{Q}_t^T\delta\mathbf{x} \qquad (9.1.24)$$

where

$$\mathbf{P}_t = \mathbf{A}_t - \tfrac{1}{4}\mathbf{B}_t^T\mathbf{C}_t^{-1}\mathbf{B}_t \qquad (9.1.25)$$

$$\mathbf{Q}_t = \tfrac{1}{2}\mathbf{D}_t^T\mathbf{C}_t^{-1} + \mathbf{E}_t^T \qquad (9.1.26)$$

These equations are necessary for the DDP backward recursion. $\boldsymbol{\alpha}_t$ and $\boldsymbol{\beta}_t$ for $1 \le t \le N$ must be stored for use in the forward sweep. The forward sweep determines the successor DDP policy by successively selecting controls according to the rule $\mathbf{u}(\mathbf{x}^*, t)$ and then calculating the successor state at each time so that $\mathbf{u}_1^* = \mathbf{u}(\mathbf{x}_1; 1)$ and $\mathbf{x}_2^* = \mathbf{g}(\mathbf{x}_1^*, \mathbf{u}_1^*, 1)$. Then the following is for $t = 2, \ldots, T$

$$\mathbf{u}_t^* = \mathbf{u}(\mathbf{x}_t^*; t) + \bar{\mathbf{u}}_t \qquad (9.1.27)$$

and

$$\mathbf{x}_{t+1}^* = \mathbf{g}(\mathbf{x}_t^*, \mathbf{u}_t^*, t) \qquad (9.1.28)$$

For the next DDP iteration, the DDP successor control is the current control sequence $\bar{\mathbf{u}}$.

9.1.3 Algorithm Description

Input for the DDP procedure consists of T, the number of decision times; m, the dimension of the control variable; n, the dimension of the state variable; $\bar{\mathbf{u}}$, a nominal policy; components of the gradients, $\partial f/\partial\mathbf{x}$ and $\partial f/\partial\mathbf{u}$; components of the Hessian, $\partial^2 f/\partial\mathbf{x}^2$, $\partial^2 f/\partial\mathbf{u}^2$, and $\partial^2 f/\partial\mathbf{x}\,\partial\mathbf{u}$; the Jacobian of $\mathbf{g}$, $\partial\mathbf{g}/\partial\mathbf{x}$ and $\partial\mathbf{g}/\partial\mathbf{u}$; and the blocks of the Hessian matrices of the coordinates of $\mathbf{g}$, $\partial^2\mathbf{g}/\partial x_i^2$, $\partial^2\mathbf{g}/\partial u_i^2$, and $\partial^2\mathbf{g}/\partial x_i\,\partial u_i$.

The program parameters are $\mathbf{A}_t$, $\mathbf{B}_t$, $\mathbf{C}_t$, $\mathbf{D}_t$, and $\mathbf{E}_t$ for $1 \le t \le T$, which are the coefficients of $\mathbf{L}(\mathbf{x}, \mathbf{u}, t)$; $\mathbf{P}_t$ and $\mathbf{Q}_t$ for $2 \le t \le T + 1$; $\boldsymbol{\alpha}_t$ and $\boldsymbol{\beta}_t$ for $1 \le t \le N$ which are the coefficients of the linear strategy function $\mathbf{u}(\mathbf{x}, t)$; and $\boldsymbol{\theta}_t$ for $1 \le t \le N + 1$, which are the parameters for the acceptance tests. The steps of the algorithm are outlined below.

Step 0: Select an initial (nominal) policy.

$$\bar{\mathbf{u}}_t, \quad t = 1, \ldots, N$$

Step 1: Initialize parameters and compute loss and trajectory for the given policy.

$$\begin{aligned}
\text{Set}\quad & \mathbf{P}_{T+1} = \mathbf{0}_{n\times n} \\
& \mathbf{Q}_{T+1} = \mathbf{0}_{1\times n} \\
& \boldsymbol{\theta}_{T+1} = \mathbf{0} \\
& \bar{\mathbf{x}}_1 \text{ is given and fixed} \\
& \bar{\mathbf{x}}_{t+1} = \mathbf{g}(\bar{\mathbf{x}}_t, \bar{\mathbf{u}}_t, t) \\
& Z(\bar{\mathbf{u}}) = \sum_{t=1}^{T} f_t(\bar{\mathbf{x}}_t, \bar{\mathbf{u}}_t, t)
\end{aligned}$$

Step 2: Backward sweep (Perform the following [(a)–(e)] for $t = T, \ldots, 1$).

(a) Compute $\mathbf{A}_t$, $\mathbf{B}_t^T$, and $\mathbf{C}_t$

$$\mathbf{A}_t = \frac{1}{2}\left(\frac{\partial^2 f}{\partial \mathbf{x}^2}\right)_t + \left(\frac{\partial \mathbf{g}}{\partial \mathbf{x}}\right)_t^T \mathbf{P}_{t+1}\left(\frac{\partial \mathbf{g}}{\partial \mathbf{x}}\right)_t + \frac{1}{2}\sum_{i=1}^{n} (\mathbf{Q}_{t+1})_i \left(\frac{\partial^2 g}{\partial x_i^2}\right) \tag{9.1.29}$$

$$\mathbf{B}_t^T = \left(\frac{\partial^2 f}{\partial \mathbf{x}\, \partial \mathbf{u}}\right)_t + 2\left(\frac{\partial \mathbf{g}}{\partial \mathbf{x}}\right)_t^T \mathbf{P}_{t+1}\left(\frac{\partial \mathbf{g}}{\partial \mathbf{x}}\right)_t + \sum_{i=1}^{n} (\mathbf{Q}_{t+1})_i \left(\frac{\partial^2 g}{\partial x_i\, \partial u_i}\right) \tag{9.1.30}$$

$$\mathbf{C}_t = \frac{1}{2}\left(\frac{\partial^2 f}{\partial \mathbf{u}^2}\right)_t + \left(\frac{\partial \mathbf{g}}{\partial \mathbf{x}}\right)_t^T \mathbf{P}_{t+1}\left(\frac{\partial \mathbf{g}}{\partial \mathbf{x}}\right)_t + \tfrac{1}{2}\sum_{i=1}^{n} (\mathbf{Q}_{t+1})_i \left(\frac{\partial^2 g}{\partial u_i^2}\right) \tag{9.1.31}$$

(b) Compute $\mathbf{D}_t^T$ and $\mathbf{E}_t^T$

$$\mathbf{D}_t^T = \left(\frac{\partial f}{\partial \mathbf{u}}\right)_t + \mathbf{Q}_{t+1}\left(\frac{\partial \mathbf{g}}{\partial \mathbf{u}}\right)_t \tag{9.1.32}$$

$$\mathbf{E}_t^T = \left(\frac{\partial f}{\partial \mathbf{x}}\right)_t + \mathbf{Q}_{t+1}\left(\frac{\partial \mathbf{g}}{\partial \mathbf{x}}\right)_t \tag{9.1.33}$$

(c) Compute $\mathbf{P}_t$ and $\mathbf{Q}_t$

$$\mathbf{P}_t = \mathbf{A}_t - \tfrac{1}{4}\mathbf{B}_t^T\mathbf{C}_t^{-1}\mathbf{B}_t \tag{9.1.34}$$

$$\mathbf{Q}_t = -\tfrac{1}{2}\mathbf{D}_t^T\mathbf{C}_t^{-1}\mathbf{B}_t + \mathbf{E}_t \tag{9.1.35}$$

(d) Compute $\boldsymbol{\alpha}_t$ and $\boldsymbol{\beta}_t$

$$\boldsymbol{\alpha}_t = -\tfrac{1}{2}\mathbf{C}_t^{-1}\mathbf{D}_t \tag{9.1.36}$$

$$\boldsymbol{\beta}_t = -\tfrac{1}{2}\mathbf{C}_t^{-1}\mathbf{B}_t \tag{9.1.37}$$

(e) Compute $\boldsymbol{\theta}_t$

$$\boldsymbol{\theta}_t = -\tfrac{1}{2}\mathbf{D}_t^T\mathbf{C}_t^{-1}\mathbf{D}_t + \boldsymbol{\theta}_{t+1} \tag{9.1.38}$$

In the above steps (a)–(e), store $\mathbf{P}_t$, $\mathbf{Q}_t$, and $\boldsymbol{\theta}_t$ and replace $\mathbf{P}_{t+1}$, $\mathbf{Q}_{t+1}$, and $\boldsymbol{\theta}_{t+1}$, respectively. Also store $\boldsymbol{\alpha}_t$ and $\boldsymbol{\beta}_t$.

Step 3: Forward sweep

(a) Set $\varepsilon = 1.0$

(b) Compute $\mathbf{u}_t(\varepsilon)$ and $\mathbf{x}_{t+1}$recursively for $t = 1, \ldots, T$ using

$$\mathbf{u}_t(\varepsilon) = \varepsilon\boldsymbol{\alpha}_t + \boldsymbol{\beta}_t(\mathbf{x}_t - \bar{\mathbf{x}}_t) + \bar{\mathbf{u}}_t \tag{9.1.39}$$

$$\mathbf{x}_{t+1} = \mathbf{g}(\mathbf{x}_t, \mathbf{u}_t(\varepsilon), t) \tag{9.1.40}$$

(c) Compute $Z(\mathbf{u}(\lambda))$.

$$Z(\mathbf{u}(\varepsilon)) = \sum_{t=1}^{T} f_t(\mathbf{x}_t, \mathbf{u}_t(\varepsilon), t) \tag{9.1.41}$$

(d) If $[Z(\mathbf{u}(\varepsilon)) - Z(\bar{\mathbf{u}})] \leq \varepsilon\ (\theta_1/2)$, set $\bar{\mathbf{u}}_t = \mathbf{u}_t(\varepsilon)$ for $t = 1, \ldots, T$ and go to step 1
Otherwise, perform a line search; set $\varepsilon = \varepsilon/2$ and go to step 3(b)

9.2 CONVERGENCE OF UNCONSTRAINED DDP

Liao and Shoemaker (1991) investigated the condition under which the DDP algorithm can be expected to converge and proposed algorithm changes to improve convergence. These important points are as follows:

1. Quadratic convergence requires that the stagewise Hessian matrices (C_t) be positive definite.

2. For linear transition equations, a sufficient condition to guarantee positive definite stagewise Hessian matrices and quadratic convergence is for the objective function to be positive definite-convex.
3. For nonlinear transition equations, the DDP algorithm will not necessarily generate positive definite stagewise Hessian matrices (C_t) even if both the objective function and transition equations are positive definite-convex.

These procedures for modifying the stagewise Hessian matrices ($\mathbf{C}_t$) to guarantee convergence when $\mathbf{C}_t$ is not positive definite were presented by Liao and Shoemaker (1991). Murray and Yakowitz (1981) and Yakowitz and Rutherford (1984) suggested that if $\mathbf{C}_t$ has any nonpositive eigenvalues, then a shift $\boldsymbol{\varepsilon}_t > 0$ is used to replace $\mathbf{C}_t$ by

$$\tilde{\mathbf{C}}_t = \mathbf{C}_t + \varepsilon_t^c \mathbf{I}_m \tag{9.2.1}$$

where $\boldsymbol{\varepsilon}_t$ is large enough to make $\tilde{\mathbf{C}}_t$ positive definite, and $\mathbf{I}_m$ is an identity matrix. If $\boldsymbol{\varepsilon}_t$ is too large or too small, the convergence will be slowed or there may be numerical difficulties, respectively. Liao and Shoemaker (1991) proposed the following two-step adaptive shift procedure:

Step 1: Apply the DDP algorithm with the modification to pick a constant shift, $\varepsilon^c > 0$ (independent of time) so that $\mathbf{C}_t^c$ is positive definite

$$\mathbf{C}_t^c = \mathbf{C}_t + \varepsilon^c \mathbf{I}_m \tag{9.2.2}$$

and replace $\mathbf{C}_t$ with $\mathbf{C}_t^c$ in the DDP algorithm.

Step 2: Define $\mathbf{C}_t^a$ with a given constant $\delta > 0$ as

$$\mathbf{C}_t^a = \mathbf{C}_t + \varepsilon_t^a \mathbf{I}_m \tag{9.2.3}$$

where $\boldsymbol{\varepsilon}_t^a(\delta)$ is defined as

$$\boldsymbol{\varepsilon}_t^a(\delta) = \begin{cases} \delta - \lambda(\mathbf{C}_t) & \text{if } \lambda(\mathbf{C}_t) < \delta \\ 0 & \text{if } \lambda(\mathbf{C}_t) \geq \delta \end{cases} \tag{9.2.4}$$

and

$$\lambda(\mathbf{C}_t) \text{ is the minimum eigenvalue of } \mathbf{C}_t$$

Liao and Shoemaker (1991) show that the DDP algorithm with the active shift will converge quadratically, whereas the constant shift method converges linearly. To apply the adaptive shift, one must determine the following:

1. How to choose $\varepsilon^c > 0$ in step 1.

2. How to choose $\delta > 0$ in step 2.
3. When to change from the constant shift to the active shift.

The computational experience of Liao and Shoemaker (1991) suggested a range of $0.001 \le \delta \le 1.0$.

$\varepsilon^c > 0$ could be chosen so that

$$\varepsilon^c = \max\left(-\mu \min_{1\le t\le N} \{\lambda(\mathbf{C}_t(\bar{\mathbf{u}})\}; 0\right)$$

where $\mu > 0$ and $\bar{\mathbf{u}}$ is the initial control value. If the ε^c cannot guarantee that $\{\lambda(\mathbf{C}_t(\bar{\mathbf{u}}))\} > 0$, then this step can be repeated with a larger value of μ.

9.3 MULTIRESERVOIR OPERATION

Murray and Yakowitz (1979) applied constrained DDP to mutireservoir control problems. Consider the four reservoir systems shown in Figure 9.1. Let $r_{i,t}$ denote the release of the *i*th reservoir during decision time t, $s_{i,t}$ denotes the beginning storage, and $q_{i,t}$ denotes the inflow to the *i*th reservoir. The basic reservoir mass balance equation is

$$\mathbf{s}_{t+1} = \mathbf{s}_t + \mathbf{q}_t + \mathbf{M}\mathbf{r}_t \tag{9.3.1}$$

where $\mathbf{q}_t = (q_{1,t}, \ldots, q_{4,t})^T$, $\mathbf{s}_t = (s_{1,t}, \ldots, s_{4,t})^T$, $\mathbf{r}_t = (r_{1,t}, \ldots, r_{4,t})^T$, and $\mathbf{M}$ is a fourth-order matrix with -1's on the diagonal and $+1$'s in the position (k, j) where reservoir k releases into reservoir j, and zero elsewhere. The mass balance equation for the four-reservoir problem is

$$\begin{bmatrix} s_{1,t+1} \\ s_{2,t+1} \\ s_{3,t+1} \\ s_{4,t+1} \end{bmatrix} = \begin{bmatrix} s_{1,t} \\ s_{2,t} \\ s_{3,t} \\ s_{4,t} \end{bmatrix} + \begin{bmatrix} q_{1,t} \\ q_{2,t} \\ 0 \\ 0 \end{bmatrix} + \begin{bmatrix} -1 & 0 & 0 & 0 \\ 0 & -1 & 0 & 0 \\ 0 & 1 & -1 & 0 \\ 1 & 0 & 1 & -1 \end{bmatrix} \begin{bmatrix} r_{1,t} \\ r_{2,t} \\ r_{3,t} \\ r_{4,t} \end{bmatrix} \tag{9.3.2}$$

Reservoir storage bound constraints are defined as

$$\begin{aligned} 0 &\le s_{1,t} \le 10 \\ 0 &\le s_{2,t} \le 10 \\ 0 &\le s_{3,t} \le 10 \\ 0 &\le s_{4,t} \le 15 \end{aligned} \tag{9.3.3}$$

The initial state is $\mathbf{s}_1 = (5, 5, 5, 5)^T$ and the terminal state is $\mathbf{s}_{13} = (5, 5, 5, 7)^T$ for 12 time periods. Inflows are $q_{1,t} = 2$ and $q_{2,t} = 3$ for $t = 1, \ldots, 12$. Release constraints are

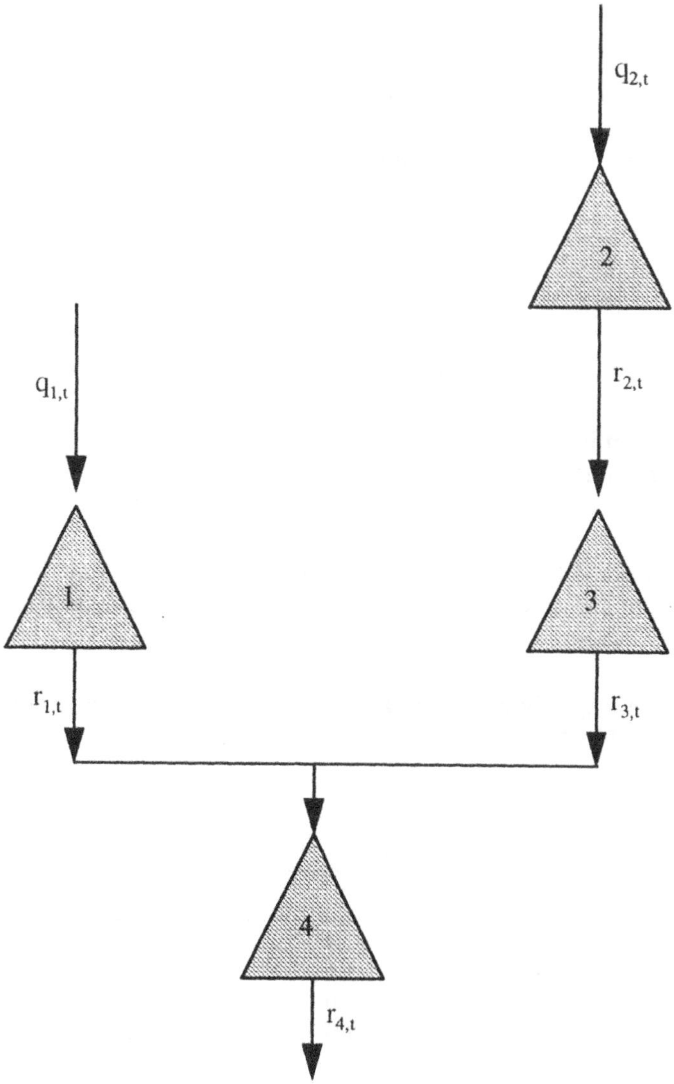

Figure 9.1 Four-reservoir configuration (adapted from Murray and Yakowitz, 1979).

$$\begin{bmatrix} 0 \\ 0 \\ 0 \\ 0 \end{bmatrix} \leq \begin{bmatrix} r_{1,t} \\ r_{2,t} \\ r_{3,t} \\ r_{4,t} \end{bmatrix} \leq \begin{bmatrix} 3 \\ 4 \\ 4 \\ 7 \end{bmatrix}, \quad 1 \leq t \leq 12 \tag{9.3.4}$$

The loss function to be minimized is

$$F(\mathbf{r}) = \sum_{t=1}^{12} f(\mathbf{s}_t, \mathbf{r}_t, t) \tag{9.3.5}$$

where

$$f(\mathbf{s}_t, \mathbf{r}_t, t) = \sum_{j=1}^{4} c_{j,t} r_{j,t} \tag{9.3.6}$$

The loss coefficients are given in Table 9.1.

The terminal constraint

$$\mathbf{s}_{13} = (5, 5, 5, 7)^T$$

can be enforced by translating this condition into constraints on the preceding state and control; for example, for $s_{1,13} = 5$, the mass balance is

$$r_{1,12} = 5 - s_{1,12} - q_{1,12} = 3 - s_{1,12}$$

The release $r_{1,12}$ is bounded by $0 \leq r_{1,12} \leq 3$, so that for the terminal state to be reachable, then

Table 9.1 Loss Function Coefficient

t	$c_{1,t}$	$c_{2,t}$	$c_{3,t}$	$c_{4,t}$
1	−1.1	−1.4	−1.0	−2.6
2	−1.0	−1.1	−1.0	−2.9
3	−1.0	−1.0	−1.2	−3.6
4	−1.2	−1.0	−1.8	−4.4
5	−1.8	−1.2	−2.5	−4.2
6	−2.5	−1.8	−2.2	−4.0
7	−2.2	−2.5	−2.0	−3.8
8	−2.0	−2.2	−1.8	−4.1
9	−1.8	−2.0	−2.2	−3.6
10	−2.2	−1.8	−1.8	−3.1
11	−1.8	−2.2	−1.4	−2.7
12	−1.4	−1.8	−1.1	−2.5

Source: Murray and Yakowitz (1979).

$$3 \leq s_{1,12} \leq 6$$

This, in turn, requires bounds on the control $r_{1,11}$ and this sets allowable ranges for $s_{1,11}$. One could then construct sequences $\{\eta_{1,t}\}$ and $\{\xi_{1,t}\}$ so that for any t, $\eta_{1,t} < s_{1,t} < \xi_{1,t}$. In general form, this can be written as

$$\eta_t < \mathbf{s}_t < \xi_t \tag{9.3.7}$$

Equation (9.3.7) is assumed to satisfy the state constraints (9.3.3), which may be rewritten in the form

$$\boldsymbol{\eta}_t < \mathbf{s}_{t-1} + \mathbf{q}_{t-1} + \mathbf{Mr}_{t-1} < \boldsymbol{\xi}_t \tag{9.3.8}$$

These linear constraints can be rewritten as

$$\boldsymbol{\eta}^c_{t-1} \leq \mathbf{r}_{t-1} \leq \boldsymbol{\xi}^c_{t-1} \tag{9.3.9}$$

where $\boldsymbol{\eta}^c_{t-1}$ and $\boldsymbol{\xi}^c_{t-1}$ are constructed from $\boldsymbol{\eta}_t$, $\boldsymbol{\xi}_t$, and $\mathbf{M}^{-1}$ $(\mathbf{s}_{t-1} + \mathbf{q}_{t-1})$. The problem would then be to require that Inequalities (9.3.4) and (9.3.9) be satisfied simultaneously.

9.4 OPTIMAL CONTROL OF GROUNDWATER HYDRAULICS

This section presents a differential dynamic programming (DDP) algorithm for solving large-scale, nonlinear groundwater management problems. The groundwater management model for the optimal control of operational costs of an unconfined aquifer was posed by Jones et al. (1987) as follows:

$$\underset{\mathbf{q}_t}{\text{Min}}\ Z = \sum_{t=1}^{T} \mathbf{q}_t^T(\mathbf{e} - \hat{\mathbf{h}}_{t+1}) \tag{9.4.1}$$

subject to

$$\mathbf{h}_{t+1} = \mathbf{g}(\mathbf{h}_t, \mathbf{q}_t, t), \quad t = 1, \ldots, T \tag{9.4.2}$$

$$\mathbf{l}^T\mathbf{q}_t \geq \mathbf{d}_t, \quad t = 1, \ldots, T \tag{9.4.3}$$

$$\underline{\mathbf{q}}_t \leq \mathbf{q}_t \leq \overline{\mathbf{q}}_t, \quad t = 1, \ldots, T \tag{9.4.4}$$

where $\mathbf{l}^T$ is a row vector of 1's, $\mathbf{e}$ is an m vector of the distance from the ground surface to the lower datum of the aquifer, $\hat{\mathbf{h}}_t$ is an m vector of hydraulic heads, $\mathbf{h}_t$ is an n vector of hydraulic heads, and $m \leq n$. Operational costs are assumed to be the product of the pumpage rate, $\mathbf{q}_t$, and the loss $(\mathbf{L} - \hat{\mathbf{h}}_{t+1})$. Equation (9.4.3) requires that the sum of pumping in each planning period satisfy the demand $\mathbf{d}_t$. Equation (9.4.4) is the capacity constraint.

The above problem (9.4.1)–(9.4.4) is a nonlinear, discrete-time, optimal control problem, in which the state equation defines the groundwater hydraulics for two-dimensional flow for confined and/or unconfined flow. These partial differential equations can be expressed in finite-difference form for unconfined conditions as

$$\alpha_1 h^2_{1,t+1} + \alpha_2 h^2_{2,t+1} + \alpha_3 h^2_{3,t+1} + \alpha_4 h^2_{4,t+1} + \alpha_0 h^2_{0,t+1} + \beta h_{0,t+1} = \beta_0 h_{0,t} + l_0 \gamma_{0,t} \quad (9.4.5)$$

where the node numbers are defined in Figure 9.2. α_0, α_1, α_2, α_3, α_4, β_0, and γ_0 are known functions of the aquifer parameters.

The DDP approach begins with nominal control for each stage ($\mathbf{q}_1$, $\mathbf{q}_2$, . . . , $\mathbf{q}_n$) to solve the state equations for the nominal states ($\mathbf{h}_1$, $\mathbf{h}_2$, . . . , $\mathbf{h}_{n+1}$). The nominal solution ($\mathbf{h}^*$, $\mathbf{q}^*$) results in a nominal objective Z^*. Next is to determine a quadratic approximation to the above optimal control problem (9.4.1)–(9.4.4).

In order to develop the quadratic approximation, a first-order Taylor series approximation (linearization) of the groundwater simulation equations about the nominal solution ($\mathbf{h}^*$, $\mathbf{q}^*$) is developed,

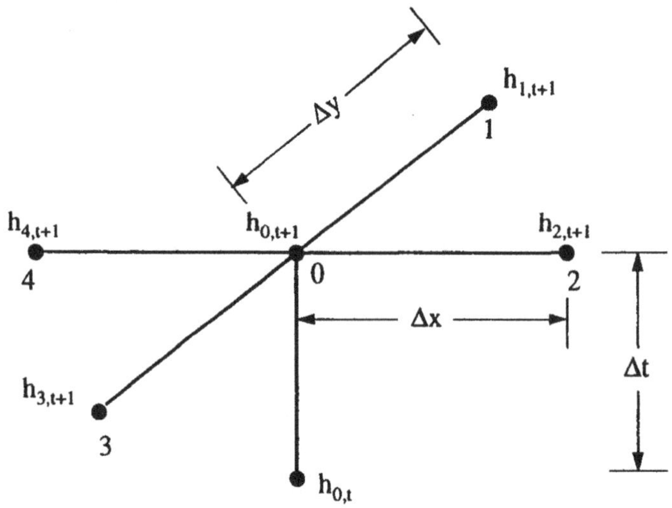

Figure 9.2 Finite-difference approximation at node 0 (Jones et al., 1987).

$$h_{i,t+1} = h^*_{i,t+1} + \sum_{j=1}^{n} \frac{\partial h_{i,t+1}}{\partial h_{i,t}} (h_{j,t} - h^*_{j,t}) + \sum_{j=1}^{n} \frac{\partial h_{i,t+1}}{\partial q_{j,t}} (q_{j,t} - q^*_{i,t})$$

where the deviations are evaluated at ($\mathbf{h}_t^*$, $\mathbf{q}_t^*$). The Taylor series approximation for the vector of heads is

$$\mathbf{h}_{t+1} = \mathbf{h}^*_{t+1} + \mathbf{R}_1(\mathbf{h}_t - \mathbf{h}_t^*) + \mathbf{R}_2(\mathbf{q}_t - \mathbf{q}_t^*) \qquad (9.4.6)$$

which is a multivariate first-order Taylor series approximation of $\mathbf{h}_{t+1}$ as a function of $\mathbf{h}_t$, where $\mathbf{R}_1$ and $\mathbf{R}_2$ are Jacobian matrices. The ith row, jth column of $\mathbf{R}_1$ is $\partial h_{i,t+1}/\partial h_{j,t}$, and the ith row, jth column of $\mathbf{R}_2$ is $\partial h_{i,t+1}/\partial q_{j,t}$ both evaluated at ($\mathbf{h}_t^*$, $\mathbf{q}_t^*$).

Determination of the Jacobian matrices are found by implicit differentiation of the simulation equation; for example, differentiation of Eq. (9.4.5) with respect to $h_{i,t}$ results in

$$2\alpha_1 h_{1,t+1} \frac{\partial h_{1,t+1}}{\partial h_{j,t}} + 2\alpha_2 h_{2,t+1} \frac{\partial h_{2,t+1}}{\partial h_{j,t}} + 2\alpha_3 h_{3,t+1} \frac{\partial h_{3,t+1}}{\partial h_{j,t}}$$

$$+ 2\alpha_4 h_{4,t+1} \frac{\partial h_{4,t+1}}{\partial h_{j,t}} + 2\alpha_0 h_{0,t+1} \frac{\partial h_{0,t+1}}{\partial h_{j,t}}$$

$$+ \beta_0 \frac{\partial h_{0,t+1}}{\partial h_{i,t}} = \beta_0 \frac{\partial h_{0,t}}{\partial h_{i,t}} + \gamma_0 \frac{\partial q_{0,t}}{\partial h_{i,t}} \qquad (9.4.7)$$

Differentiating each simulation equation with respect to $\mathbf{h}_{i,t}$ yields a simultaneous system of n equations for the jth columns of the Jacobian matrix $\mathbf{R}_1$. The Jacobian matrix is evaluated at ($\mathbf{h}_t^*$, $\mathbf{q}_t^*$), $\mathbf{h}^*_{t+1}$ is used instead of $\mathbf{h}_{t+1}$, resulting in a system of n simultaneous linear equations for the jth column of the Jacobian matrix $\mathbf{R}_1$ evaluated at ($\mathbf{h}_t^*$, $\mathbf{q}_t^*$). A system of n simultaneous linear equations is found for each $h_{j,t}$ and $q_{1,t}$ by the same procedure. The left-hand-side coefficient matrix of the n simultaneous linear equations for the jth column is the same, so that for n nodes and m wells, the Jacobian matrices for Eq. (9.4.6) are found from the solution of n simultaneous linear equations with $n + m$ right-hand sides.

The Taylor series approximation is made for each finite-difference time step, dividing the time between stages into a number of finite-difference time steps. The linear approximation for each finite-difference time step allows simple substitution and matrix algebra to relate $\mathbf{h}_{t+1}$ to $\mathbf{h}_t$ and $\mathbf{q}_t$ to $\mathbf{q}_{t+1}$. Through the use of the first-order approximation to the nonlinear simulation equation, $\mathbf{h}_{t+1}$ can be expressed as

$$\mathbf{h}_{t+1} = \mathbf{A}_t\mathbf{h}_t + \mathbf{C}_t\mathbf{q}_t + \mathbf{z}_t \qquad (9.4.8)$$

where $\mathbf{A}_t$ is an $n \times n$ matrix, $\mathbf{C}_t$ is an $n \times n$ matrix, and $\mathbf{z}_t$ is an n vector.

The original problem (9.4.1)–(9.4.4) can now be expressed as

$$\underset{\mathbf{q}_t}{\text{Min}}\ Z = \sum_{t=1}^{T} \mathbf{h}_t^T\mathbf{U}_t\mathbf{h}_t + \mathbf{q}_t^T\mathbf{V}_t\mathbf{h}_t + \mathbf{q}_t^T\mathbf{W}_t\mathbf{q}_t + \mathbf{X}_t\mathbf{q}_t + \mathbf{Y}_t^T\mathbf{h}_t \quad (9.4.9)$$

subject to

$$\mathbf{h}_{t+1} = \mathbf{A}_t\mathbf{h}_t + \mathbf{C}_t\mathbf{q}_t + \mathbf{z}_t \quad (9.4.10)$$

$$\mathbf{l}^T\mathbf{q}_t \le \mathbf{d}_t \quad (9.4.11)$$

$$\underline{\mathbf{q}}_t \le \mathbf{q}_t \le \overline{\mathbf{q}}_t \quad (9.4.12)$$

where the matrices in Eq. (9.4.9) are $\mathbf{U}_t = \mathbf{0}$, $\mathbf{V}_t = -\hat{\mathbf{A}}_t$, $\mathbf{W}_t = -\frac{1}{2}(\hat{\mathbf{B}} + \hat{\mathbf{B}}^1)$, $\mathbf{X}_t = \mathbf{L} - \hat{\mathbf{z}}_t$, and $\mathbf{Y}_t = \mathbf{0}$.

The values of $\hat{\mathbf{h}}_{t+1}$ in Eq. (9.4.1) for nodes with wells are determined from

$$\hat{\mathbf{h}}_{t+1} = \hat{\mathbf{A}}_t\mathbf{h}_t + \hat{\mathbf{B}}_t\mathbf{q}_t + \hat{\mathbf{z}}_t \quad (9.4.13)$$

The optimization problem (9.4.9)–(9.4.12) can be referred to as the nominal control problem, which is an approximation to the original control problem (9.4.1)–(9.4.4) about the nominal solution. This problem would be a linear quadratic programming (LQP) problem if the two inequalities were not present. They can be placed in the objective through the use of penalty terms. Jones et al. (1987) suggest incorporating Eq. (9.4.13) into the objective of the nominal control problem and solving it as an LQP problem using dynamic programming as described by Murray and Yakowitz (1979). The resulting recursive equation is

$$f_t(\mathbf{h}_t) = \underset{\mathbf{q}_t}{\text{Min}}\ \{\mathbf{h}_t^T\hat{\mathbf{U}}_t\mathbf{h}_t + \mathbf{q}_t^T\hat{\mathbf{V}}_t\mathbf{h}_t + \mathbf{q}_t^T\hat{\mathbf{W}}_t\mathbf{q}_t + \hat{\mathbf{X}}_t^T\mathbf{q}_t + \hat{\mathbf{Y}}^T\mathbf{h}_t + f_{t+1}(\mathbf{h}_{t+1}) - \lambda^T(\mathbf{l}^T\mathbf{q}_t - \mathbf{d}_t)\} \quad (9.4.14)$$

where λ is a Lagrange multiplier estimate and $f_{t+1}(\mathbf{h}_{t+1})$ is the optimal value that gives the optimal return for stages $t + 1$ to T, which is a scalar for this problem. Using the nominal stage $\mathbf{h}_t^*$ in place of $\mathbf{h}_t$ in Eq. (9.4.14) results in a QP problem which is a subproblem in the backward sweep.

9.5 GROUNDWATER RECLAMATION MODELS

The groundwater reclamation problem can be stated using the general form of the optimal control problem as

$$\underset{\mathbf{u}}{\text{Min}} = \sum_{t=1}^{T} f_t(\mathbf{x}_t, \mathbf{u}_t, t) \tag{9.5.1}$$

subject to

$$\mathbf{x}_{t+1} = \mathbf{g}_t(\mathbf{x}_t, \mathbf{u}_t\, t), \quad t = 1, \ldots, T \tag{9.5.2}$$

$$\mathbf{w}(\mathbf{x}_t, \mathbf{u}_t, t) \leq \mathbf{0}, \quad t = 1, \ldots, T \tag{9.5.3}$$

where the vector of state variables, $\mathbf{x}_t$, is the vector of hydraulic heads, $\mathbf{h}_t$, and concentrations, $\mathbf{c}_t$,

$$\mathbf{x}_t = \begin{bmatrix} \mathbf{h}_t \\ \mathbf{c}_t \end{bmatrix} \tag{9.5.4}$$

The vector of control variables are the pumping rates at a set of m possible and or existing well locations. The state equation (9.5.2) defines the groundwater flow and contaminant transport for the aquifer. Constraint equation (9.5.3) includes constraints on both the state and control. As an example, constraints on the state variable can include water quality requirements on the concentration levels and pumpage limitations on the control variable.

Culver (1990) and Shoemaker (1992) present a groundwater remediation model based on a DDP approach that they refer to as a successive approximation linear quadratic regulator (SALQR). Their definition of SALQR is that it "differs from DDP only in that the nonlinear simulation equations are linearized in the optimization step." SALQR and DDP are identical if the simulation equations are linear. The optimal control problem solved by Culver and Shoemaker (1992) with flexible management periods is defined as

$$\text{Min } Z = \sum_{k=1}^{K} f_k(\mathbf{x}_k, \mathbf{u}_k, k) \tag{9.5.5}$$

subject to

$$\mathbf{x}_{k+1} = \mathbf{g}(\mathbf{x}_k, \mathbf{u}_k, k) \quad k = 1, \ldots, K \tag{9.5.6}$$

$$\mathbf{w}(\mathbf{x}_k, \mathbf{u}_k, k) \leq \mathbf{0} \quad k = 1, \ldots, K \tag{9.5.7}$$

where k refers to the management period, K is the total number of management periods defined as $K = T/d$, T is the total number of simulation steps, and d is the number of simulation periods per management period. The above problem (9.5.5)–(9.5.7) is directly analogous to problem (9.5.1)–(9.5.4). The state equation (9.5.6) describes the change in $\mathbf{x}$ over a management period instead of over a simulation step. The algorithm defined by Culver and Shoemaker (1992) to solve this problem is presented in Figure (9.3).

The cost function used by Culver and Shoemaker (1992) defines the total operating costs of pumping and treatment during each time period given as

$$f_k(\mathbf{x}_k, \mathbf{u}_k, k) = \sum_{i=1}^{m} a_{1i}u_{ki} + \sum_{i=1}^{m} a_{2i}u_{ki}(\bar{\mathbf{h}} - \mathbf{h}_{k+1}) \tag{9.5.8}$$

in which the first term defines water treatment costs as a linear function of the amount of water pumped (extracted) and the second term defines the pumping costs related to the product of the extraction rate and the lift $(\bar{\mathbf{h}} - \mathbf{h}_{k+1})$. $\bar{\mathbf{h}}$ is the vector of distances from the ground surface to the lower boundary of the aquifer.

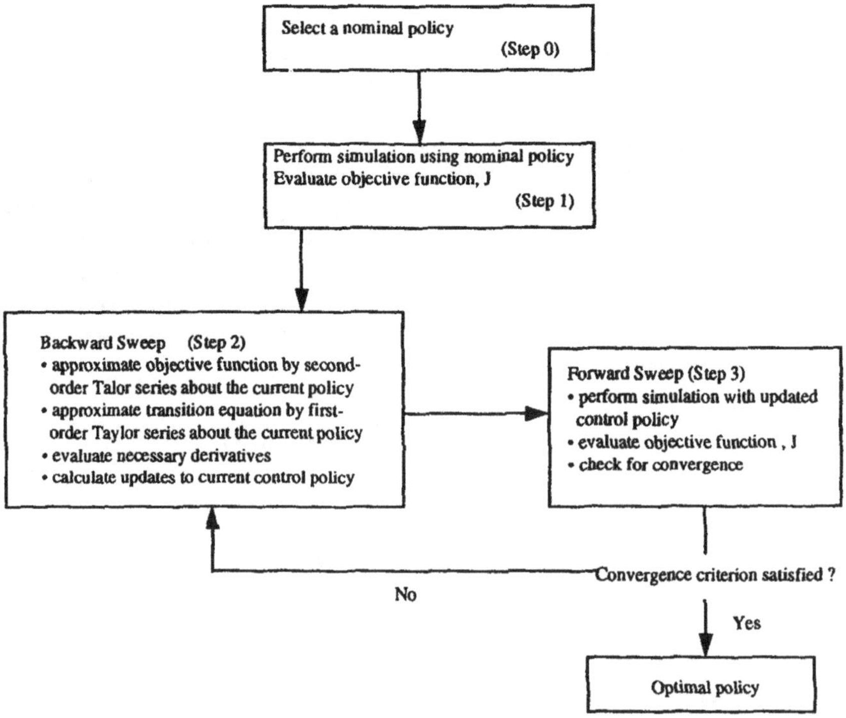

Figure 9.3 Diagram of optimization algorithm for SALQR. Step numbers correspond to the mathematical description in the appendix (from Culver and Shoemaker, 1992).

The constraint equation (9.5.7) on the control and state variables are incorporated into the objective function (9.5.8) utilizing a penalty function so that

$$\hat{f}_k(\mathbf{x}_k, \mathbf{u}_k, k) = \sum_{i=1}^{m} a_{1i}u_{ki} + \sum_{i=1}^{m} a_{2i}u_{ki}(\mathbf{h} - \mathbf{h}_{k+1}) + \sum_{j=1}^{J} \eta_{ki}(r_{ki}) \qquad (9.5.9)$$

where η_{ki} ( ) is the penalty (cost) associated with the violation k_i of constraint i in period k. Culver and Shoemaker (1992) used the following penalty function:

$$\eta_{ki}(r_{ki}) = \varepsilon_{ki}, \qquad \varepsilon_{ki} \leq 1 \qquad (9.5.10)$$

$$\eta_{ki}(r_{ki}) = a\varepsilon_{ki}^2 + b(\varepsilon_{ki})^{1/2} + c, \quad \varepsilon_{ki} > 1 \qquad (9.5.11)$$

gives the following hyperbolic penalty function by Lin (1990):

$$\varepsilon_{ki} = (\alpha_{ki}^2 r_{ki}^2 + \beta_{ki}^2)^{1/2} + \alpha_{ki} r_{ki} \qquad (9.5.12)$$

where α_{ki} is the weighting coefficient of the ith constraint, β_{ki} is a shape parameter of the hyperbolic function ε_{ki}, and a, b, and c are constant coefficients.

Culver and Shoemaker (1992) point out that with the inclusion of equations (9.5.10) and (9.5.11) the objective function (9.5.9) is not quadratic. During the backward sweep (refer to Fig. 9.3), the objective (9.5.9) is approximated as a quadratic by a second-order Taylor series expansion about the current policy as suggested by Yakowitz and Rutherford (1984). The first derivatives of the state equation over a management period requires the first derivatives of the state equation over a simulation period:

$$\left(\frac{\partial g}{\partial x}\right) = \prod_{t=p}^{p+d-1} \left(\frac{\partial g}{\partial x}\right)_t \qquad (9.5.13)$$

$$\begin{aligned}\left(\frac{\partial g}{\partial u}\right)_k &= \left(\frac{\partial g}{\partial u}\right)_{p+d-1} + \left(\frac{\partial g}{\partial x}\right)_{p+d-1}\left(\frac{\partial g}{\partial u}\right)_{p+d-2} \\ &\quad + \left(\frac{\partial g}{\partial x}\right)_{p+d-1}\left(\frac{\partial g}{\partial x}\right)_{p+d-2}\left(\frac{\partial g}{\partial u}\right)_{p+d-3} \qquad (9.5.14) \\ &\quad + \cdots + \prod_{t=p+1}^{p+d-1}\left[\left(\frac{\partial g}{\partial x}\right)_t\right]\left(\frac{\partial g}{\partial u}\right)_p\end{aligned}$$

where $p = (k - 1)\, d + 1$. These derivatives for $(\partial g/\partial x)_k$ and $(\partial g/\partial u)_k$ above are based on the product rule of differentiation.

The derivatives can be determined analytically over a simulation period as shown by Chang (1990), who computed the derivatives from the equations of the finite element model ISOQUAD by Pinder (1979). The algorithm used by Culver and Shoemaker (1992) is basically the same as the one described in Sec. 9.1 with the exception of Eqs. (9.1.15)–(9.1.17) in which

$$\frac{1}{2}\sum_{i=1}^{n} (Q_{t+1})_i \left(\frac{\partial^2 \mathbf{g}}{\partial \mathbf{x}_i^2}\right) = 0 \tag{9.5.15}$$

$$\sum_{i=1}^{n} (Q_{t+1})_i \left(\frac{\partial^2 \mathbf{g}}{\partial \mathbf{x}_i \partial \mathbf{u}_i}\right) = 0 \tag{9.5.16}$$

$$\frac{1}{2}\sum_{i=1}^{n} (Q_{t+1})_i \left(\frac{\partial^2 \mathbf{g}}{\partial \mathbf{u}_i^2}\right) = 0 \tag{9.5.17}$$

Control theory algorithms such as DDP and SALQR have a major advantage in that the computational effort to solve for a nonsteady control policy increases only linearly with the number of time steps, T (Liao and Shoemaker, 1991.)

APPENDIX: DERIVATION OF COEFFICIENTS FOR QUADRATIC

The following derivation of the terms $\mathbf{A}_t$, $\mathbf{B}_t$, $\mathbf{C}_t$, $\mathbf{D}_t$, and $\mathbf{E}_t$ follows that presented by Mayne (1966). A nonlinear system is defined by the difference equation

$$\mathbf{x}_t = \mathbf{g}_{t-1}(\mathbf{x}_{t-1}, \mathbf{u}_{t-1}) \tag{9.A.1}$$

with the performance function

$$Z_r = \sum_{t=r}^{T-1} f_t(\mathbf{x}_t, \mathbf{u}_t) + f_T(\mathbf{x}_T) \tag{9.A.2}$$

The optimization problem is to determine the sequence of controls to minimize Z, given the initial condition $\mathbf{x}_I$. A nominal control sequence is used to determine a state variable sequence by the state equation. The optimal incremental control law, $\boldsymbol{\delta}\mathbf{u}_t$, is determined in reverse time ($t = T, \ldots, 1$) for a small region about the initial trajectory, which enables improved control and state variable sequences to be calculated:

$$\delta\mathbf{u}_t = \mathbf{g}_t(\delta\mathbf{x}_t) \tag{9.A.3}$$

$\delta\mathbf{x}_{t+r}$ for $r \geq 1$ is a function of $\delta\mathbf{x}_t$, $\delta\mathbf{u}_t$, . . . , $\delta\mathbf{u}_{t+r-1}$; therefore ΔZ_r can be expressed as a function of $\delta\mathbf{x}_t, \delta\mathbf{u}_t, \ldots, \delta\mathbf{u}_{t+r-1}$. If ΔZ_t is optimized with respect to $\delta\mathbf{u}_t, \ldots, \delta\mathbf{u}_{t-1}$, the resultant $\Delta\mathbf{Z}_t^\circ$is a function of $\delta\mathbf{x}_t$ only:

$$\Delta Z_t^\circ(\delta\mathbf{x}_t) = \underset{\delta\mathbf{u}_t,\ldots,\delta\mathbf{u}_{T-1}}{\mathrm{Min}} \Delta Z_t(\delta\mathbf{x}_t, \delta\mathbf{u}_t, \ldots, \delta\mathbf{u}_{T-1}) \tag{9.A.4}$$

ΔZ_t° is also a function of $\mathbf{x}_1$ and the initial control sequence. Assuming that $\delta\mathbf{x}_t$ is sufficiently small and the ΔZ_t° is sufficiently smooth, then a second-order power series expansion is used

$$\Delta Z_t^\circ = \tfrac{1}{2}\delta\mathbf{x}_t^T Z_{\mathbf{xx}_t}\delta\mathbf{x}_t + Z_{\mathbf{x}_t}^T\delta\mathbf{x}_t + \mathbf{a}_t \tag{9.A.5}$$

where

$$Z_{\mathbf{xx}_t} = \left(\frac{\partial^2 Z}{\partial\mathbf{x}^2}\right)_t \tag{9.A.6}$$

$$Z_{\mathbf{x}_t} = \left(\frac{\partial Z}{\partial\mathbf{x}}\right)_t \tag{9.A.7}$$

Based on dynamic programming, a recursive function can be expressed as

$$\Delta Z_t^\circ(\delta\mathbf{x}_t) = \underset{\delta\mathbf{u}_t,\ldots,\delta\mathbf{u}_{T-1}}{\min} [\Delta f_t(\delta\mathbf{x}_t, \delta\mathbf{u}_t) + \Delta Z_{t+1}(\delta\mathbf{x}_{t+1}, \delta\mathbf{u}_{t+1}, \ldots, \delta\mathbf{u}_{T-1})] \tag{9.A.8}$$

where $\delta\mathbf{x}_{t+1}$ is by equation (9.A.1) a function of $\delta\mathbf{x}_t$ and $\delta\mathbf{u}_t$:

$$\delta\mathbf{x}_{t+1} = \Delta\mathbf{g}_t(\delta\mathbf{x}_t, \delta\mathbf{u}_t) \tag{9.A.9}$$

Minimizing with respect to $\delta\mathbf{u}_{t+1}, \ldots, \delta\mathbf{u}_{T-1}$ yields

$$\Delta Z_t^\circ(\delta\mathbf{x}_t) = \underset{\delta\mathbf{u}_t}{\min} [\Delta f_t(\delta\mathbf{x}_t, \delta\mathbf{u}_t) + \Delta Z_{t+1}^\circ(\delta\mathbf{x}_{t+1})] \tag{9.A.10}$$

where $\delta\mathbf{x}_{t+1}$ is given by Eq. (9.A.9).

For the sake of simplicity, assume that $\mathbf{x}_t$ and $\mathbf{u}_t$ are scalars. Expand Δf_t and Δg_t up to second-order terms

$$\Delta f_t(\delta x_t, \delta u_t) = \delta f(\delta x_t, \delta u_t) + \tfrac{1}{2}\delta^2 f(\delta x_t, \delta u_t) \tag{9.A.11}$$

where

$$\delta f(\delta x_t, \delta u_t) = f_{x_t}\delta x_t + f_{u_t}\delta u_t \tag{9.A.12}$$

$$\delta^2 f(\delta x_t, \delta u_t) = f_{xx_t}\delta x_t^2 + 2f_{xu_t}\delta x_t\delta u_t + f_{uu_t}\delta u_t^2 \tag{9.A.13}$$

and

$$f_{x_t} = \frac{\partial f_t(x_t, u_t)}{\partial x_t} \quad \text{and} \quad f_{xu_t} = \frac{\partial^2 f_t(x_t, u_t)}{\partial x_t, \partial u_t}$$

Substituting the above expansion for δf and $\delta^2 f$ into Eq. (9.A.11) results in

$$\Delta f_t(\delta x_t, \delta u_t) = f_{x_t}\delta x_t + f_{u_t}\delta u_t + f_{xx_t}\delta x_t^2 + 2f_{xu_t}\delta x_t\delta u_t + f_{uu_t}\delta u_t^2 \quad (9.A.14)$$

The expansion for $\Delta Z^\circ_{t+1}(\delta x_{t+1})$ in Eq. (9.A.10) is also needed

$$\Delta Z^\circ_{t+1}(\delta x_{t+1}) = \tfrac{1}{2}Z_{xx_{t+1}} + Z_{x_{t+1}}\delta x_{t+1} + a_{t+1} \quad (9.A.15)$$

δx_{t+1} is determined using Eq. (9.A.9) so that

$$\Delta g_t(\delta x_t, \delta u_t) = \delta g_t(\delta x_t, \delta u_t) + \tfrac{1}{2}\delta^2 g_t(\delta x_t, \delta u_t) \quad (9.A.16)$$

where

$$\delta g_t(\delta x_t, \delta u_t) = g_{x_t}\delta x_t + g_{u_t}\delta u_t \quad (9.A.17)$$

$$\delta^2 g_t(\delta x_t, \delta u_t) = g_{xx_t}\delta x_t^2 + 2g_{xu_t}\delta x_t\delta u_t + g_{uu_t}\delta u_t^2 \quad (9.A.18)$$

Substituting the above expansion for δg_t and $\delta^2 g_t$ into Eq. (9.A.16) results in

$$\Delta g_t = g_{x_t}\delta x_t + g_{u_t}\delta u_t + g_{xx_t}\delta x_t^2 + 2g_{xu_t}\delta x_t\delta u_t + g_{uu_t}\delta u_t^2 \quad (9.A.19)$$

This expression for Δg_t is then substituted into Eq. (9.A.15) for $\delta x_{t+1} = \Delta g_t(\delta x_t, \delta_t)$ so that

$$\Delta Z^\circ_{t+1}(\delta x_{t+1}) = \tfrac{1}{2}Z_{xx_{t+1}}[\Delta g_t(\delta x_t, \delta u_t)]^2 + Z_{x_{t+1}}[\Delta g_t(\delta x_t, \delta u_t)] + a_{t+1} \quad (9.A.20)$$

Then the above expressions for $\Delta Z^\circ_{t+1}(\delta x_{t+1})$ and $\Delta f_t(\delta x_t, \delta u_t)$ from Eq. (9.A.14) are substituted into Eq. (9.A.10) and rearranged to obtain

$$\begin{aligned}\Delta Z^\circ_t(\delta x_t) = \min_{\delta u_t} [&(f_{x_t} + Z_{x_{t+1}}g_{x_t})\delta x_t + (f_{u_t} + Z_{x_{t+1}}g_{u_t})\delta u_t \\ &+ \tfrac{1}{2}(f_{xx_t} + Z_{x_{t+1}}g_{xx_t})\delta x_t^2 + (f_{xu_t} + Z_{x_{t+1}}g_{xu_t})\delta x_t\delta u_t \\ &+ \tfrac{1}{2}(f_{uu_t} + Z_{x_{t+1}}g_{uu_t})\delta u_t^2 + \tfrac{1}{2}Z_{xx_{t+1}}g_{x_t}^2\delta x_t^2 \\ &+ Z_{xx_{t+1}}g_{x_t}g_{u_t}\delta x_t\delta u_t + \tfrac{1}{2}Z_{xx_{t+1}}g_{u_t}^2\delta u_t^2] + a_{t+1}\end{aligned} \quad (9.A.21)$$

The Hamiltonian, H_t, can be defined as

$$H_t(x_t, u_t, Z_{x_{t+1}}) = f_t(x_t, u_t) + Z_{x_{t+1}}g_t(x_t, u_t) \quad (9.A.22)$$

and

$$H_{x_t} = f_{x_t} + Z_{x_{t+1}} g_{x_t} \tag{9.A.23}$$

$$H_{u_t} = f_{u_t} + Z_{u_{t+1}} g_{u_t} \tag{9.A.24}$$

$$H_{xx_t} = f_{xx_t} + Z_{x_{t+1}} g_{xx_t} \tag{9.A.25}$$

$$H_{xu_t} = f_{xu_t} + Z_{x_{t+1}} g_{xu_t} \tag{9.A.26}$$

$$H_{uu_t} = f_{uu_t} + Z_{x_{t+1}} g_{uu_t} \tag{9.A.27}$$

Using the above definition for the scalar case,

$$\begin{aligned} \Delta Z_t^\circ(\delta x_t) = \min_{\delta u_t} [&H_{x_t}\delta x_t + H_{u_t}\delta u_t + \tfrac{1}{2}H_{xx_t}\delta x_t^2 \\ &+ H_{xu_t}\delta x_t \delta u_t + \tfrac{1}{2}H_{uu_t}\delta u_t^2 + \tfrac{1}{2}Z_{xx_{t+1}} g_{x_t}^2 \delta x_t^2 \\ &+ Z_{xx_{t+1}} g_{x_t} g_{u_t} \delta x_t \delta u_t + \tfrac{1}{2}Z_{xx_{t+1}} g_{u_t}^2 \delta u_t^2] + a_{t+1} \end{aligned} \tag{9.A.28}$$

For the vector case with $\mathbf{x}_t$ and $\mathbf{u}_t$ being vectors, then

$$\mathbf{H}_t(\mathbf{x}_t, \mathbf{u}_t, \mathbf{Z}_{x_{t+1}}) = f_t(\mathbf{x}_t, \mathbf{u}_t) + \mathbf{Z}_{x_{t+1}}^T f_t(\mathbf{x}_t, \mathbf{u}_t) \tag{9.A.29}$$

$$\begin{aligned} \Delta \mathbf{Z}_t^\circ(\delta \mathbf{x}_t) = \min_{\delta \mathbf{u}_t} [&\mathbf{H}_{x_t}^T \delta \mathbf{x}_t + \mathbf{H}_{u_t}^T \delta \mathbf{u}_t + \tfrac{1}{2}\delta \mathbf{x}_t^T \mathbf{A}_t \delta \mathbf{x}_t \\ &+ \tfrac{1}{2}\delta \mathbf{u}_t^T \mathbf{B}_t \delta \mathbf{x}_t + \tfrac{1}{2}\delta \mathbf{x}_t^T \mathbf{B}_t^T \delta \mathbf{x} \\ &+ \tfrac{1}{2}\delta \mathbf{u}_t \mathbf{C}_t \delta \mathbf{u}_t] + \mathbf{a}_{t+1} \end{aligned} \tag{9.A.30}$$

where

$$\mathbf{A}_t = \mathbf{H}_{xx_t} \mathbf{g}_{x_t}^T \mathbf{Z}_{xx_{t+1}} \mathbf{g}_{x_t} \tag{9.A.31}$$

$$\mathbf{B}_t = \mathbf{H}_{ux_t} \mathbf{g}_{u_t}^T \mathbf{Z}_{xx_{t+1}} \mathbf{g}_{x_t} \tag{9.A.32}$$

$$\mathbf{C}_t = \mathbf{H}_{uu_t} \mathbf{g}_{u_t}^T \mathbf{Z}_{xx_{t+1}} \mathbf{g}_{u_t} \tag{9.A.33}$$

$$\mathbf{D}_t = \mathbf{H}_{u_t} \tag{9.A.34}$$

$$\mathbf{E}_t = \mathbf{H}_{x_t} \tag{9.A.35}$$

$$\mathbf{H}_{x_t} = \begin{bmatrix} \dfrac{\partial \mathbf{H}_t}{(\partial \mathbf{x}_t)_1} \\ \vdots \\ \dfrac{\partial \mathbf{H}_t}{(\partial \mathbf{x}_t)_n} \end{bmatrix}$$

$$\mathbf{H}_{ux_t} = \begin{bmatrix} \dfrac{\partial^2 \mathbf{H}_t}{(\partial \mathbf{u}_t)_1 (\partial \mathbf{x}_t)_1} & \cdots & \dfrac{\partial^2 \mathbf{H}_t}{(\partial \mathbf{u}_t)_1 (\partial \mathbf{x}_t)_n} \\ \vdots & & \vdots \\ \dfrac{\partial^2 \mathbf{H}_t}{(\partial \mathbf{u}_t)_m (\partial \mathbf{u}_t)_1} & \cdots & \dfrac{\partial^2 \mathbf{H}_t}{(\partial \mathbf{u}_t)_m (\partial \mathbf{x}_t)_n} \end{bmatrix}$$

Minimizing the right-hand side of (9.A.30) with respect to $\delta\mathbf{u}_t$ results in

$$\delta \mathbf{u}_t = \mathbf{a}_t + \boldsymbol{\beta}_t \delta \mathbf{x}_t \tag{9.A.36}$$

where

$$\boldsymbol{\alpha}_t = -\mathbf{C}_t^{-1} \mathbf{H}_{u_t} \tag{9.A.37}$$

$$\boldsymbol{\beta}_t = -\mathbf{C}_t^{-1} \mathbf{B}_t \tag{9.A.38}$$

Substituting Eqs. (9.A.36), (9.A.37), and (9.A.38) into Eq. (9.A.30) results in

$$\begin{aligned} \Delta Z_t^\circ(\delta \mathbf{x}_t) &= (\mathbf{H}_{x_t} + \boldsymbol{\beta}_t^T \mathbf{H}_{u_t})^T \delta \mathbf{x}_t \\ &\quad + \tfrac{1}{2} \delta \mathbf{x}_t^T (\mathbf{A}_t - \mathbf{B}_t^T \mathbf{C}_t^{-1} \mathbf{B}_t) \delta \mathbf{x}_t \\ &\quad - \tfrac{1}{2} \mathbf{H}_{u_t}^T \mathbf{C}_t^{-1} \mathbf{H}_{u_t} + \mathbf{a}_{t+1} \end{aligned} \tag{9.A.39}$$

then

$$\mathbf{Z}_{xx_t} = \mathbf{A}_t - \mathbf{B}_t^T \mathbf{C}_t^{-1} \mathbf{B}_t = \mathbf{P}_t \tag{9.A.40}$$

$$\mathbf{Z}_{x_t} = \mathbf{H}_{x_t} + \boldsymbol{\beta}_t^T \mathbf{H}_{u_t} = \mathbf{Q}_t \tag{9.A.41}$$

$$\mathbf{a}_t = \mathbf{a}_{t+1} - \tfrac{1}{2} \mathbf{H}_{u_t}^T \mathbf{C}_t^{-1} \mathbf{H}_{u_t} = \boldsymbol{\theta}_t \tag{9.A.42}$$

Because

$$\Delta \mathbf{Z}_T^\circ(\delta \mathbf{x}_t) = \mathbf{f}_{x_T}^T \delta \mathbf{x}_T + \tfrac{1}{2} \delta \mathbf{x}_T^T \mathbf{f}_{xx_T} \delta \mathbf{x}_T$$

the boundary conditions for Eqs. (9.A.40), (9.A.41), and (9.A.42) are

$$\mathbf{Z}_{xx_T} = \mathbf{f}_{xx_T} \tag{9.A.43}$$

$$\mathbf{Z}_{x_T} = \mathbf{f}_{x_T} \tag{9.A.44}$$

$$\mathbf{a}_t = \mathbf{0} \tag{9.A.45}$$

The algorithm can be summarized as follows:

(i) A nominal control sequence $\mathbf{u}_t$ is selected to determine the state variable sequence $\mathbf{x}_t$, using the state equation (9.A.1). $\mathbf{u}_t$ and $\mathbf{x}_t$ are stored.

(ii) The sequences, $\mathbf{z}_{xxT}$, $\mathbf{z}_{xT}$ and $\mathbf{a}_t$ (which are respectively $\mathbf{P}_t$, $\mathbf{Q}_t$ and $\boldsymbol{\theta}_t$) are computed in reverse time $t = T, \ldots, 1$ using Eqs. (9.A.40)–(9.A.42) along with the boundary conditions (9.A.43)–(9.A.45). The incremental control law variables $\boldsymbol{\alpha}_t$ and $\boldsymbol{\beta}_t$ are calculated. Only $\boldsymbol{\alpha}_t$ and $\boldsymbol{\beta}_t$ are stored.

(iii) Improved $\mathbf{x}_t$ and $\mathbf{u}_t$ sequences are calculated and stored using Eq. (9.A.1)

$$\delta\mathbf{u}_t = \boldsymbol{\alpha}_t + \boldsymbol{\beta}_t[(\mathbf{x}_t)_{\text{new}} - (\mathbf{x}_t)_{\text{old}}]$$

$$(\mathbf{u}_t)_{\text{new}} = (\mathbf{u}_t)_{\text{old}} + \delta\mathbf{u}_t$$

(iv) Repeat steps (ii) and (iii).

The improvement obtained for one iteration is $\Delta Z_1^\circ(0) = a$, since $\delta\mathbf{x}_1 = \mathbf{0}$ if $\delta\mathbf{u}_t$ and $\delta\mathbf{x}_t$ are sufficiently small for all t.

REFERENCES

Bellman, R. and Dreyfus, S., *Applied Dynamic Programming*, Princeton, University Press., Princeton, NJ, 1962.

Chang, L. C., The Application of Constrained Optimal Control Algorithms to Groundwater Remediation, Ph.D. Dissertation, Cornell University, Ithaca, NY, 1990.

Culver, T. B., Dynamic Optimal Control of Groundwater Remediation with Management Periods: Linearized and Quasi-Newton Approaches, Ph.D. Dissertation, Cornell University, Ithaca, NY, 1991.

Culver, T. and Shoemaker, C., Dynamic Optimal Control for Groundwater Remediation with Flexible Management Periods, *Water Resources Research*, Vol. 28, No. 3, pp. 629–641, 1992.

Dyer P. and McReynolds, S., *The Computational Theory of Optimal Control*, Academic Press, New York, 1970.

Gershwin, S. and Jacobson, D., A Discrete-Time Differential Dynamic Programming Algorithm with Application to Optimal Orbit Transfer, *AIAA Journal*, Vol. 8, pp. 1616–1626, 1970.

Jacobson, D. and Mayne, D., *Differential Dynamic Programming*, Elsevier, New York, 1970.

Jones, L. C., Willis, R., and Yeh, W. W., Optimal Control of Nonlinear Groundwater Hydraulics Using Differential Dynamic Programming, *Water Resources Research*, Vol. 23, No. 11, pp. 2097–2217, 1987.

Liao, L.-Z., and Shoemaker, C. A., Convergence in Unconstrained Discrete-Time Differential Dynamic Programming, *IEEE Transactions on Automatic Control*, Vol. 4C-36, No. 6, pp. 692–706, 1991.

Lin, T-W., Well-Behaved Penalty Functions for Constrained Optimization, *Journal of the Chinese Institute of Engineers*, Vol. 13, No. 2, pp. 157–166, 1990.

Mayne D., A Second-Order Gradient Method for Determining Optimal Trajectories of Non-linear Discrete-Time Systems, *International Journal on Control*, Vol. 3, pp. 85–95, 1966.

Murray M. and Yakowitz, S. J., The Application of Optimal Control Methodology to Nonlinear Programming Problems, *Mathematical Programming*, Vol. 21, pp. 331–347, 1981.

Murray, D. M., and Yakowitz, S. J., Constrained Differential Dynamic Programming and Its Application to Multireservoir Control, *Water Resources Research*, Vol. 15, No. 5, pp. 1017–1027, 1979.

Ohno, K., A New Approach of Differential Dynamic Programming for Discrete Time Systems, *IEEE Transactions on Automatic Control*, Vol. AC-23, pp. 37–47, 1978.

Pinder, G. F., Galerkin Finite Element Models for Aquifer Simulation, Report 76-WR-5, Dept of Civil Eng., Princeton University, Princeton, NJ, 1979.

Yakowitz, S., Algorithms and Computational Techniques in Differential Dynamic Programming, *Control and Dynamic Systems*, Vol. 31, pp. 75–91, 1989.

Yakowitz, S. and Rutherford, B., Computational Aspects of Discrete-Time Optimal Control, *Applied Mathematics and Computation*, Vol. 15, pp. 29–45, 1984.

Chapter 10
Estuarine Management Model Using SALQR

10.1 PROBLEM FORMULATION

The estuarine management problem can be formulated with different management objectives or even as a multiobjective problem (see Chapter 7; Bao and Mays, 1994a,b; LeBlanc, 1993; Mao and Mays, 1994; Martin, 1987; Shi, 1992; Siebert, 1993; and Tung et al., 1990). One of the objectives is the enhancement of fishery harvest of selected fish species while meeting viability limits for salinity and satisfying monthly and seasonal freshwater inflow needs. The mathematical model can be expressed as follows:

$$\text{Max} \sum H_i \tag{10.1.1}$$

$$\text{S.T.} \quad S_{t,j} = f(Q_{t,j}, t) \tag{10.1.2}$$

$$\underline{Q}_{t,j} \le Q_{t,j} \le \overline{Q}_{t,j} \tag{10.1.3}$$

$$\underline{S}_{t,j} \le S_{t,j} \le \overline{S}_{t,j} \tag{10.1.4}$$

where

H_i is the fishery harvest for the ith species (1000 pounds)
$Q_{t,j}$ is the tth monthly inflows from the jth river (cfs)
$S_{t,j}$ is the tth monthly average salinity at a specified location in the estuary, for river j (ppt)
$\overline{S}_{t,j}$ and $\underline{S}_{t,j}$ are the upper and lower limits, respectively, on monthly

average salinity at a specified location in the estuary, for river j (ppt)

$\overline{Q}_{t,j}$ and $\underline{Q}_{t,j}$ are the upper and lower limits, respectively, of monthly freshwater inflow (cfs)

The objective function [Eq. (10.1.1)] is the summation of the regression equations (Table 10.1) on fishery harvest as a function of freshwater inflow. Constraint (10.1.2) defines the relation of the state variable (salinity) and the control variable (freshwater inflow) at time t. Constraint (10.1.3) defines bounds on the monthly freshwater inflows. Constraint (10.1.4) defines bounds on the monthly salinity.

In this chapter, the mathematical programming model on estuarine management is reformulated as an optimal control model. To formulate the optimal control problem, the mathematical programming problem must have an objective function that is separable. In the estuarine management model, the objective function is to maximize the summation of fishery harvest. Fishery harvest is expressed in terms of mean monthly freshwater inflow per season (1000 acre-ft). Table 10.1 lists some regression equations on fishery harvests for the Lavaca-Tres Palacios Estuary in Texas. These regression equations were developed by the Texas Department of Water Resources (1980) and modified by Bao (1992). A separable objective function can be obtained by replacing monthly freshwater inflow in each season with monthly freshwater inflow (cfs) as

Table 10.1 Regression Equations of Fishery Harvest and Freshwater Inflow Relation

Species	Equations	Inflow used in regression equation
All shellfish	$H_1 = 3109.5 - 3.782QS_1 + 2.553QS_2 - 12.14QS_3$	a
All penaeid shrimp	$H_2 = 1888.6 - 1.061QS_1 + 1.088QS_2 - 1.071QS_5$	c
Blue crab	$H_3 = 289.5 + 1.725QS_3 + 0.429QS_4 + 0.202QS_5$	c

Notes: a: Freshwater inflow at the Lavaca Bay.
c: Combined freshwater inflows from all contributing rivers and coastal drainage basins.
H_k: The commercial harvest in thousands of pounds.
QS_j: The mean monthly freshwater inflow during the season (1000 acre–ft) (1 acre-ft = 1233.5 m^3).
QS_1: January–March; QS_2: April–June; QS_3: July–August; QS_4: September–October; QS_5: November–December.

$$\text{Max } J = \sum_{t=1}^{N} \sum_{j=1}^{m} g(Q_{t,j}, t) \tag{10.1.5}$$

where $g(Q_{t,j}, t) = (a_{t,j} + b_{t,j}Q_{t,j})$ are regression equations expressed in terms of monthly freshwater inflow, $a_{t,j}$ and $b_{t,j}$ are coefficients that are given in Table 10.2, and N is time period and m is the dimension of freshwater inflow. It is convenient to transform the maximization into a minimization. Thus, the objective function for the estuarine management problem becomes

$$\text{Min } \tilde{J} = -\sum_{t=1}^{N} \sum_{j=1}^{m} g(Q_{t,j}, t) \tag{10.1.6}$$

For optimal control problems, a state transition equation to represent the relationship of the state variable at time t and at time $t + 1$ is required. In the estuarine management problem, salinity at the beginning of each month is chosen as the state variable, and monthly freshwater inflow is chosen as the control variable. Equation (10.1.2) is modified to the form

$$S_{t+1,k} = T(S_{t,k}, Q_{t,j}, t) \tag{10.1.7}$$

where k is the dimension of salinity.

Table 10.2 Coefficients $a_{t,j}$ and $b_{t,j}$ in the Objective Function for Fishery Harvest Regression Equations

	$a_{t,j}$		$b_{t,j}$	
Month	Lavaca Bay	Matagorda Bay	Lavaca Bay	Matagorda Bay
January	506.73	118.04	−0.099	−0.022
February	506.73	118.04	−0.090	−0.020
March	506.73	118.04	−0.096	−0.021
April	506.73	118.04	0.072	0.022
May	506.73	118.04	0.075	0.022
June	506.73	118.04	0.072	0.022
July	412.82	24.13	−0.320	0.053
August	412.82	24.13	−0.320	0.053
September	24.13	24.13	0.013	0.013
October	24.13	24.13	0.013	0.013
November	142.17	142.17	−0.026	−0.026
December	142.17	142.17	−0.026	−0.026

From Eq. (10.1.7) it is known that $S_{t+1,k}$ is determined by $S_{t,k}$ and $Q_{t,j}$; meanwhile, $S_{t+1,k}$ should also satisfy its bound constraint. Therefore, Eq. (10.1.7) is modified to

$$\underline{S}_{t+1,k} \leq S_{t+1,k} \leq \overline{S}_{t+1,k} \qquad (10.1.8)$$

This means that control variable $Q_{t,j}$ is determined such that $S_{t+1,j}$ is also within its bound. The constrained discrete-time optimal control model on estuarine management consists of Eqs. (10.1.6), (10.1.7), (10.1.3), and (10.1.8).

Using the bracket penalty function (Reklaitis, 1983; Li and Mays, 1995), the penalty term associated with bound constraints on the control variable (freshwater inflow) is

$$P_1(Q_{t,j}, R_i, t) = R_i\{\min[0, \min((Q_{t,j} - \underline{Q}_{t,j}), (\overline{Q}_{t,j} - Q_{t,j}))]\}^2 \qquad (10.1.9)$$

The penalty term associated with the bound constraints on the state variable (salinity) is

$$P_2(S_{t,j}, Q_{t,j}, R_2, t)$$
$$= R_2\{\min[0, \min((S_{t+1,j} - \underline{S}_{t+1,j}), (\overline{S}_{t+1,j} - S_{t+1,j}))]\}^2 \qquad (10.1.10)$$

After introducing the penalty function into the objective function, Eq. (10.1.6) becomes

$$\text{Min } \hat{J} = \sum_{t=1}^{N} \sum_{j=1}^{m} \hat{G}(Q_{t,j}, S_{t,j}, R_1, R_2, t) \qquad (10.1.11)$$

where $\hat{G}(S_{t,j}, Q_{t,j}, R_1, R_2, t) = -g(Q_{t,j}, t) + P_1(Q_{t,j}, R_1, t) + P_2(S_{t,j}, Q_{t,j}, R_2, t)$. $P_1(\ )$ and $P_2(\ )$ are determined by Eqs. (10.1.9) and (10.1.10), respectively. The augmented estuarine management model consists of Eqs. (10.1.11) and (10.1.2).

10.2 PROBLEM SOLUTION

10.2.1 SALQR Algorithm with Constraints

A system with nonlinear dynamics can be addressed by linearizing the simulation dynamics in the optimization stage and then updating the control and the state vectors through simulation of the full nonlinear model. Culver and Shoemaker (1992) refer to this as the successive approximation linear quadratic regulator (SALQR) method. SALQR differs from differential dynamic programming (DDP) only in that the nonlinear simulation equations are linearized in the optimization step. Therefore, if the simulation equations are linear, SALQR and DDP are identical. The mathematical details of the SALQR algorithm are given below.

A constrained discrete-time optimal problem can be described as

$$\text{Min } J = \sum_{t=1}^{N} g(\mathbf{x}_t, \mathbf{u}_t, t) \tag{10.2.1}$$

$$\text{S.T.} \quad \mathbf{x}_{t+1} = \mathbf{T}(\mathbf{x}_t, \mathbf{u}_t, t) \tag{10.2.2}$$

$$\mathbf{L}(\mathbf{x}_t, \mathbf{u}_t, t) \geq \mathbf{0} \tag{10.2.3}$$

where $t = 1, 2, \ldots, N - 1, N$, $\mathbf{x}_1 = \mathbf{x}_1^c$ is given and fixed, $\mathbf{x}_t$ is the state variable, $\mathbf{u}_t$ is the control variable, $\mathbf{g}(\mathbf{x}_t, \mathbf{u}_t, t)$ is called the loss function, and $\mathbf{T}(\mathbf{x}_t, \mathbf{u}_t, t)$ is called the transition function.

Several techniques can be used to handle the constraints (Jones et al., 1987; Yakowitz, 1989; Andricevic and Kitanidis, 1990; Chang et al., 1992). In this chapter, one technique is through the use of a penalty function in which the penalty function $P(x_t, u_t, R, t)$ is associated with the violation of constraints and is added to the objective function. There are many kinds of functions that can be used. Some functions are complicated and include the estimation of many parameters. Here, a very simple penalty function called a bracket penalty function (Reklaitis, 1983; Li and Mays, 1995) is selected in the algorithm description and model formulation. The bracket function has the following form:

$$P(\mathbf{x}_t, \mathbf{u}_t, R, t) = R[\mathbf{L}(\mathbf{x}_t, \mathbf{u}_t, t)]^2 \tag{10.2.4}$$

where R is a penalty parameter that is a numerical value that must be assigned; $[\mathbf{L}(\mathbf{x}_t, \mathbf{u}_t, t)]$ is the violation of constraints;

$$\begin{aligned} [\mathbf{L}(\mathbf{x}_t, \mathbf{u}_t, t)] &= \mathbf{0} && \text{if } \mathbf{L}(\mathbf{x}_t, \mathbf{u}_t, t) \geq \mathbf{0} \\ [\mathbf{L}[\mathbf{x}_t, \mathbf{u}_t, t)] &= \mathbf{L}(\mathbf{x}_t, \mathbf{u}_t, t) && \text{if } \mathbf{L}(\mathbf{x}_t, \mathbf{u}_t, t) \leq \mathbf{0} \end{aligned} \tag{10.2.5}$$

The penalty function is added to Eq. (10.1.1). Thus, a constrained, discrete-time, optimal control problem becomes an unconstrained formulation which has the form

$$\text{Min } \hat{J} = \sum_{t=1}^{N} \hat{G}(\mathbf{x}_t, \mathbf{u}_t, R, t) \tag{10.2.6}$$

$$\text{S.T.} \quad \mathbf{x}_{t+1} = \mathbf{T}(\mathbf{x}_t, \mathbf{u}_t, t) \tag{10.2.7}$$

where

$$\hat{G}(\mathbf{x}_t, \mathbf{u}_t, R, t) = g(\mathbf{x}_t, \mathbf{u}_t, t) + P(\mathbf{x}_t, \mathbf{u}_t, R, t) \tag{10.2.8}$$

The iteration process of the SALQR consists of five steps:

Step 1: Initialize the penalty parameter R.

Step 2: Initialize parameters and compute the state and objective functions associated with the given policy $\mathbf{u}$.

(i) Select a starting guess $\mathbf{u}^c = (\mathbf{u}_1^c, \mathbf{u}_2^c, \ldots, \mathbf{u}_N^c)$. Select ε_1 as the stopping criterion for the inner optimal and $\theta_{\min}$ as the stopping criterion for θ_1;

(ii) Compute $\mathbf{x}_{t+1}^c = \mathbf{T}(\mathbf{x}_t^c, \mathbf{u}_t^c, t)$ recursively for $t = 1, 2, \ldots, N - 1$, $\mathbf{x}_1^c$ is given.

(iii) Initialize $\mathbf{P}_{N+1} = \mathbf{0}$, $\mathbf{Q}_{N+1} = \mathbf{0}$, $\boldsymbol{\theta}_{N+1} = \mathbf{0}$.

Step 3: Perform the backward sweep: perform steps (i)–(iv) below recursively for $t = N, N - 1, \ldots, 1$.

(i) Compute $\mathbf{A}_t$, $\mathbf{B}_t$, $\mathbf{C}_t$, $\mathbf{D}_t$, and $\mathbf{E}_t$ according to

$$\begin{aligned}
\mathbf{A}_t &= \frac{1}{2}\left[\hat{G}_{xx} + 2\left(\frac{\partial \mathbf{T}}{\partial \mathbf{x}}\right)^T \mathbf{P}_{t+1}\left(\frac{\partial \mathbf{T}}{\partial \mathbf{x}}\right)\right] \\
\mathbf{B}_t &= \hat{\mathbf{G}}_{xu} + 2\left(\frac{\partial \mathbf{T}}{\partial \mathbf{x}}\right)^T \mathbf{P}_{t+1}\left(\frac{\partial \mathbf{T}}{\partial \mathbf{u}}\right) \\
\mathbf{C}_t &= \frac{1}{2}\left[\hat{\mathbf{G}}_{uu} + 2\left(\frac{\partial \mathbf{T}}{\partial \mathbf{u}}\right)^T \mathbf{P}_{t+1}\left(\frac{\partial \mathbf{T}}{\partial \mathbf{u}}\right)\right] \\
\mathbf{D}_t^T &= \nabla_u \hat{\mathbf{G}} + \mathbf{Q}_{t+1}\left(\frac{\partial \mathbf{T}}{\partial \mathbf{u}}\right) \\
\mathbf{E}_t^T &= \nabla_x \hat{\mathbf{G}} + \mathbf{Q}_{t+1}\left(\frac{\partial \mathbf{T}}{\partial \mathbf{x}}\right)
\end{aligned} \tag{10.2.9}$$

(ii) Compute $\mathbf{P}_t$ and $\mathbf{Q}_t$ according to

$$\begin{aligned}
\mathbf{P}_t &= \mathbf{A}_t - \mathbf{B}_t^T \mathbf{C}_t^{-1} \mathbf{B}_t \\
\mathbf{Q}_t &= -\tfrac{1}{2}\, \mathbf{D}_t^T \mathbf{C}_t^{-1} \mathbf{B}_t + \mathbf{E}_t^T
\end{aligned} \tag{10.2.10}$$

and store $\mathbf{P}_t$ and $\mathbf{Q}_t$ in memory, replacing $\mathbf{P}_{t+1}$ and $\mathbf{Q}_{t+1}$.

(iii) Compute $\boldsymbol{\alpha}_t$ and $\boldsymbol{\beta}_t$ according to

$$\begin{aligned}
\boldsymbol{\alpha}_t &= -\tfrac{1}{2}\, \mathbf{C}_t^{-1} \mathbf{D}_t^T \\
\boldsymbol{\beta}_t &= -\tfrac{1}{2}\, \mathbf{C}_t^{-1} \mathbf{B}_t
\end{aligned} \tag{10.2.11}$$

and store in memory.

(iv) Compute

$$\boldsymbol{\theta}_t = -\tfrac{1}{2}\, \mathbf{D}_t^T\, \mathbf{C}_t^{-1} \mathbf{D}_t + \boldsymbol{\theta}_{t+1} \tag{10.2.12}$$

and store $\boldsymbol{\theta}_t$ in place of $\boldsymbol{\theta}_{t+1}$ in memory.

Step 4: Perform the forward sweep: compute the successor policy and trajectory.

(i) Compute $\mathbf{u}_t$ and $\mathbf{x}_t$ recursively for $t = 1, 2, \ldots, N - 1$ according to

$$\mathbf{u}_t = \mathbf{u}_t^c + \boldsymbol{\alpha}_t + \boldsymbol{\beta}_t(\mathbf{x}_t - \mathbf{x}_t^c)$$
$$\mathbf{x}_{t+1} = \mathbf{T}(\mathbf{x}_t, \mathbf{u}_t, t) \tag{10.2.13}$$

(ii) If $\max_t |\mathbf{u}_t - \mathbf{u}_t^c| \leq \varepsilon_1$, then the inner optimal is reached, so go step 5. Otherwise, set $\mathbf{u}_t^c = \mathbf{u}_t$ and $\mathbf{x}_t^c = \mathbf{x}_t$ and go to step 2.

Step 5: If $\theta_1 > \theta_{\min}$, update the parameter R and go to step 1. If $\theta_1 \leq \theta_{\min}$ and R converges, then the overall optimal is reached and the stopping criteria are satisfied, so stop.

The flowchart of the SALQR method is shown in Figure 10.1. The SALQR algorithm can be selected where the second derivatives of the transition equation are difficult to calculate.

10.2.2 SALQR Interfaced with HYD-SAL

The purpose of the estuarine hydrodynamic transport model is to simulate the flow circulation in the bay system and to be able to compute the spatial distribution of salinity in the bay for the time period of interest for given freshwater inflows and other boundary conditions. In the past two decades, especially in the last 10 years, numerous estuarine hydrodynamic transport models have been developed. Most of these models have been applied to real-world situations such as to an estuary or a river delta system with various degrees of success. Bao and Mays (1994a, 1994b) have successfully adapted the Texas Department of Water Resources (TDWR) finite-difference model HYD-SAL into an estuarine optimal management model (see Chapter 7, Figure 7.5 and Appendix 7.A).

The basic tidal hydrodynamic equations are nonlinear partial differential equations. The transport equation is a linear second-order partial differential equation. Even for the most ideal situations, analytical solutions of these equations are a formidable undertaking. This, compounded with complex geometry, intricate interior features, and variable boundary conditions, makes purely analytical approaches unsuitable for the bays under study. For these reasons, numerical methods (Masch and Associates, 1971) are utilized to obtain the solution of equations. In the numerical approach, the bay is discretized into computational elements. The elements are arranged in time and space so that the output from one element becomes the

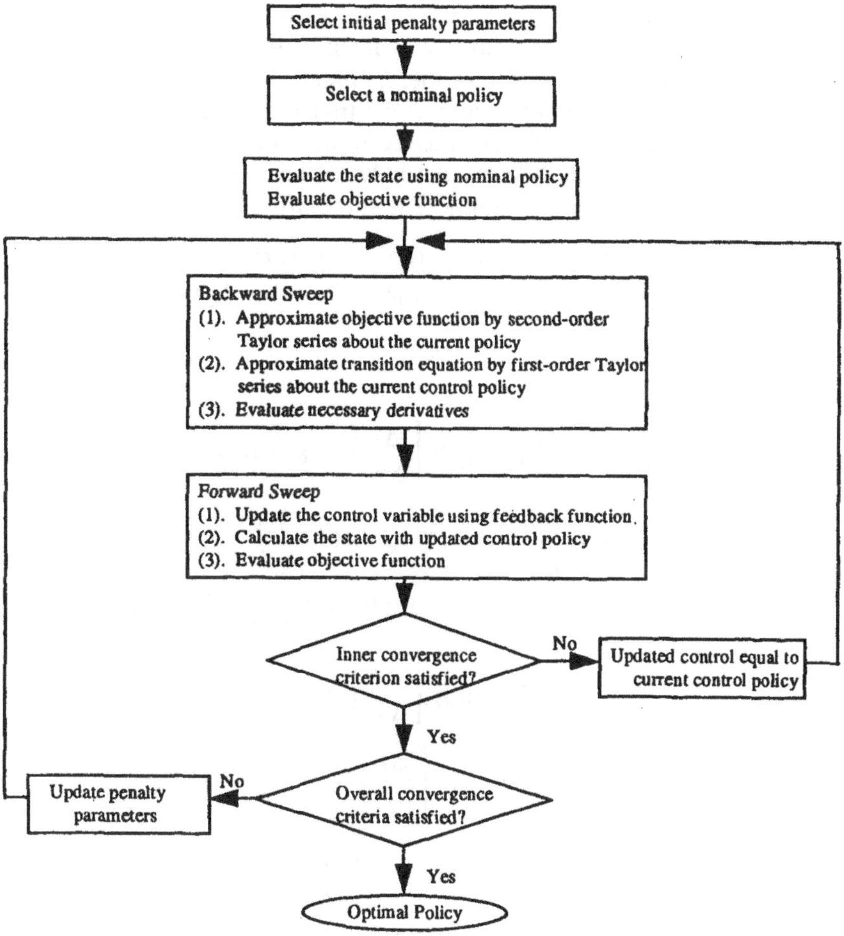

Figure 10.1 Flowchart of the SALQR method (Li, 1994).

input to the next and so on. Each input is operated on by the transfer function for the element, and through an advancing series of spatial and time steps, the functional behavior of the entire bay system is determined. The selection of these spatial and time steps is controlled by mathematical considerations involving stability, convergence, and compatibility. Underlying these considerations is the further requirement that the hydrodynamic and transport equations effectively interface with one another so that the computed tidal amplitudes and net velocities integrated over successive tidal cycles can be used directly as input to the transport model.

In the backward sweep of the optimal control method, it is required to calculate the partial derivatives of the transition equation. This is the most difficult part of HYD-SAL interfaced with the optimizer in the optimal control method. The SAL model is based on a linear second-order partial difference equation on salinity. The finite-difference form is

$$\alpha_1 \mathbf{S}_{t+1}(i-1, j) + \alpha_2 \mathbf{S}_{t+1}(i, j) + \alpha_3 \mathbf{S}_{t+1}(i+1, j) = \beta_1 \mathbf{S}_t(i, j-1) + \beta_2 \mathbf{S}_t(i, j) + \beta_3 \mathbf{S}_t(i, j+1) \tag{10.2.14}$$

where

$$\alpha_1 = -E_x^{t+1}(i-1, j)\left[\frac{\Delta t}{\Delta x^2}\right] - U^{t+1}(i-1, j)\left[\frac{\Delta t}{2\Delta x}\right]$$

$$\alpha_2 = 1 + E_x^{t+1}(i, j)\left[\frac{\Delta t}{\Delta x^2}\right] + E_x^{t+1}(i-1, j)\left[\frac{\Delta t}{\Delta x^2}\right] - U^{t+1}(i-1, j)\left[\frac{\Delta t}{2\Delta x}\right] + U^{t+1}(i, j)\left[\frac{\Delta t}{2\Delta x}\right]$$

$$\alpha_3 = -E_x^{t+1}(i, j)\left[\frac{\Delta t}{\Delta x^2}\right] + U^{t+1}(i, j)\left[\frac{\Delta t}{2\Delta x}\right]$$

$$\beta_1 = E_y^t(i, j-1)\left[\frac{\Delta t}{\Delta y^2}\right] + V^t(i, j-1)\left[\frac{\Delta t}{2\Delta y}\right]$$

$$\beta_2 = 1 - E_y^t(i, j)\left[\frac{\Delta t}{\Delta y^2}\right] - E_y(i, j-1)\left[\frac{\Delta t}{\Delta y^2}\right] + V^t(i, j-1)\left[\frac{\Delta t}{2\Delta y}\right] - V^t(i, j)\left[\frac{\Delta t}{2\Delta y}\right] + K_1\Delta t$$

$$\beta_3 = E_y^t(i, j)\left[\frac{\Delta t}{\Delta y^2}\right] - V^t(i, j)\left[\frac{\Delta t}{2\Delta y}\right]$$

where the variables are defined as in Section 7.2. From Eq. (10.2.14), it is easy to obtain $\partial \mathbf{S}_{t+1}/\partial \mathbf{S}_t$ analytically.

The spatial and temporal salinities in the bay system are computed by solving a set of partial differential equations (PDEs) numerically. The freshwater inflows, **Q**, are part of the boundary conditions for the hydrodynamic model (i.e., **Q** does not appear explicitly in the governing partial differential equation for the hydrodynamic model). In order to compute the derivatives $\partial \mathbf{S}_{t+1}/\partial \mathbf{Q}_t$ analytically, a new set of PDEs needs to be derived and an analytical solution may be very difficult, if not impossible (Bao, 1992). Therefore, the computation of $\partial \mathbf{S}_{t+1}/\partial \mathbf{Q}_t$ is performed by a finite-difference approximation method by the perturbation of Q and running the hydrodynamic transport simulator repeatedly for each perturbation of $\mathbf{Q}$. The Jacobian matrix of $\partial \mathbf{S}_{t+1}/\partial \mathbf{Q}_t$ is

$$J\left(\frac{\partial \mathbf{S}_{t+1}}{\partial \mathbf{Q}_t}\right) = \begin{bmatrix} \frac{\partial S_{t+1,1}}{\partial Q_{t,1}} & \frac{\partial S_{t+1,2}}{\partial Q_{t,1}} & \cdots & \frac{\partial S_{t+1,k}}{\partial Q_{t,1}} & \frac{\partial S_{t+1,n}}{\partial Q_{t,1}} \\ & & \vdots & & \\ \frac{\partial S_{t+1,1}}{\partial Q_{t,j}} & \frac{\partial S_{t+1,2}}{\partial Q_{t,j}} & \cdots & \frac{\partial S_{t+1,k}}{\partial Q_{t,j}} & \frac{\partial S_{t+1,n}}{\partial Q_{t,j}} \\ & & \vdots & & \\ \frac{\partial S_{t+1,1}}{\partial Q_{t,m}} & \frac{\partial S_{t+1,2}}{\partial Q_{t,m}} & \cdots & \frac{\partial S_{t+1,k}}{\partial Q_{t,m}} & \frac{\partial S_{t+1,n}}{\partial Q_{t,m}} \end{bmatrix} \tag{10.2.15}$$

where

$$\frac{\partial S_{t+1,k}}{\partial Q_{t,j}} = [S_{t+1,k}(S_{t,1}, S_{t,2}, \ldots, S_{t,n}, Q_{t,1} + \Delta Q_{t,1}, \ldots, Q_{t,j} + \Delta Q_{t,j}, \ldots, Q_{t,m} + \Delta Q_{t,m}, t) - S_{t+1,k}(S_{t,1}, S_{t,2}, \ldots, S_{t,n}, Q_{t,1}, \ldots, Q_{t,m}, t)]\,(\Delta Q_{t,j})^{-1}$$

Running HYD-SAL for simulation of the salinity for a 1-year period for the Lavaca-Tres Palacios Estuary requires 140 sec on a Cray YM-P8/864 (Bao, 1992; Bao and Mays, 1994b). Calculating the derivatives of transition equations requires many calls to HYD-SAL. Therefore, it is necessary to select a suitable solution technique in order to reduce the calls to HYD-SAL and to save CPU time. Three methods are considered: the DDP method, the SALQR method, and the quasi-Newton method. If the derivatives of the transition equations are easy to obtain, using the DDP method has a great advantage (Li and Mays, 1995). But the DDP method requires the second derivatives of the transition equation. The hydrodynamic transport model HYD-SAL consists of four partial differential equations, It is not possible to determine the derivative $\partial \mathbf{S}_{t+1}/\partial \mathbf{Q}_t$, $\partial^2 \mathbf{S}_{t+1}/\partial \mathbf{Q}_t^2$, and $\partial^2 \mathbf{S}_{t+1}/\partial \mathbf{S}_t\, \partial \mathbf{Q}_t$ from HYD-SAL directly. A finite-difference approximation method must be used. Thus, to obtain these derivatives requires many calls to HYD-SAL for each iteration, and the CPU time will increase rapidly. The DDP method loses its advantage for the estuarine management model, with HYD-SAL as the transition equation. Therefore, it is not suitable in the DDP method to interface HYD-SAL with the optimization procedure.

The main idea of the SALQR method is that in the backward sweep, $\partial^2 \mathbf{S}_{t+1}/\partial \mathbf{Q}_t^2$ and $\partial^2 \mathbf{S}_{t+1}/\partial \mathbf{S}_t \, \partial \mathbf{Q}_t$ are set to zero (i.e., linearizing the transition equation in the backward sweep). In the forward sweep, use the full nonlinear transition equation to update the state variables. Thus, only the derivative $\partial \mathbf{S}_{t+1}/\partial \mathbf{Q}_t$ needs to be calculated using the finite-difference approximation method. The number of calls to HYD-SAL reduces rapidly for each iteration.

The SALQR method differs from DDP in that the nonlinear simulation equation is linearized in the optimization step. Hence, if the transition equations are linear, SALQR and DDP are identical. The SALQR technique is useful for dynamic optimal control when the second derivatives of the simulation model are difficult to calculate.

Sen and Yakowitz (1987) suggested utilizing the quasi-Newton method in conjunction with the DDP algorithm for application when the second derivatives of the transition equation are difficult to calculate. The quasi-Newton technique approximates the second derivatives of the transition equation from the values of the first derivatives. The mathematical details of the quasi-Newton DDP (QNDDP) algorithm is equivalent to DDP with the addition of the quasi-Newton approximations. This method converges faster than the SALQR method. But the calculations of the second derivatives of the transition equation are much more complicated than those of the SALQR method.

Table 10.3 lists the number of calls to HYD-SAL for each method. The DDP method has the largest number of calls to HYD-SAL. The SALQR and QNDDP methods have the same number of calls to HYD-SAL. But the SALQR method is much simpler than the QNDDP method. Therefore, the SALQR method is chosen to interface with HYD-SAL. The flowchart SALQR interfaced with HYD-SAL is shown in Figure 10.2, which illustrates that there are two loops for interfacing SALQR with HYD-SAL. The outer loop determines the penalty parameters R_1 and R_2. The inner loop solves the augmented optimal control model using the SALQR method with R_1 and R_2 fixed at the values determined by the outer loop. Once the inner loop is completed, the convergence criterion is checked. If the convergence criterion is satisfied, the procedure terminates; otherwise, the procedure returns to the outer loop and updates the penalty parameter and then goes to the inner loop, and the augmented optimal control model is solved again with the updated R_1 and R_2 from the outer loop. This process continues until an optimal solution of the overall problem is found.

Table 10.3 Calls to HYD-SAL for Different Methods in Each Iteration

Method	$\mathbf{S}_{t+1}\ \frac{\partial \mathbf{S}_{t+1}}{\partial \mathbf{S}_t}$	$\frac{\partial \mathbf{S}_{t+1}}{\partial \mathbf{Q}_t}$	$\frac{\partial^2 \mathbf{S}_{t+1}}{\partial \mathbf{Q}_t^2}$	$\frac{\partial^2 \mathbf{S}_{t+1}}{\partial \mathbf{S}\ \partial \mathbf{Q}_t}$	Total calls to HYD-SAL
DDP	N	mN	$0.5(nm^2 + nm)N$	n^2mN	$(0.5nm^2 + 0.5nm + n^2m + 1)N$
SALQR	N	mN	0	0	$(m + 1)N$
QNDDP	N	mN	0	0	$(m + 1)N$

Notes: N is the number of months in the simulation period; m is the number of rivers; and n is the number of salinity test stations.

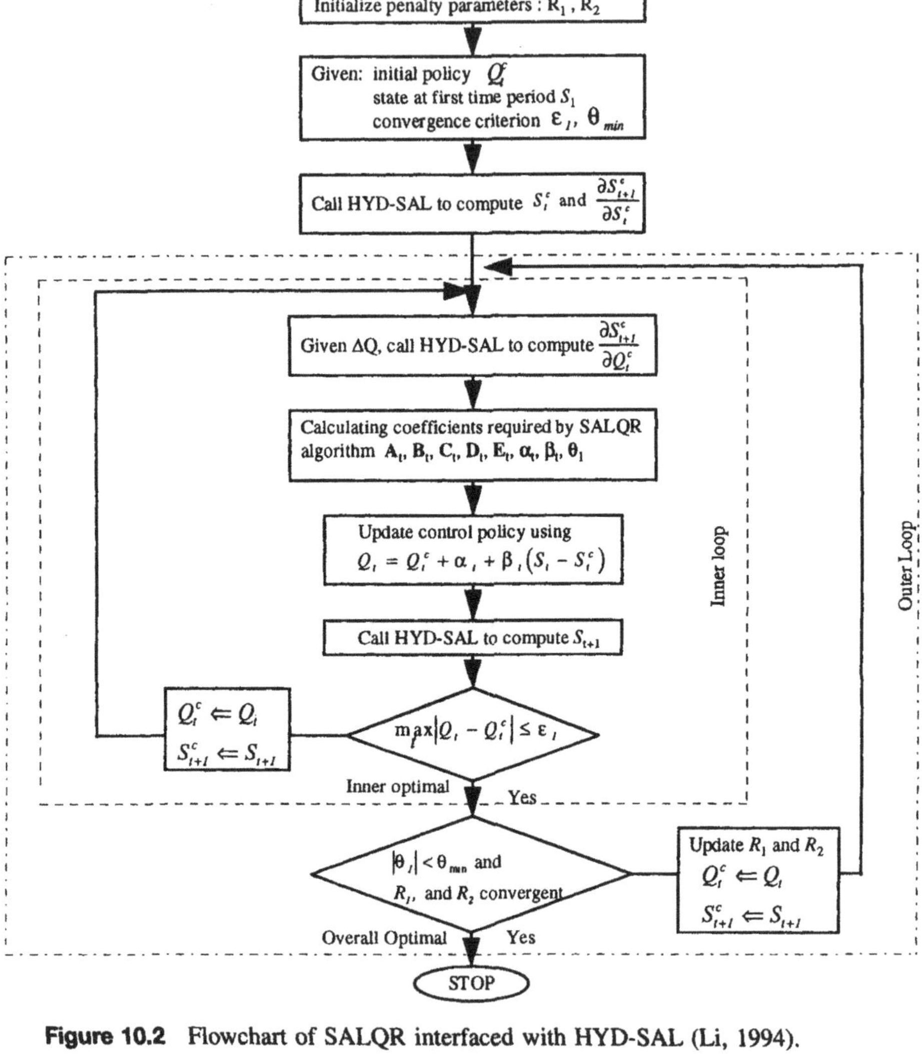

Figure 10.2 Flowchart of SALQR interfaced with HYD-SAL (Li, 1994).

10.3 EXAMPLE APPLICATION

The SALQR method is applied to the Lavaca-Tres Palacios Estuary in Texas. The estuarine management model has been run for various scenarios which focus on three key issues: (1) different adaptive shift parameters to evaluate the impact of the parameters on the convergence of the model;

(2) different initial penalty parameters R_1 and R_2 to test the effect of the penalty parameters on the solution process; and (3) various starting points to explore the global optimal issue.

When the SALQR method is used to interface the hydrodynamic model, an adaptive shift procedure by Liao and Shoemaker (1991) is used to obtain positive definite Hessian matrices. Parameters ε^c, δ, and M_1 have some effects on the convergence rate of the SALQR method. ε^c and δ depend on the first derivatives of the objective function with respect to the control variable. If the derivatives are small, the values of ε^c and δ should be small. M_1 controls the number of iterations for the constant shift step which should be a value to make the model converge fast. For the Lavaca-Tres Palacios Estuary management problem, the effects of ε^c, δ, and M_1 on the convergence rate of the SALQR method are shown in Tables 10.4, 10.5, and 10.6, respectively.

Table 10.4 indicates that when ε^c is very small, there is a numerical problem. When ε^c is much larger, the convergence rate is reduced. The appropriate value is $\varepsilon^c = 0.00001$ (for the derivative) and $\varepsilon^c = 0.0001$ [for the derivative $\hat{G}_Q(Q_{t,j}, S_{t,k}, R_1, R_2, t) > 0.1$]. Table 10.5 shows that the effect of δ on the convergence rata of the model is similar as that of ε^c. δ is a value not too small and not too large. $\delta = 0.001$ [for the derivative $\hat{G}_Q(Q_{t,j}, S_{t,k}, R_1, R_2, t) < 0.1$] and $\delta = 0.002$ (for the derivative $\hat{G}_Q(Q_{t,j}, S_{t,k}, R_1, R_2/, t) > 0.1$) are suggested for the Lavaca-Tres Estuary management model. Table 10.6 shows the effect of M_1 on the convergence rate of the SALQR method. $M_1 = 10$ is recommended for the estuarine management problem.

The two parameters considered for the bracket penalty function are R_1 (parameter for the control variable) and R_2 (parameter for the state variable). The test runs of the computer model for penalty parameters are designed to examine whether the model converges faster (with fewer iterations) for different initial values of R_1 and R_2. The results are summarized in Tables 10.7 and 10.8, respectively. Table 10.7 indicates that higher initial values of R_1 do not necessarily reduce the number of iterations. The reason is that the higher penalty parameter R_1 results in greater changes in the optimal monthly freshwater inflow, which cause more frequent switches of the salinity constraint violation. In other words, because of the great changes in optimal monthly freshwater inflow from iteration k to $k + 1$, some salinity constraints which are satisfied in iteration k will probably be violated in iteration $k + 1$. An appropriate initial value of R_1 for the Lavaca-Tres Palacios Estuary problem is $R_1 = 10$.

Table 10.8 indicates that as R_2 increases, the convergence rate of the model increases. This is because salinity values are much smaller compared to freshwater inflows. Only a large penalty parameter R_2 can make the

Table 10.4 Effect of ε^c on the Convergence Rate of the SALQR Method

Parameter ε^c	Optimal objective value (1000 lbs)	θ_1	No. of iterations	CPU time (hr:min:sec)
$\varepsilon^c = \begin{cases} 10^{-4} & \text{if } \hat{G}_Q(Q_{t,j}, S_{t,j}, R_1, R_2, t) \le 0.1 \\ 10^{-3} & \text{if } \hat{G}_Q(Q_{t,j}, S_{t,j}, R_1, R_2, t) > 0.1 \end{cases}$	6740	−0.6554	45	12:40:42.06
$\varepsilon^c = \begin{cases} 10^{-5} & \text{if } \hat{G}_Q(Q_{t,j}, S_{t,j}, R_1, R_2, t) \le 0.1 \\ 10^{-4} & \text{if } \hat{G}_Q(Q_{t,j}, S_{t,j}, R_1, R_2, t) > 0.1 \end{cases}$	6817	−0.009838	21	4:15:5.42
$\varepsilon^c = \begin{cases} 10^{-6} & \text{if } \hat{G}_Q(Q_{t,j}, S_{t,j}, R_1, R_2, t) \le 0.1 \\ 10^{-5} & \text{if } \hat{G}_Q(Q_{t,j}, S_{t,j}, R_1, R_2, t) > 0.1 \end{cases}$	There is a numerical problem			

Note: 1 lb = 0.4536 kg.

Table 10.5 Effect of δ on the Convergence Rate of the SALQR Method

Parameter δ	Optimal objective value (1000 lbs)	θ_1	No. of iterations	CPU time (hr:min:sec)
$\delta = \begin{cases} 1 \times 10^{-4} & \text{if } \hat{G}_Q(Q_{t,j}, S_{t,j}, R_1, R_2, t) \le 0.1 \\ 2 \times 10^{-4} & \text{if } \hat{G}_Q(Q_{t,j}, S_{t,j}, R_1, R_2, t) > 0.1 \end{cases}$	6817	−0.01077	29	4:40:26.17
$\delta = \begin{cases} 1 \times 10^{-3} & \text{if } \hat{G}_Q(Q_{t,j}, S_{t,j}, R_1, R_2, t) \le 0.1 \\ 2 \times 10^{-3} & \text{if } \hat{G}_Q(Q_{t,j}, S_{t,j}, R_1, R_2, t) > 0.1 \end{cases}$	6817	−0.009836	21	4:15:5.42
$\delta = \begin{cases} 1 \times 10^{-2} & \text{if } \hat{G}_Q(Q_{t,j}, S_{t,j}, R_1, R_2, t) \le 0.1 \\ 2 \times 10^{-2} & \text{if } \hat{G}_Q(Q_{t,j}, S_{t,j}, R_1, R_2, t) > 0.1 \end{cases}$	6720	−0.1200	30	5:29:42.77

Note: 1 lb = 0.4536 kg.

Table 10.6 Effect of M_1 on the Convergence Rate of the SALQR Method

Parameter M_1	Optimal objective value (1000 lbs)	θ_1	No. of iterations	CPU time (hr:min:sec)
5	6818	−0.04395	41	6:34:20.38
10	6817	−0.009836	21	4:15:5.42

Note: 1 lb = 0.4536 kg.

penalty term on salinity have the same level effect for the objective function as that of the penalty term on the freshwater inflow. R_2 = 10000 is suggested for the Lavaca-Tres Palacios Estuary management model.

Although the SALQR method cannot guarantee a global solution, a variety of starting points can help decide the local–global issue. If all starting points yield approximately the same final solution, then the solution point is more likely to be a global optimal (Lasdon and Waren, 1989).

The results for model runs using different initial values are listed in Table 10.9. In this table, Case 1 represents that initial monthly freshwater

Table 10.7 Effect of R_1 on the Convergence Rate of the SALQR Method

Parameter R_1	Optimal objective value (1000 lbs)	θ_1	No. of iterations	CPU time (hr:min:sec)
0.1	6718	−0.8772	36	5:36:52.12
1	6789	−0.4706	31	5:25:50.70
10	6817	−0.009836	21	4:15:5.42

Note: 1 lb = 0.4536 kg.

Table 10.8 Effect of R_2 on the Convergence Rate of the SALQR Method

Parameter R_2	Optimal objective value (1000 lbs)	θ_1	No. of iterations	CPU time (hr:min:sec)
100	6,719	−0.1084	34	5:15:32.40
1,000	6,819	−0.05623	23	4:06:56.17
10,000	6,817	−0.009838	21	4:15:5.42

Note: 1 lb = 0.4536 kg.

Table 10.9 Results of Using the SALQR Method with Different Initial Values

Month	Case 1 $\underline{Q} < Q_{\text{initial}} < \overline{Q}$		Case 2 $Q_{\text{initial}} = \underline{Q}$		Case 3 $Q_{\text{initial}} = \overline{Q}$	
	Lavaca River (cfs)	Colorado River (cfs)	Lavaca River (cfs)	Colorado River (cfs)	Lavaca River (cfs)	Colorado River (cfs)
Jan.	37	293	37	293	37	293
Feb.	41	324	41	324	41	324
Mar.	37	293	37	293	37	293
Apr.	1752	7600	1737	7600	1752	7600
May	2125	8863	2125	8863	2131	8863
June	1758	9209	1758	9209	1770	9209
July	114	4130	114	6017	114	4124
Aug.	114	2900	114	2900	114	2900
Sep.	949	5042	976	5042	949	5042
Oct.	474	4879	468	4879	499	4879
Nov.	114	573	114	573	114	573
Dec.	111	496	111	496	111	496
Fishery harvest (1000 lbs)	6718		6817		6719	
No. of iterations	21		21		34	
CPU time (hr:min:sec)	5:05:46.57		4:15:5.42		4:16:29.82	

Note: 1 lb = 0.4536 kg, 1 cfs = 0.0283 m^3/sec.

Table 10.10 Results for Different Transition Equations

Month	Linear regression equation		Nonlinear regression equation		Simulation model HYD-SAL	
	Freshwater inflow from Lavaca River (cfs)	Freshwater inflow from Colorado River (cfs)	Freshwater inflow from Lavaca River (cfs)	Freshwater inflow from Colorado River (cfs)	Freshwater inflow from Lavaca River (cfs)	Freshwater inflow from Colorado River (cfs)
Jan.	37	293	37	424	37	293
Feb.	1112	324	41	1934	41	324
Mar.	1283	293	37	837	37	293
Apr.	3784	6070	2442	1619	1734	7600
May	2256	2429	2172	3179	2125	8863
June	2257	2429	2172	3180	1785	9209
July	656	6017	114	3308	114	6017
Aug.	732	2900	114	2900	114	2900
Sep.	3272	653	2184	1966	976	5042
Oct.	2569	2429	2171	3179	468	4879
Nov.	114	573	114	1087	114	573
Dec.	111	496	111	844	111	496
Fishery harvest (1000 lbs)	6695		7294		6817	
No. of iterations	17		164		21	
CPU time (hr:min:sec)	0:0:2.19		0:0:3.49		4:15:5.42	

Note: 1 cfs = 0.0283 m^3/sec, 1 lb = 0.4536 kg.

inflows are equal to the mean of the lower bounds and the upper bounds; for Case 2, the initial monthly freshwater inflows are the lower bounds; and for Case 3, the initial monthly freshwater inflows are equal to the upper bounds. The results are similar to those for using the nonlinear transition equation. For Case 2, the fishery harvest is the largest. The optimal fishery harvest values for Cases 1 and 3 are almost the same. The relative differences in the annual fishery harvest are consistently less than 100 (1000 lbs) for all cases. That means the difference in the optimal solution of the total annual fishery harvest is considered to be acceptable and the global optimal is likely reached at 6817 (1000 lbs).

For comparison, the results of using regression equations as the transition (Li and Mays, 1995) and those of using HYD-SAL as the transition are listed in Table 10.10. Table 10.10 shows that for different transition equations, the optimal freshwater inflow and fishery harvest are very similar. For example, the two monthly peak flows all take place during April to May and September to October. Using the linear transition equation, the optimal fishery harvest is 6695 (1000 lbs), and using the nonlinear transition equation, it is 7294 (1000 lbs). When HYD-SAL is used as the transition equation, the optimal fishery harvest is 6817 (1000 lbs), which is smaller than that of using the nonlinear transition equation and greater than that of using the linear transition equation. Because the hydrodynamic transport model incorporates more variables for the complex ecosystem such as tidal mixing, wind-induced mixing, precipitation, evaporation, freshwater inflow, and salinity, the results of using HYD-SAL as the transition are more reliable than that of using regression equations as the transition.

REFERENCES

Andricevic, R. and Kitanidis, P. K., Optimization of the Pumping Schedule in Aquifer Remediation Under Uncertainty, *Water Resources Research*, Vol. 26, No. 5, pp. 875–885, 1990.

Bao, Y. X., Methodology for Determining the Optimal Freshwater Inflows into Bays and Estuaries, Ph.D. Dissertation, Department of Civil Engineering, The University of Texas at Austin, 1992.

Bao, Y. and Mays, L. W., New Methodology for Optimization of Freshwater Inflows to Estuaries, *Journal of Water Resources Planning and Management*, Vol. 120, No. 2, pp. 199–217, 1994a.

Bao, Y. and Mays, L. W., Optimization of Freshwater Inflows to the Lavaca-Tres Palacios, Texas, Estuary, *Journal of Water Resources Planning and Management*, Vol. 120, No. 2, pp. 218–236, 1994b.

Chang, L-C, Shoemaker, C. A., and Liu, P. L.-F., Optimal Time-Varying Pumping Rates for Groundwater Remediation: Application of a Constrained Optimal Control Algorithm, *Water Resources Research*, Vol. 28, No. 12, pp. 3157–3173, 1992.

Culver, T. B. and Shoemaker, C. A., Dynamic Optimal Control for Groundwater Remediation with Flexible Management Periods, *Water Resources Research*, Vol. 28, No. 3, pp. 629–641, 1992.

Jones, L. C., Willis, R., and Yeh, W. W., Optimal Control of Nonlinear Groundwater Hydraulics Using Differential Dynamic Programming, *Water Resources Research*, Vol. 20, No. 4, pp. 415–427, 1984.

Lasdon, L. S. and Waren, A. D., *GRG2 User's Guide*, Department of General Business, The University of Texas at Austin, Austin, TX, 1989.

LeBlanc, L., Epsilon-Constrain Method for Determining Freshwater Inflows into Bays and Estuaries, Master Thesis, Department of Civil Engineering, Arizona State University, Tempe, 1993.

Li, G. L., Differential Dynamic Programming for Estuarine Management, Ph.D. Dissertation, Department of Civil and Environmental Engineering, Arizona State University, Tempe, 1994.

Li, G. L. and Mays, L. W., Differential Dynamic Programming for Estuarine Management, *Journal of Water Resources Planning and Management*, Vol. 121, No. 6, pp. 455–462, 1995.

Liao, L. and Shoemaker, C. A., Convergence in Unconstrained Discrete-Time Differential Dynamic Programming, *IEEE Transitions on Automatic Control*, Vol. AC-36, No. 6, pp. 692–706, 1991.

Mao, N. and Mays, L. W., Goal Programming Models for Determining Freshwater Inflows to Estuaries, *Journal of Water Resource Planning and Management*, Vol. 120, No. 3, pp. 316–329, 1994.

Masch, F. D. and Associates, Tidal Hydrodynamic and Salinity Model for San Antonio and Matagorda Bays, Texas, Report to Texas Water Development Board, Austin, 1971.

Martin, Q. W., Estimating Freshwater Inflows Needs for Texas Estuaries by Mathematical Programming, *Water Resources Research*, Vol. 23, No. 2, pp. 230–238, 1987.

Reklaitis, G. V., Ravindran, A., and Ragsdell, K. M., *Engineering Optimization: Methods and Applications*, Wiley–Interscience, New York, 1983.

Sen, S., and Yakowitz, S. J., Constrained Differential Dynamic Programming Algorithm for Discrete-Time Optimal Control, *Automatika*, Vol. 23, No. 6, pp. 749–752, 1987.

Shi, W., Multiobjective Optimization of Freshwater Inflows into Estuaries Using the Surrogate Worth Tradeoff Method, Master Thesis, Department of Civil Engineering, Arizona State University, Tempe, 1992.

Siebert, J., Multiobjective Optimization Involving Freshwater Inflows into Bays and Estuaries Using Utility Function Assessments, Master Thesis, Department of Civil Engineering, Arizona State University, Tempe, 1993.

Texas Department of Water Resources, Lavaca-Tres Palacios Estuary: A Study of the Influence of Freshwater Flows, Report LP-106, TDWR, 1980.

Tung, Y. K., Bao, Y., Mays L., and Ward, G., Optimization of Freshwater Inflow to Estuaries, *Journal of Water Resources Planning and Management*, Vol. 116, No. 4, pp. 567–584, 1990.

Yakowitz, S., Algorithm and Computational Techniques in Differential Dynamic Programming, *Control and Dynamic Systems*, Vol. 31, pp. 75–91, 1989.

Chapter 11
Sediment Control in Rivers and Reservoirs

11.1 PROBLEM IDENTIFICATION

Sedimentation occurring in reservoirs and rivers are always associated with floods being stored and conveyed. In rivers, sedimentation, which refers to either deposition or scouring of a riverbed, is caused principally by the passage of flood events. Continued sedimentation in the river over time physically diminishes the channel's capability to contain flows, posing a major threat to the economics of the system. For instance, the rise of the channel bed profile due to aggradation reduces the conveyance capacity of the channel. As aggradation processes become chronic, flood encroachment into the flood plain follows, leading to property damages, probable loss of lives, and endangerment in the means of livelihood of the community.

Scouring (or degradation) on the other hand, threatens instream and bank structures like flood levees, bridge piers, as well as underground utility lines. Either way, if sedimentation processes are left unchecked or uncontrolled in the system, there could be serious economic consequences. Interestingly, under a given flood event or series of flood events, both aggradation and degradation phenomena occur side-by-side across the channel and simultaneously in the same river. Any bed change, however, as a result of sediment movements due to floods, is a sign of channel instability, and efforts must be pursued to minimize such instability in the channel. Because flood events are known to be the principal instigator of bed mobility, focus must be made on how such flood events could be altered or modified into a series of events that permits the least amount of bed changes in the river.

There has been minimal effort in the past to develop an optimization procedure to determine the optimal control of sedimentation occurring in alluvial rivers. On the other hand, there has been a major concentration in modeling efforts in the last two decades to develop simulation codes to predict the bed material movement in alluvial streams. Since the development of HEC-6 by the Hydrologic Engineering Center of the U. S. Army Corps of Engineers (1977, 1991), over 50 other sedimentation codes have been developed.

A simple system described by the interaction of a reservoir and a river as shown in Figure 11.1 is used to describe the mathematical model. More complex systems with multiple reservoirs and downstream river channels can be analyzed by the new methodology. Because the mobility and transport of sediments in the downstream river are directly associated with the reservoir releases, the determination of the optimal releases are critical in the analysis because they have a direct bearing on the hydraulics and sedimentation processes in the river downstream. The optimization model that addresses such sedimentation control problems is expressed as follows:

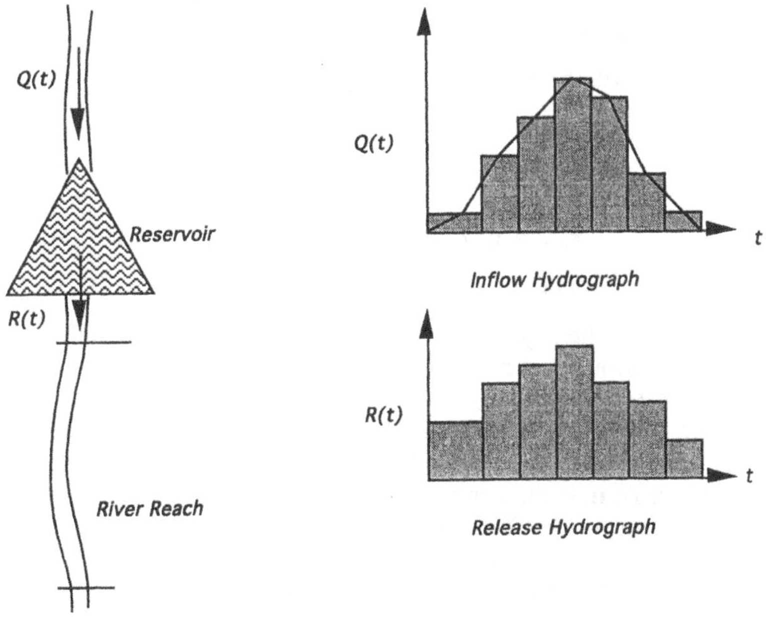

Figure 11.1 Sketch of the system for the sedimentation problem (Carriaga, 1993).

Minimize [the sum of aggradation and degradation depths at key locations along the river reach]
Subject to: the system's governing physical equations;
the boundary conditions;
the operating, budgetary, and design constraints.

Carriaga and Mays (1995a) developed nonlinear programming models to determine reservoir releases for sediment control. Uncertainties were also considered on some sediment transport parameters identified—like the sediment load and mean grain size—leading to the formulation of three chance-constrained models. The optimization problem was also formulated in a multistage decision process and solved using the conventional dynamic programming (DP) methodology. In this DP problem, the approach was extended to the discrete differential dynamic programming (DDDP) approach. The optimization problem was also formulated and solved using a simplified differential dynamic programming (DDP) methodology.

11.2 PROBLEM FORMULATION

11.2.1 Optimal Control Problem

The optimization problem is a discrete-time optimal control problem with the following form:

$$\text{Min } Z(\mathbf{u}) = \sum_t f_t(\mathbf{x}_t, \mathbf{u}_t, t) = \sum_t (\mathbf{a}_t + \mathbf{b}_t), \quad t = 1, 2, \ldots, T \tag{11.2.1}$$

subject to

$$\mathbf{x}_{t+1} = \mathbf{g}_t(\mathbf{x}_t, \mathbf{u}_t, t), \quad t = 1, 2, \ldots, T - 1 \tag{11.2.2}$$

$$\mathbf{h}_t(\mathbf{x}_t, \mathbf{u}_t, t) \leq \mathbf{0}, \quad t = 1, 2, \ldots, T \tag{11.2.3}$$

where $Z(\mathbf{u})$ is the objective function (sum of aggradation and degradation depths in downstream river) or the overall cost associated with the control policy (reservoir releases), $\mathbf{u}$; t is the simulation time step; T is the total number of simulation time steps; $\mathbf{x}_t$ is the state vector (reservoir storage level and downstream riverbed elevations) at the starting time t (having dimension I); $\mathbf{u}_t$ is the control vector during time t (having dimension m); f_t is the cost or loss function during time step t, given $\mathbf{x}_t$, $\mathbf{u}_t$, and t; $\mathbf{a}_t$ is the aggradation depth at time t (having dimension $I - 1$); $\mathbf{b}_t$ is the degradation depth at time t (having dimension $I - 1$); $\mathbf{g}_t$ is the transition equation or the governing equation for the simulation model (reservoir operation model and sediment transport routing model), given $\mathbf{x}_t$, $\mathbf{u}_t$, and

t; and $\mathbf{h}_t$ is the set of constraints or boundary conditions on the control and state vectors.

The loss function, $f_t(\mathbf{x}_t, \mathbf{u}_t, t)$, defined by Eq. (11.2.1), can be approximated by using the following quadratic form:

$$f_t(\mathbf{x}_t, \mathbf{u}_t, t) = \tfrac{1}{2}\, \delta\mathbf{x}^T\mathbf{A}_t\delta\mathbf{x} + \delta\mathbf{x}^T\mathbf{B}_t\delta\mathbf{u} + \tfrac{1}{2}\delta\mathbf{u}^T\mathbf{C}_t\delta\mathbf{u} + \mathbf{D}_t\delta\mathbf{u} + \mathbf{E}_t\delta\mathbf{x} \qquad \text{for } t = 1, 2, \ldots, T \quad (11.2.4)$$

where $\mathbf{A}_t$, $\mathbf{B}_t$, $\mathbf{C}_t$, $\mathbf{D}_t$, and $\mathbf{E}_t$ are the parameters at time period t determined from the first- and second-order derivatives of the loss function, $f_t(\mathbf{x}_t, \mathbf{u}_t, t)$, and the state transition function, $\mathbf{g}_t(\mathbf{x}_t, \mathbf{u}_t, t)$; $\delta\mathbf{x}$ and $\delta\mathbf{u}$ are the increment changes between the current and the previous values of the state $\mathbf{x}$ and control $\mathbf{u}$. The DDP concept is to minimize the quadratic approximation as presented in Eq. (11.2.4). Estimation of parameters $\mathbf{A}_t$, $\mathbf{B}_t$, $\mathbf{C}_t$, $\mathbf{D}_t$, and $\mathbf{E}_t$ in the work of Mayne (1966) was based on an unconstrained problem. See Section 9.1 for definition. A reduced problem is, therefore, essential in order to use the original formula for the above parameters.

The DDP algorithm requires an initial state vector (reservoir storage and bed elevation at time $t = 1$) designated as $\mathbf{x}(1)$ and a nominal control policy, $\mathbf{u}^c$, expressed as

$$\mathbf{u}^c = (\mathbf{u}^c(1), \mathbf{u}^c(2), \ldots, \mathbf{u}^c(T)) \quad (11.2.5)$$

From these initial values, $\mathbf{x}^c$ could be determined using the transition equation (simulator) in Eq. (11.2.2) expressed as follows:

$$\mathbf{x}^c = (\mathbf{x}^c(1), \mathbf{x}^c(2), \ldots, \mathbf{x}^c(T + 1)) \quad (11.2.6)$$

The state transition functions generally expressed by Eq. (11.2.2) for the system defined include (i) the hydrologic routing for the reservoir described by the mass balance equation and (ii) the hydraulic and sediment routing for the alluvial river performed by the HEC-6 code. In addition to the transition equations are the bound constraints for the control and state variables plus the boundary condition for the state variable.

The system model is described basically by the relations associated with the reservoir operation and the river hydraulics and sedimentation. Before presenting the governing equations involved, the state and control variables must be defined. The state variable, $\mathbf{x}$, represents the storage state, S, of the reservoir and the bed elevations, E, at various sections of the river, namely

$$\mathbf{x}_t = \begin{bmatrix} \text{Bed elevation, } E \\ \text{Storage state, } S \end{bmatrix} = \begin{bmatrix} x_{i,t};\ i = 1, 2, \ldots, I - 1 \\ x_{I,t} \end{bmatrix} \qquad \text{for } t = 1, 2, \ldots, T \quad (11.2.7)$$

Thus, the state of the reservoir–river system at the start of any period t can be described by the vector of riverbed elevations and the storage state. The control variable, **u**, represents the reservoir release expressed as

$$\mathbf{u}_t = [\text{Reservoir release}, R], \quad \text{for } t = 1, 2, \ldots, T \qquad (11.2.8)$$

11.2.2 Reservoir Operation Constraints

The following equations for the reservoir operation include the mass balance equation (11.2.9), the release and storage state bounds (11.2.10) and (11.2.11), and the boundary condition (11.2.12) that defines the end-of-the-operation storage state of the reservoir:

$$S_{t+1} = S_t + Q_t - R_t - L_t + P_t - D_t\, t = 1, 2, \ldots, T \qquad (11.2.9)$$

$$R_{\min,t} \le R_t \le R_{\max,t}, \qquad t = 1, 2, \ldots, T \qquad (11.2.10)$$

$$S_{\min,t} \le S_t \le S_{\max,t}, \qquad t = 1, 2, \ldots, T \qquad (11.2.11)$$

$$S_T = S_{\max,T} \qquad (11.2.12)$$

where S_t and S_{t+1} are the beginning and ending storage states, respectively, of the reservoir during time (t), Q_t is the inflow, R_t is the release, L_t is the seepage flow, P_t is the excess rainfall/precipitation, D_t is the evaporation loss, $R_{\min,t}$ and $R_{\max,t}$ are the minimum and maximum releases, respectively, at time period t, $S_{\min,t}$ and $S_{\max,t}$ are the minimum and maximum storages of the reservoir at time period t, and $S_{\max,T}$ is the target storage state at the end of reservoir operation.

11.2.3 River Hydraulics and Sedimentation Constraints

The HEC-6 code (U. S. Army Corps of Engineers, 1991) is used to solve the hydraulic and sediment transport routing constraints each time the optimizer requires these solutions. The one-dimensional physical governing equations are the conservation of mass and energy, and some established empirical relations. The steady-state flow continuity is expressed as

$$\frac{\partial R_t}{\partial l_c} - q_t = 0, \quad t = 1, 2, 3, \ldots, T \qquad (11.2.13)$$

where R_t is the release discharge from the reservoir, q_t is the lateral inflow rate per unit length, and l_c is the distance along the channel. The energy equation is expressed as

$$E_{i+1,t} + d_{i+1,t} + \frac{\alpha_{i+1,t} V_{i+1,t}^2}{2g} = E_{i,t} + d_{i,t} + \frac{\alpha_{i,t} V_{i+1,t}^2}{2g} + h_{i,t},$$

$$t = 1, 2, 3, \ldots, T, i = 1, 2, 3, \ldots, I - 1 \quad (11.2.14)$$

where $E_{i,t}$ is the bed elevation at station i and at time t measured from a designated reference datum, $d_{i,t}$ is the depth of flow at station i during time t, α_t is the energy distribution coefficient associated with station i at time t, $h_{i,t}$ is the energy loss from station $i + 1$ to station i during time t, $V_{i,t}$ is the flow velocity at station i during time t, and g is the gravitational constant. The sediment continuity (routing) equation (Exner equation) is expressed as

$$\frac{\partial G_{i,t}}{\partial l_c} + \gamma_s (1 - \eta) W_i \frac{\partial E_{i,t}}{\partial t} - q_{s_t} = 0,$$
$$i = 1, 2, \ldots, I - 1, t = 1, 2, 3, \ldots \quad (11.2.15)$$

where $G_{i,t}$ is the sediment discharge at section i and at time period t, q_{s_t} is the lateral or local sediment input during time t from bank or tributaries per unit length, g_s is the specific weight of bed material, h is the porosity of the bed sediment material, and W_i is the average width of the movable bed between sections i and $i + 1$. The sediment transport equation is expressed in general form as (see American Society of Civil Engineers, 1977)

$$G_{i,t} = F_{i,t}(R, C_w, W, d, r, V, S_e, f, n, \nu, r_s, m_g,$$
$$s_g, w, g, l, S_s, S_c, f_s, C_T, C_F) \quad (11.2.16)$$

where the parameters inside the parenthesis are the variables that define the sediment load $G_{i,t}$; $F_{i,t}$ indicates that the parameters enumerated vary in time t and in space i; C_w is the concentration of wash load; d is the flow depth; r is the hydraulic radius of the channel; S_e is the energy gradient; f is the Darcy–Weisbach friction factor; n is the fluid kinematic viscosity; ν is the fluid density; r_s is the particle density; m_g is the geometric mean size of the bed material sediments; s_g is the geometric standard deviation of the bed material; w is the mean settling velocity of the sediment particle; S_p is the plan-form geometry; l is the apparent dynamic viscosity; S_s is the shape factor of the sediment particles; S_c is the shape factor of the channel reach; f_s is the seepage force in the channel bed; C_T is the concentration of bed material discharge; and C_F is the fine material concentration. Equations (11.2.10)–(11.2.12) cause the problem to be constrained. In order to follow the algorithm proposed by Yakowitz and Rutherford (1984), the constrained optimization problem must be converted into an unconstrained format using a penalty function method.

11.2.4 Penalty Function Method

A penalty function method is used for the purpose of converting a constrained problem into an unconstrained problem by adding to the objective function a term that prescribes a high cost (penalty) for violation of the constraint. Because Eqs. (11.2.10)–(11.2.12) cause the problem to be constrained, a reduced problem that fits the original DDP problem must be derived. The procedure would result in a reduced problem expressed as

$$\underset{\mathbf{u}}{\text{Min}}\ Z = \sum_t F_t(\mathbf{x}_t, \mathbf{u}_t, t) = \sum_t f_t(\mathbf{x}_t, \mathbf{u}_t, t) + \sum_t \phi_t Y_t(\mathbf{x}_t, \mathbf{u}_t),$$

$$t = 1, 2, 3, \ldots, T \quad (11.2.17)$$

subject to the transition equations defined by the mass balance equation [i.e., Eq. (11.2.6)] and the governing equations for the river hydraulics and sedimentation [i.e., Eqs. (11.2.13)–(11.2.16)] being solved by the HEC-6 code. In Eq. (11.2.17), $f_t(\mathbf{x}_t, \mathbf{u}_t, t)$ is the original objective function of the constrained DDP problem, f_t is the penalty weight (or cost) associated with the penalty function $Y_t(\mathbf{x}_t, \mathbf{u}_t)$, and $F_t(\mathbf{x}_t, \mathbf{u}_t, t)$ is the objective function of the reduced problem during time period t.

The hyperbolic penalty function introduced by Lin (1990) is used for the constrained sedimentation control problem because of its analytical characteristics as described by Culver and Shoemaker (1992, 1993) in their groundwater remediation studies.

The basic penalty function can be expressed as

$$\mathbf{y}_t = \sqrt{p^2\mathbf{H}_t^2(\mathbf{x}_t, \mathbf{u}_t) + w^2} - p\,\mathbf{H}_t(\mathbf{x}_t, \mathbf{u}_t), \quad t = 1, 2, 3, \ldots, T \quad (11.2.18)$$

where p is the scale factor and $\mathbf{H}_t(\mathbf{x}_t, \mathbf{u}_t)$ is the constraint violation involving the control $\mathbf{u}$ and/or state variable $\mathbf{x}$, w is the shape factor of the parabola which is the ordinate intercept distance from the origin. The scale factor, p, takes a positive value ($p > 0$) if the constraints being considered are either $\mathbf{H}_t(\mathbf{x}_t, \mathbf{u}_t) \leq \mathbf{0}$ or $\mathbf{H}_t(\mathbf{x}_t, \mathbf{u}_t) = \mathbf{0}$. Conversely, p takes a negative value ($p < 0$) if the constraints considered are either $\mathbf{H}_t(\mathbf{x}_t, \mathbf{u}_t) \geq \mathbf{0}$ or $\mathbf{H}_t(\mathbf{x}_t, \mathbf{u}_t) = \mathbf{0}$.

For a penalty function that dominates, a function $\mathrm{Y}_t(\mathbf{x}_t, \mathbf{u}_t)$ which is an expression of the control $\mathbf{u}$ and/or state $\mathbf{x}$ could be expressed in general form as

$$Y_t(\mathbf{x}_t, \mathbf{u}_t) = \begin{cases} y_t & \text{for elements of } y_t \le y_0, \\ & \text{where } t = 1, 2, 3, \ldots, T \quad (11.2.19a) \\ Ay_t^n + By_t^m + C & \text{for elements of } y_t \ge y_0, \\ & \text{where } t = 1, 2, 3, \ldots, T \quad (11.2.19b) \end{cases}$$

where

$$A = \frac{m-1}{n(m-n)y_0^{n-1}} \tag{11.2.20}$$

$$B = \frac{n-1}{m(n-m)y_0^{m-1}} \tag{11.2.21}$$

$$C = \left(1 - \frac{m-1}{n(m-n)} - \frac{n-1}{m(n-m)}\right) y_0 \tag{11.2.22}$$

The coefficients A, B, and C in Eq. (11.2.19b) can be determined by the continuous of functions $Y_t(\mathbf{x}_t, \mathbf{u}_t)$, $Y_t'(\mathbf{x}_t, \mathbf{u}_t)$, and $Y_t''(\mathbf{x}_t, \mathbf{u}_t)$ for all the elements of $y = y_0$. To have positive penalty values, the coefficient A in Eq. (11.2.19b) must be positive. This then requires m to be less than 1 ($m < 1$) for any positive value of y_0. For $y_0 = 1$ and $m = 0.5$, Eqs. (11.2.19a) and (11.2.19b) could be simplified as

$$Y_t(\mathbf{x}_t, \mathbf{u}_t) = y_t \quad \text{for all elements of } y_t \le 1, \quad \text{where } t = 1, 2, 3, \ldots, T \tag{11.2.23}$$

and

$$Y_t(\mathbf{x}_t, \mathbf{u}_t) = \frac{y_t^n + 4\,n\,(n-1)\sqrt{y_t} - (n-1)(2n-1)}{n\,(2n-1)}$$
$$\text{for all elements of } y_t \ge 1, \text{ where } t = 1, 2, 3, \ldots, T \tag{11.2.24}$$

Equations (11.2.23) and (11.2.24) are both continuous at least up to the second derivative. The parameter y_0 in Eqs. (11.2.19a) and (11.2.19b) was given a value of 1.0 for purposes of simplicity in evaluating Eq. (11.2.19b), although y_0 can take any value. The shape factor w which represents the intercept distance from the origin, takes a value from 0.001 to 0.01, from which the value of p could be adjusted for the required accuracy. The n in Eq. (11.2.24) is chosen such that the penalty function can dominate the negative objective function, if one exists.

The penalty functions in Eqs. (11.2.23) and (11.2.24) have the following advantages:

(i) It is defined for all real values of the product of p and $\mathbf{H}_t(\mathbf{x}_t, \mathbf{u}_t)$.
(ii) It is continuous and differentiable, for which the first and second derivatives of the function, $Y_t(\mathbf{x}_t, \mathbf{u}_t)$, are also continuous.
(iii) It is not flat at the infeasible side near the constraint boundary.

11.3 PROBLEM SOLUTION

11.3.1 Evaluation of Derivatives

The DDP method requires the computation of the first- and second-order derivatives of the objective function in Eq. (11.2.17) defined as

$$\frac{\partial F}{\partial \mathbf{x}}, \frac{\partial F}{\partial \mathbf{u}}, \frac{\partial^2 F}{\partial \mathbf{x}^2}, \frac{\partial^2 F}{\partial \mathbf{u}^2}, \text{ and } \frac{\partial^2 F}{\partial \mathbf{x}\, \partial \mathbf{u}} \tag{11.3.1}$$

and the derivatives of the state transition equations, expressed as

$$\frac{\partial \mathbf{g}}{\partial \mathbf{x}}, \frac{\partial \mathbf{g}}{\partial \mathbf{u}}, \frac{\partial^2 \mathbf{g}}{\partial \mathbf{x}^2}, \frac{\partial^2 \mathbf{g}}{\partial \mathbf{u}^2}, \frac{\partial^2 \mathbf{g}}{\partial \mathbf{x}\, \partial \mathbf{u}} \tag{11.3.2}$$

For the evaluation of the above derivatives [i.e., Eqs. (11.3.1) and (11.3.2)], a combination of numerical and analytical approaches was used. The derivatives of the state transition equation for the river model which is solved by the HEC-6 code represented by Eqs. (11.2.13)–(11.2.16) and the original objective function, $f_t(\mathbf{x}_t, \mathbf{u}_t, t)$ in Eq. (11.2.1), were evaluated numerically by a forward difference scheme.

The derivatives of the mass balance equation for the reservoir defined by Eq. (11.2.9) and the penalty terms associated with the violation of the bound constraints [in Eqs. (11.2.10) and (11.2.11)] and boundary condition [in Eq. (11.2.12)] were evaluated analytically. This is because the penalty function generally represented by $Y_t(\mathbf{x}_t, \mathbf{u}_t)$, which is a term in the objective function of the reduced problem, is explicitly defined. Thus, the gradients and the Hessian matrices could be evaluated from the function expressions. However, the gradient and Hessian matrices of the original objective function, $f_t(\mathbf{x}_t, \mathbf{u}_t, t)$, which defines the summation of bed changes (i.e., $\mathbf{a}_t$ + $\mathbf{b}_t$) along the river at time period t, is solved by a forward difference scheme. An example of the hyperbolic penalty function associated with the violation of storage bound constraint is presented in Appendix 11.A.

The Jacobian matrices and the second derivatives of the state transition equations, defined by the mass balance and the bed change equations, generally expressed as $\mathbf{g}_t(\mathbf{x}_t, \mathbf{u}_t, t)$ are evaluated analytically and numerically. Because the mass balance equation for the reservoir model is defined explicitly, the Jacobian matrix and the second derivatives are evaluated analytically. For the bed change equation, which is solved by the HEC-6 code, the forward difference scheme is used to numerically evaluate the Jacobian and the second derivatives.

11.3.2 DDP Algorithm

The interfacing approach of the optimizer and simulation models is simply represented by Figure 11.2. However, Figure 11.3, which was modified from the flowchart developed by Culver and Shoemaker (1992), describes in detail the steps involved in the solution approach. As presented in Figure 11.3, an initial or "nominal" control (or release) policy must be specified for the first simulation in order to calculate the current state (or condition) of the system, $\mathbf{x}_t$, in each time period t, and for the value of the objective function, Z [Steps (0) and (1)].

Each iteration could be divided into two parts: the backward and forward sweeps. In the backward sweep [Step (2)], an update to the current control policy is calculated through a series of matrix computations that involve derivative information as listed in Eqs. (11.2.20) and (11.2.21). In the forward sweep [Step (3)], the transition equations (or simulation models) are used to recalculate the state of the system and the value of the

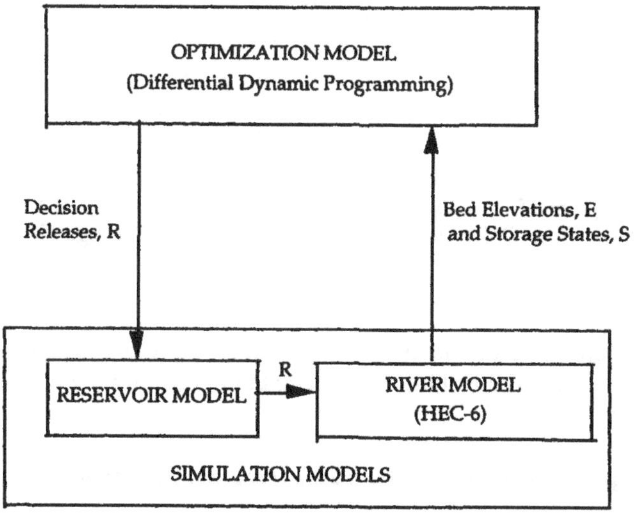

Figure 11.2 Schematic of the optimal control approach for the sedimentation problem (Carriaga, 1993).

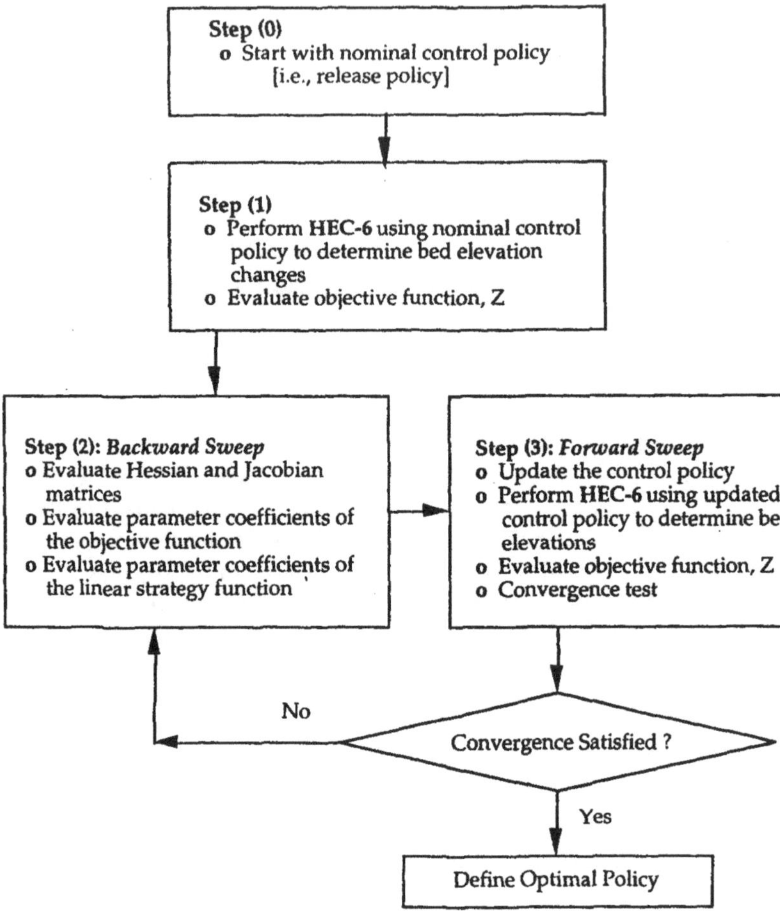

Figure 11.3 Diagram of the optimization algorithm for the sedimentation control problem. [Modified from Culver and Schoemaker (1992) and Carriaga (1993).]

objective function using the updated control policy. The two-part iteration [Steps (2) and (3)] continues until the calculated control policy converges.

11.4 EXAMPLE APPLICATION TO AGUA FRIA RIVER

The application (Carriaga, 1993; Carriaga and Mays, 1995b) was to the existing New Waddell Dam and Agua Fria River in Arizona as shown in

Figure 11.4 and whose hydraulic and numerical input parameters are given in Table 11.1. Three scenarios identified in Figure 11.5 were considered under the second application which cover different problem scenarios of the Agua Fria River. The New Waddell Dam on the Agua Fria River in Arizona are used to demonstrate the capability of the developed model (see Fig. 11.4). Table 11.1 lists the hydraulic and numerical input parameters. For this large application study, the entire 34-mile reach of the Agua Fria River has been considered. The geometry of the river model is described by 96 stations (Carriaga, 1993; Carriaga et al., 1994; Carriaga and Mays, 1995b). The river stretches from the most upstream station near the New Waddell Dam in the north to the most downstream station at the confluence with the Gila River in the south. In order to evaluate the capability of the optimal control model developed, three problem scenarios for the Agua Fria River are considered. Figure 11.5 shows the configuration sketches of these case studies. These studies evaluate the extent of (i) sedimentation for the entire river consisting of 96 cross-section stations (Case I); (ii) sedimentation at the most downstream reaches consisting of 50 cross-section stations (Case II); and (iii) sedimentation at selected stations consisting of 20 stations (Case III).

Although the analysis under Case II as shown in Figure 11.5b is only focused on the evaluation of sedimentation of the downstream reaches of the river which covers a length of 16 miles, the upstream stations are still included in the model because they provide the inputs essential for the evaluation of sediment movement and transport at the locations of interest. This scenario, however, does not consider how much sediment transport dynamics are permitted upstream as long as their impact do not have significant bearing on the sedimentation processes at the downstream reaches. This is because only the bed changes at the downstream stations are evaluated for the objective function.

Critical areas along the river that require special consideration are addressed by Case III. Because these stations are identified as critical stations due to excessive bed sedimentation (i.e., scouring or deposition), they could be taken to comprise the list of stations that the model should specifically address. These stations, however, may be identified at different segments of the river as shown in Figure 11.4.2c. Locations that are considered critical are stations at and around bridge locations, or stations where channel widths change significantly. For the application example, four bridge locations and their vicinities are considered to be analyzed by the model. This means that the vicinity around each bridge location selected is represented by five cross-section stations—two stations defining the channel geometries of the immediate upstream, two stations downstream, and one station defining the channel geometry at the bridge site.

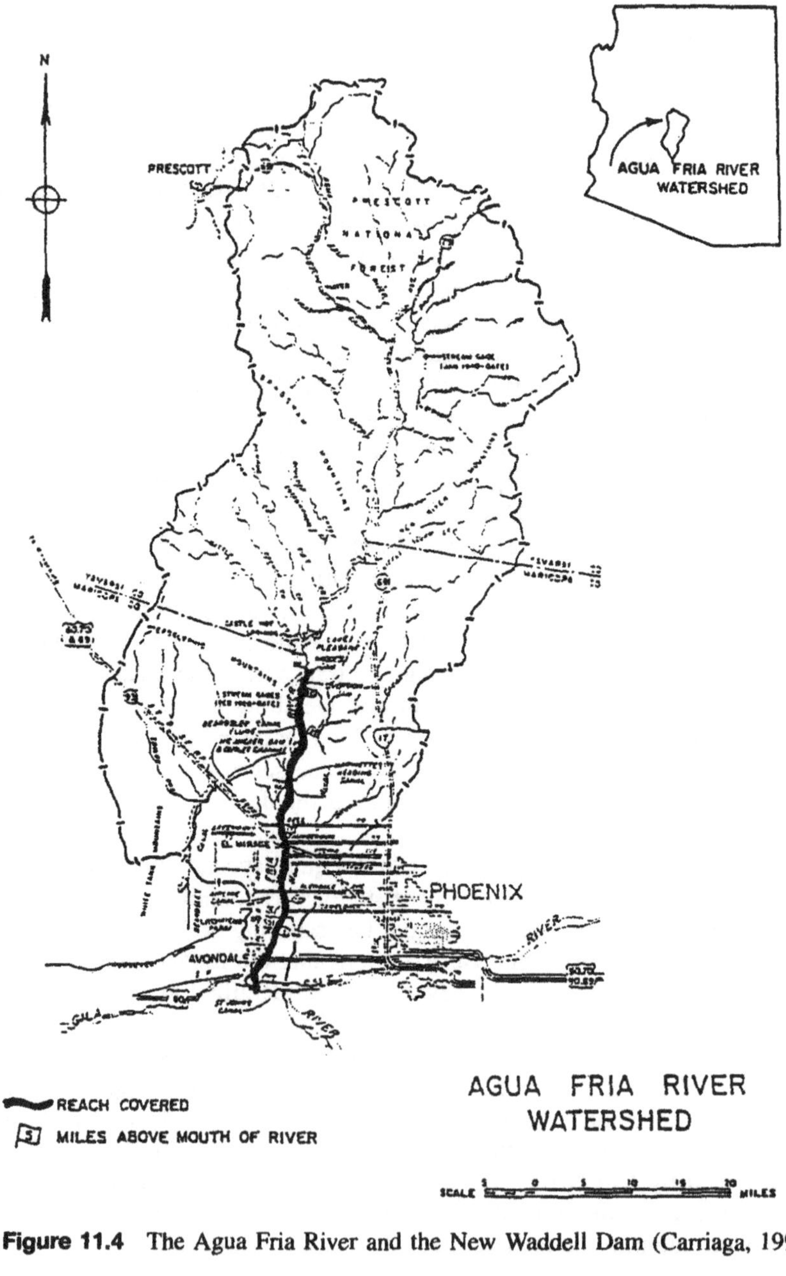

Figure 11.4 The Agua Fria River and the New Waddell Dam (Carriaga, 1993).

Table 11.1 Hydraulic and Numerical Input Parameters for the New Waddell Dam and Agua Fria River

Parameters				Value	
I. New Waddell Dam					
(a) Maximum storage capacity (flood storage) (acre-ft)				86,100	
(b) Minimum (dead) storage capacity (acre-ft)				40,500	
(c) Maximum release (cfs)				105,000	
(d) Minimum release (cfs)				20,000	
(e) Beginning-of-operation storage (acre-ft)				86,100	
(f) End-of-operation storage (acre-ft)				86,100	
(g) Seepage and evaporation losses (acre-ft)				0.0	
(h) Number of simulation time steps				5	
Time index (t)	$t = 1$	$t = 2$	$t = 3$	$t = 4$	$t = 5$
(i) Inflows, Q_t (cfs)	30,000	80,000	120,000	70,000	30,000
(j) Time duration per period (hr)				12.0	
II. Agua Fria River					
2.1 Geometric and Hydraulic Data					
(a) Number of channel cross sections					
(i) Case I—Entire river				96	
(ii) Case II—Lower reaches				50	
(iii) Case III—Selected channel sections				20	
(b) Other data				(see Carriaga, 1993)	
2.2 Sediment Data					
(a) Transport Function (MTC No. 9)				Toffaleti/Schoklitsch	
(b) Specific weight of water sediment (lb/ft³)				93	
(c) Discharge(R)–sediment load (G_s) relationship:					

No.	Discharge, R (cfs)	Sed. load, G_s (ton/day)
1	0.0	0.1
2	4,000	3,082
3	20,000	26,604
4	40,000	56,358
5	60,000	97,056
6	85,000	155,824
7	115,000	200,000

Parameters				Value	
(d) Gradation Data				(see Carriaga, 1993)	
2.3 Hydrologic Data					
Time Index (t)	$t = 1$	$t = 2$	$t = 3$	$t = 4$	$t = 5$
(a) Nominal policy, R_t	15,000	95,000	110,000	95,000	15,000
(b) Rating data[a] $\text{WSE} = 913.91 + 1.5376\times10^{-4}R - 6.6746\times10^{-10}R^2$					

[a]WSE is the water surface elevation, ft; and R is the discharge event, cfs

Conversions: 1 cfs = 0.0283 m³/sec; 1 acre-ft = 1234.4 m³; 1 lb/sec = 4.453 N/sec; 1 ft = 0.3049 m; 1 lb/ft³ = 156.966 N/m³

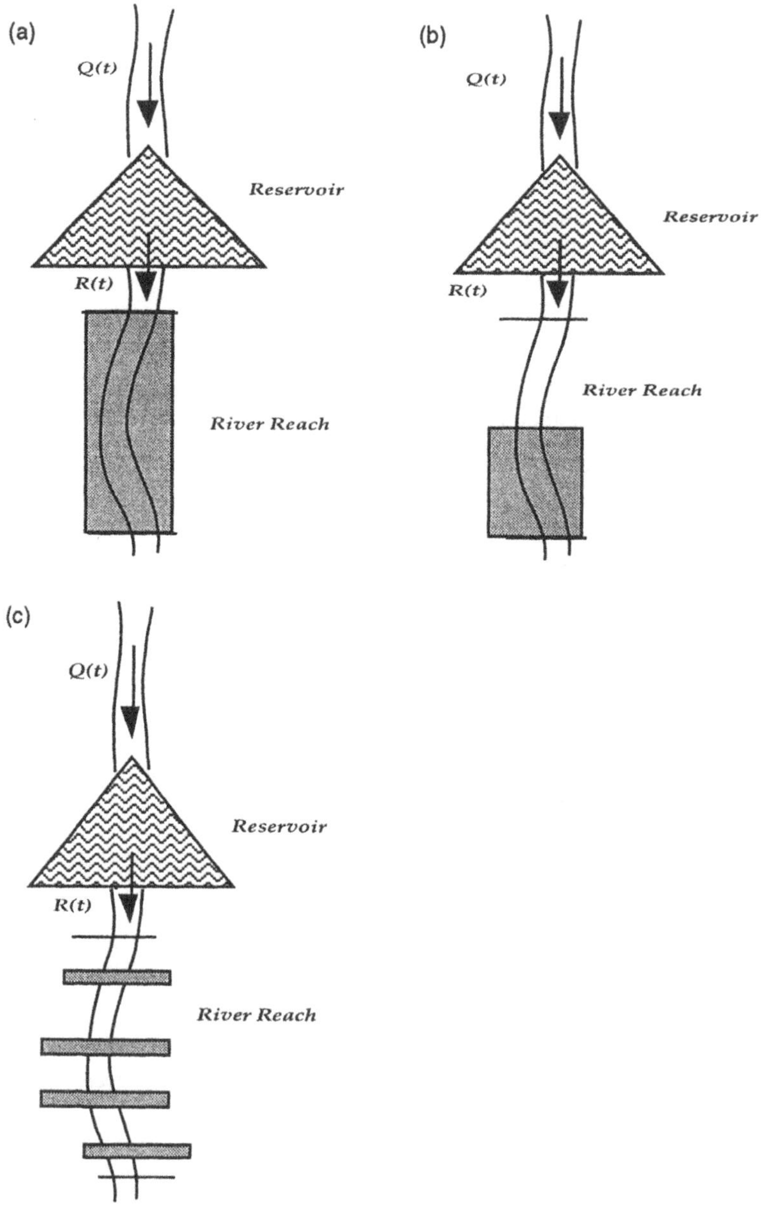

Figure 11.5 Case studies for the optimal control model. (a) Case I—entire river. (b) Case II—lower reaches. (c) Case III—selected stations (Carriaga, 1993).

Table 11.2 lists the optimum results of the five-section hypothetical model for four different sediment transport functions used in the optimization. As could be observed in columns 5, 6 and 7, the extent of sedimentation is different for different sediment transport functions. Despite this difference in sedimentation response, the optimum release policies, R_t^*, obtained for all the four sediment transport functions as listed in column 3 are almost identical.

For the large application model comprising the New Waddell Dam and Agua Fria River, Table 11.3 lists the optimum results for the three case studies considered. Common to these studies was the nominal release policy, R_t, used which was in the infeasible space violating the release bounds defined by Eq. (11.2.10). Here, the minimum and maximum releases are R_{min} = 20,000 cfs and R_{max} = 105,000 cfs, respectively. Associated with this release policy is also a violation of the state bound constraint (i.e., $S_{min} \leq S_t \leq S_{max}$, where S_{max} = 86,100 acre-ft and S_{min} = 40,500 acre-ft). The initial solution, therefore, of the DDP problem starts in the infeasible region of the solution space.

Case I considers aggradation and degradation effect in the entire Agua Fria River. The optimum release policy, R_t^*, satisfied both the control and state bounds and the target storage state at the end of operation. These results are listed in Table 11.3a.

Case II considers aggradation and degradation effects in the 50 most downstream stations in the river. Similarly, a release policy, R_t^*, like that obtained in Case I, was determined (see Table 11.3b). Although, there is close matching of the optimum release policy, R_t^*, the evaluated total aggradation and degradation processes are different. For example at stage 5 (i.e., t = 5), the total evaluated aggradation value of 25.29 ft in Case II is much larger than the aggradation value of 17.45 ft obtained in Case I. Also, at stage 4 (i.e., t = 4), the total evaluated degradation value of 27.15 ft in Case II is almost twice as much as the evaluated degradation value of 14.59 ft obtained in Case I. Furthermore, at stage 1 (i.e., t = 1), the degradation value of 17.395 ft in Case III is slightly larger than the degradation value of 17.23 ft in Case II, although Case III only considered 20 stations compared to 50 stations for Case II. In all these instances, the evaluated degradation and aggradation values in Case I should provide higher values of sedimentation than the other cases by virtue of the larger coverage considered in Case I.

Computationally, the analysis made for the hypothetical reservoir–river system for the four different transport functions have provided an identical number of iterations as listed in Table 11.4a. Except for the analysis made using the Toffaleti and Schoklitsch formula, the other three analyses spent almost similar CPU times. For the large model, significant differences are

Table 11.2 Optimum Results from Four Sediment Transport Functions

	Inflow data	Reservoir operation optimum values		River sediment hydraulics		
	Q_t	R_t^*	S_t^*	Cumulative aggradation	Cumulative degradation	Total
t	(cfs)	(cfs)	(acre-ft)	(ft)	(ft)	(ft)
(1)	(2)	(3)	(4)	(5)	(6)	(7)
			(a) Tofaletti Formula			
1	500	2468.8	2047.5	1.338	0.039	1.377
2	5000	3031.9	4000.0	3.704	0.000	3.704
3	3000	3744.7	3260.8	3.836	0.000	3.836
4	1000	324.5	3930.6	0.080	1.624	1.704
5	200	127.7	4000.0	0.249	1.307	1.556
Cumulative				9.207	2.970	12.177
			(b) Yang's Streampower Function			
1	500	2468.9	2047.3	0.000	1.751	1.751
2	5000	3031.7	4000.0	0.039	0.003	0.042
3	3000	3744.6	3260.8	0.043	0.002	0.045
4	1000	324.6	3930.7	0.027	0.003	0.030
5	200	127.7	4000.0	0.002	0.017	0.019
Cumulative				0.111	1.776	1.887
			(c) Ackers and White Formula			
1	500	2469.0	2047.3	0.370	2.453	2.823
2	5000	3031.7	4000.0	0.116	2.692	2.808
3	3000	3744.5	3261.0	0.117	3.413	3.530
4	1000	324.7	3930.7	0.000	0.914	0.914
5	200	127.7	4000.0	0.000	0.307	0.307
Cumulative				0.603	9.779	10.382
			(d) Tofaletti/Schoklitsch Formula			
1	500	2469.6	2046.7	1.824	6.022	7.846
2	5000	3031.1	3999.3	1.476	8.563	10.039
3	3000	3743.7	3261.8	0.008	5.766	5.774
4	1000	323.5	3932.7	0.000	4.071	4.071
5	200	129.7	4000.0	0.032	0.614	0.646
Cumulative				3.240	25.036	28.376

Note: Conversions: 1 cfs = 0.0283 m^3/sec 1 acre-ft = 1234.4 m^3
1 lb/sec = 4.453 N/sec 1 ft = 0.3049 m

Table 11.3 Optimum Results for the New Waddell Dam and Agua Fria River System

	Reservoir Operation					River Sediment Hydraulics		
	Inflow	Initial values		Optimum		Aggradation	Degradation	Total
t	Q_t	R_t	S_t	R_t^*	S_t^*	a_t	b_t	Z_t
	(cfs)	(cfs)	(ac-ft)	(cfs)	(ac-ft)	(ft)	(ft)	(ft)
(1)	(2)	(3)	(4)	(5)	(6)	(7)	(8)	(9)
Case I—Sedimentation Study for the Entire River (96 Stations)								
1	30,000	15,000	100,976	35,456	80,689	22.89	43.32	66.21
2	80,000	95,000	86,100	105,000	55,896	24.63	53.46	78.09
3	120,000	110,000	96,017	89,543	86,100	14.45	42.97	57.42
4	70,000	95,000	71,224	79,567	76,613	32.61	14.59	47.20
5	30,000	15,000	86,100	20,433	86,100	17.45	16.77	34.22
Cumulative						112.03	171.11	283.14
Case II—Sedimentation Study for the Lower Reaches (50 Stations)								
1	30,000	15,000	100,976	35,865	80,283	16.82	17.23	34.05
2	80,000	95,000	86,100	105,000	55,490	16.32	20.24	36.56
3	120,000	110,000	96,017	89,142	86,092	11.56	16.52	28.08
4	70,000	95,000	71,224	79,538	76,634	18.79	27.15	45.94
5	30,000	15,000	86,100	20,451	86,100	25.29	8.06	33.35
Cumulative						88.78	89.20	177.98
Case III—Sedimentation Study for Selected Stations (20 Stations)								
1	30,000	15,000	100,976	35,473	80,672	5.466	17.395	22.861
2	80,000	95,000	86,100	104,998	55,880	5.739	14.232	19.971
3	120,000	110,000	96,017	89,529	86,100	4.889	2.818	7.707
4	70,000	95,000	71,224	79,563	76,616	10.637	3.183	13.820
5	30,000	15,000	86,100	20,432	86,100	4.009	3.026	7.035
Cumulative						30.740	40.654	71.394

Note: Conversions: 1 cfs = 0.0283 m^3/sec 1 acre-ft = 1234.4 m^3
1 lb/sec = 4.453 N/sec 1 ft = 0.3049 m

observed on the CPU time used and the number of iterations involved for the three case studies (see Table 11.4b). Case III which required the largest number of iterations (265) is the result of an unstable or dynamic riverbed. Although Case III have only considered 20 stations in the evaluation of the aggradation and degradation, these stations are among the critical stations since they include bridge site locations.

Table 11.4 CPU Time and Number of Iterations for the Optimal Control Studies

MTC No.	Sediment transport function	Number of stations	CPU time (sec)	No. of iterations
	(a) Hypothetical Reservoir–River System			
1	Toffaleti formula	5	788.58	130
4	Yang's streampower	5	770.43	130
7	Ackers and White formula	5	780.67	130
9	Tofaletti and Schoklitsch	5	856.75	130
	(b) New Waddell Dam–Agua Fria River			
9	Toffaleti and Schoklitsch	96[a]	9865.34	255
9	Toffaleti and Schoklitsch	50[b]	8513.87	190
9	Toffaleti and Schoklitsch	20[c]	—	165

[a]Case I.
[b]Case II.
[c]Case III.

APPENDIX: HYPERBOLIC PENALTY FUNCTION FOR THE VIOLATION OF STORAGE CONSTRAINT

The hyperbolic penalty function $Y_t(\mathbf{x}_t, \mathbf{u}_t)$ associated with the violation of the storage bound constraint could be defined as follows: Let the violation of the storage state bound at time t be generally expressed as $H_t(S_t)$ which could be defined either as

$$H_t(S_t) = S_{min} - S_t, \quad t = 1, 2, 3, \ldots, T \qquad (11.A.1)$$

for the violation of the minimum storage, S_{min}, or

$$H_t(S_t) = S_t - S_{max}, \quad t = 1, 2, 3, \ldots, T \qquad (11.A.2)$$

for the violation of the maximum storage, S_{max}. [Note that violation $H_t(S_t)$ is a scalar expression indicating one reservoir.] For either case, the hyperbolic function is expressed as

$$y_t = y = \sqrt{p^2H_t^2(S_t) + w^2} - p\,H_t(S_t), \qquad t = 1, 2, 3, \ldots T \quad (11.A.3)$$

For purposes of illustration, the function y_t and its first and second derivatives (i.e., y_t' and y_t'') could be evaluated by using any of the above violation, $H_t(S_t)$, say, Eq. (11.A.1) (i.e., violation of the minimum storage, S_{min}). Thus, the function in Eq. (11.A.3) could be rewritten as

$$y_t = y = \sqrt{p^2(S_{\min} - S_t)^2 + w^2} - p(S_{min} - S_t), \quad t = 1, 2, 3, \ldots, T \quad (11.A.4)$$

so that the penalty function defined as $Y_t(x_t, u_t)$ could be expressed as

$$Y_t(S_t) = y_t \quad \text{if } y_t \le 1 \text{ for } t = 1, 2, 3, \ldots, T \quad (11.A.5)$$

or

$$Y_t(S_t) = \frac{y_t^n + 4n\,(n - 1)\sqrt{y_t} - (n - 1)(2n - 1)}{n\,(2n - 1)} \quad \text{if } y_t \ge 1 \text{ for } t = 1, 2, 3, \ldots, T \quad (11.A.6)$$

Having defined the functions, the first and second derivatives of Eq. (11.A.4) are expressed as

$$\frac{dy_t}{dS_t} = \frac{p^2(S_{\min} - S_t)}{\sqrt{p^2(S_{\min} - S_t)^2 + w^2}} - p, \quad t = 1, 2, 3, \ldots, T \quad (11.A.7)$$

$$\frac{d^2y_t}{dS_t^2} = \frac{p^2w^2}{(p^2(S_{\min} - S_t)^2 + w^2)^{3/2}}, \quad t = 1, 2, 3, \ldots, T \quad (11.A.8)$$

Using the relations in Eqs. (11.A.7) and (11.A.8), the derivatives of the hyperbolic function $Y_t(\mathbf{x}_t, \mathbf{u}_t)$ are as follows:

For the case where $y_t \le 1$,

$$\frac{dY_t}{dS_t} = \frac{dy_t}{dS_t}, \quad t = 1, 2, 3, \ldots, T \quad (11.A.9)$$

and

$$\frac{d^2Y_t}{dS_t^2} = \frac{d^2y_t}{dS_t^2}, \quad t = 1, 2, 3, \ldots, T \quad (11.A.10)$$

For the case where $y_t \ge 1$,

$$\frac{dY_t}{dS_t} = \left(nAy_t^{n-1} + \frac{0.5B}{\sqrt{y_t}}\right)\left(\frac{dy_t}{dS_t}\right), \quad t = 1, 2, 3, \ldots, T \quad (11.A.11)$$

$$\frac{d^2Y_t}{dS_t^2} = \left(nAy_t^{n-1} + \frac{0.5B}{\sqrt{y_t}}\right)\left(\frac{d^2y_t}{dS_t^2}\right) + \left(n(n - 1)Ay_t^{n-2} - \frac{0.25B}{y_t^{3/2}}\right)\left(\frac{dy_t}{dS_t}\right)^2$$

$$\text{for } t = 1, 2, 3, \ldots, T \qquad (11.A.12)$$

where A and B are coefficients which could be evaluated from Eqs. (11.2.20) and (11.2.21), where $m = 0.5$ and $y_0 = 1$.

A similar procedure could be followed for the violation of the maximum storage, S_{max}, or for the bound constraint violations for the releases, R_t. For the equality constraint defined by the boundary condition in Eq. (11.2.12), a similar procedure could be used except that the function y_t is expressed in the form

$$y_T = \sqrt{p^2(S_T - S_{max})^2 + w^2} \qquad (11.A.13)$$

REFERENCES

American Society of Civil Engineers, *Sedimentation Engineering: Manual and Reports on Engineering Practice*, V. A. Vanoni (ed.), ASCE, New York, 1977.

Brooke, A., Kendrick, D., and Meeraus, A., *GAMS: A User's Guide*, The Scientific Press, South San Francisco, California, 1988.

Carriaga, C. C., A Model for Determining Optimal Reservoir Releases to Minimize Aggradation and Degradation in Alluvial Rivers, Ph.D. Dissertation, Civil Engineering Department, Arizona State University, Tempe, 1993.

Carriaga, C. C., and Mays, L. W., Optimization Modeling for Sedimentation in Alluvial Rivers, *Journal of the Water Resources, Planning and Management*, Vol. 121, No. 3, pp. 251–259, 1995a.

Carriaga, C. C. and Mays, L. W., Optimal Control Approach for Sedimentation Control in Alluvial Rivers, *Journal of Water Resources Planning and Management Division*, Vol. 121, No. 5, pp. 408–417, 1995b.

Carriaga, C. C., Mays, L. W., and Ruff, P. F., Agua Fria River Sediment Transport Study, Final Report, to the Flood Control District of Maricopa County, Phoenix, Arizona, 1994.

Culver, T. B., and Shoemaker, C. A., Dynamic Optimal Control for Groundwater Remediation with Flexible Management Periods, *Water Resources Research*, Vol. 28, No. 3, pp. 629–641, 1992.

Culver, T. B. and Shoemaker, C. A., Optimal Control for Groundwater Remediation by Differential Dynamic Programming with Quasi-Newton Approximations, *Water Resources Research*, Vol. 29, No. 4, pp. 823–831, 1993.

Lin, T-W., Well-Behaved Penalty Functions for Constrained Optimization, *Journal of the Chinese Institute of Engineers*, Vol. 13, No. 2, pp. 157–166, 1990.

Mayne, D. A., A Second-Order Gradient Method for Determining Optimal Trajectories of Non-Linear Discrete-Time Systems, *International Journal on Control*, Vol. 3, pp. 85–95, 1966.

Simons, Li & Associates, Inc., *Engineering Analysis of Fluvial Systems*, Fort Collins, CO, 1982.

U.S. Army Corps of Engineers, *HEC-6: Scour and Deposition in Rivers and Reservoirs, User's Manual*, Hydrologic Engineering Center, Davis, CA, 1977 Version, 1977.

U.S. Army Corps of Engineers, *HEC-6: Scour and Deposition in Rivers and Reservoirs, User's Manual, CPD-6*, Hydrologic Engineering Center, Davis, CA, June 1991 Version, 1991.

Yakowitz, S. and Rutherford, B., Computational Aspects of Discrete-Time Optimal Control, *Applied Mathematics and Computation*, Vol. 15, pp. 29–45, 1984.

Chapter 12
Optimal Operation of Soil Aquifer Treatment Systems Using SALQR

12.1 PROBLEM IDENTIFICATION

Soil aquifer treatment (SAT) is an economical method used for groundwater recharge, in which surface ponded reclaimed wastewater percolates through the vadose zone and water quality improves as a series of biological and physical interactions occur. Generally, recharge facilities are operated (loading schedules) in two ways: wet–dry cycles or continuous operations (Cournoyer and Kriege, 1988; U.S. Environmental Protection Agency, 1992). Of the recharge facilities reported (U.S. Environmental Protection Agency, 1992), over 70% of the facilities reported wet–dry cycle operation. Wet–dry cycle operations consist of filling the pond to a certain depth, stopping the inflow (loading), and allowing the water to infiltrate into the ground. After all the water has infiltrated into the soil, the pond is left to dry for a period so that natural aeration can take place and the pond can reach an aerobic state. During the drying period, water percolates which increases the infiltration potential for the next application period. When clogging of the recharge basin occurs, it can be cleaned and possibly restored to its original capacity by draining, drying, and scraping. Another method of wet–dry cycle operation maintains a full pond and the influent water is maintained at a rate equal to the recharge rate. When the recharge rate reaches an unacceptable value, the operation is stopped so that the clogging layer can be removed.

These typical operation procedures are done on an ad hoc basis relying on limited judgment of the design engineer and/or operator. The U.S. Environmental Protection Agency (1992) reported that the facility in Whittier, California, fills for 7 days (4-ft. depth), then dries for 7 days. The facility in St. Croix, Virgin Islands fills for 18 days, then dries for 30 days. The facility in Hemet, California is filled for 1 day (2.5-ft. depth), drained for 2 days, and then dried for 1 day. No fundamental explanation for the wide variation of wet–dry cycles has been presented. The design and operation of SAT facilities rely primarily on judgment and experience. In this chapter, the successive approximation linear quadratic regulator (SALQR) method is used for determining the optimal operation of SAT systems, thus facilitating preliminary engineering analyses and reducing the cost of pilot studies. The methodology is based on mathematically describing the problem as a discrete-time, optimal control problem. Mathematically, the problem statement is to maximize the infiltration subject to constraints describing the following:

- Water content distribution
- Infiltration process
- Draining process
- SAT operation

The resulting mathematical problem is a large-scale, nonlinear programming problem formulated as a discrete-time, optimal control problem. Solution of the problem is accomplished by interfacing a nonlinear programming optimizer and a simulator together.

The one-dimensional, finite element, simulation model HYDRUS (Kool and van Genuchten, 1991) has been modified to simulate the water content distribution, the infiltration process, and the draining process. An optimal control approach has been developed by using a SALQR, which is interfaced with the simulator, a modified HYDRUS. The SALQR algorithm is similar to that used by Chang et al. (1992) and described in Chapter 9 and applied in Chapters 10 and 11. A computer model SATOM (SAT Operation Model) has been developed for application of the methodology for optimal SAT operation.

12.2 PROBLEM FORMULATION

Few techniques have been developed to assist in the detailed design or operation of SAT systems. Investments have been made in experimental site-specific analyses of the infiltration process. These efforts have resulted in several empirically and physically based infiltration models. However, the extension to design and operation of an SAT system using these models has not been made.

Mushtaq et al. (1994) developed the only previous optimization model for the optimal operation of infiltration systems. The objective of this nonlinear programming problem was to maximize infiltration subject to the following:

- Continuity equations for basins

  Mass balance equations during application time
  Mass balance equations during wetting time

- Soil moisture redistribution equation

  Subsurface flow equation (kinematic wave)

- System operation equations

  Operation time constraint Cycle time constraint
  Ponding water depth constraint

This nonlinear optimization problem was solved using a generalized reduced gradient approach. The model developed by Mushtaq et al. (1994) simplifies the treatment of the unsaturated flow mechanics and clogging effects. In this chapter, the simulator used is a modified HYDRUS which describes the water content distribution, the infiltration process, and the draining process. The clogging effect is also modeled through the affected hydraulic conductivity in the modified HYDRUS. The water quality aspect is not considered in this chapter, whereas the methodology demonstrates the potential of the SALQR to interface a water quality model that has been developed (Tang et al., 1995).

The operation model is used to determine optimal values of the control variables (independent variable): the application time (X) and the drying time (Z) in order to maximize the infiltration. The simulator determines the state of the SAT system for these decisions. The state variable (dependent variable) is the average water content (ω). The draining time (Y) and infiltration volume (F) are functions of the control variables (X and Z) and the state variable (ω). The infiltration rate and the ponding depth versus the operation schedule are shown conceptually in Figure 12.1. An operation period CT (cycle time) is divided into three periods: the application time X, the draining time Y, and the drying time Z.

A mathematical formulation of the overall optimization model for one pond is stated as follows:

$$\text{Maximize } V = \sum_{t=1}^{N} F_t(\omega_t, X_t, Z_t, t) \tag{12.2.1}$$

subject to the following constraints:

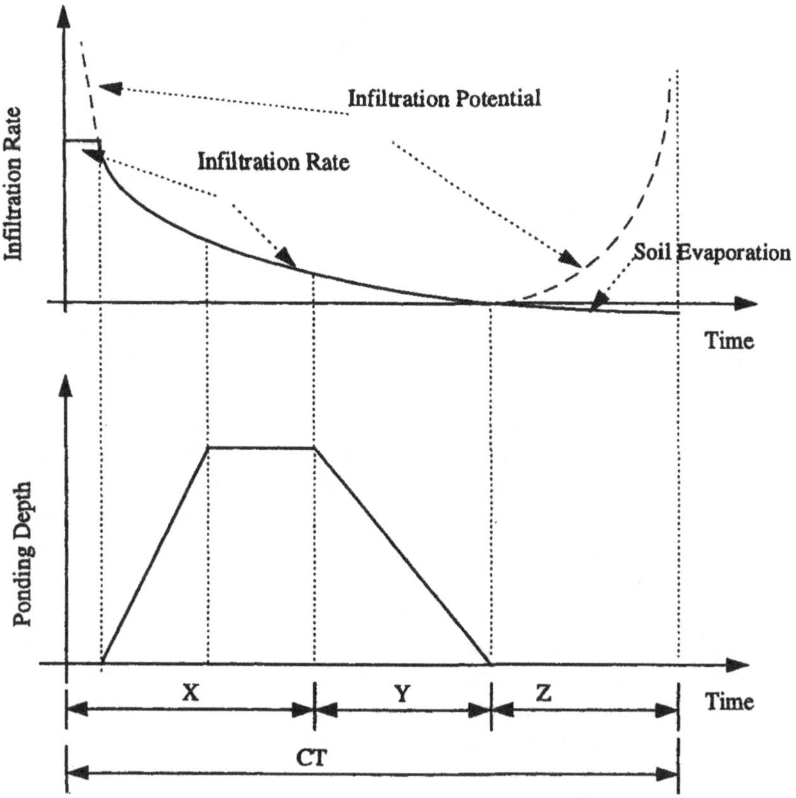

Figure 12.1 Infiltration rate and ponding depth versus operation schedule.

1. Simulator equations to describe the water content distribution, the infiltration process, and the draining process are, respectively,

$$\omega_{t+1} = T_1(\omega_t, X_t, Z_t, t) \quad (12.2.2)$$

$$F_t = T_2(\omega_t, X_t, Z_t, t) \quad (12.2.3)$$

$$Y_t = T_3(\omega_t, X_t, t) \quad (12.2.4)$$

2. The infiltration potential recovery constraint

$$\omega_{t+1} \le \omega_{\max} \quad (12.2.5)$$

3. The operation time constraint

$$X_t + Y_t + Z_t \le \mathrm{CT}_{\max} \quad (12.2.6)$$

where N is the number of cycles, ω_{max}, and CT_{max} is the upper bounds for ω_{t+1} and cycle time, and V is the total infiltration volume during N cycles.

The objective function, Eq. (12.2.1), maximizes the infiltration volume over N cycles. Equation (12.2.2) is the transition equation which relates the average water content ω_{t+1} to the average water content ω_t, given the application time X_t and the drying time Z_t. Equations (12.2.3) and (12.2.4) represent the relations of the infiltration volume and the draining time with respect to the water content, the application time, and the drying time. Equations (12.2.2)–(12.2.4) are solved by the modified HYDRUS. Constraints (12.2.5) and (12.2.6) are the bound constraints on the average water content and cycle time. Equations (12.2.1)–(12.2.6) constitute an optimization model for SAT system operation with the structure of a discrete-time optimal control problem which is solved by the optimal control algorithm, SALQR. Solution of this model will yield the optimal application time X and drying time Z which maximize the total infiltration volume over N cycles. The maximum cycle time CT_{max} and the maximum water content ω_{max} are assumed present.

The SALQR algorithm is a modification to differential dynamic programming [DDP, developed by Jacobson and Mayne (1970)]. With SALQR, nonlinear dynamics are addressed by linearizing the simulation dynamics in the optimization stage and then updating the state vector through the full nonlinear model. The SALQR algorithm is presented in Chapter 9. An important advantage of the SALQR method is that it does not require the calculation of the second derivatives of the transition equation with respect to the state and control variables, thereby significantly reducing the necessary computational effort (Li, 1994).

12.3 PROBLEM SOLUTION

12.3.1 Interface of the SALQR with the Modified HYDRUS

In the optimization model, ω_t, Y_t, and F_t are computed by the simulator, and X_t and Z_t are determined by the SALQR. In other words, $\omega_{t+1}(\omega_t, X_t, Z_t, t)$, $Y(\omega_t, X_t, t)$ and $F(\omega_t, X_t, Z_t, t)$ are implicit to the optimizer and are solved by the simulator. As a result, satisfaction of the bound constraints on ω_t and the operation time is forced through the use of a penalty function incorporated in the objective function. This results in the following reduced form of the original problem:

$$\text{Maximize } \hat{V} = \sum_{t=1}^{N} G_t(\omega_t, X_t, Z_t, t) \tag{12.3.1}$$

subject to

$$\omega_{t+1} = T_1(\omega_t, X_t, Z_t, t) \tag{12.3.2}$$

$$F_t = T_2(\omega_t, X_t, Z_t, t) \tag{12.3.3}$$

$$Y_t = T_3(\omega_t, X_t, t) \tag{12.3.4}$$

where $\hat{V}$ is the augmented objective function value, and

$$G_t(\omega_t, X_t, Z_t, t) = F_t(\omega_t, X_t, Z_t, t) - R_1[\min(0, \omega_{\max} - \omega_{t+1})]^2 - R_2[\min(0, \mathrm{CT}_{\max} - (X_t + Y_t + Z_t))]^2 \tag{12.3.5}$$

where R_1 and R_2 are the penalty parameters. The updating formula used for R_1 and R_2 are

$$R_1^{k+1} = \Delta R_1 R_1^k \tag{12.3.6}$$

$$R_2^{k+1} = \Delta R_2 R_2^k \tag{12.3.7}$$

where k is the outer loop number and ΔR_1 and ΔR_2 are determined by the following:

$$\begin{aligned} \Delta R_1 &= 1 \quad \text{if } \max_t |\omega_{\max} - \omega_{t+1}| \leq \varepsilon_\omega \\ \Delta R_1 &= 10 \quad \text{if } \max_t |\omega_{\max} - \omega_{t+1}| > \varepsilon_\omega \end{aligned} \tag{12.3.8}$$

$$\begin{aligned} \Delta R_2 &= 1 \quad \text{if } \max_t |\mathrm{CT}_{\max} - (X_t + Y_t + Z_t)| \leq \varepsilon_{\mathrm{CT}} \\ \Delta R_2 &= 10 \quad \text{if } \max_t |\mathrm{CT}_{\max} - (X_t + Y_t + Z_t)| > \varepsilon_{\mathrm{CT}} \end{aligned} \tag{12.3.9}$$

where ε_ω and $\varepsilon_{\mathrm{CT}}$ are the criteria for checking the bound violations on the average water content and the cycle time. The reduced problem (12.3.1)–(12.3.4) is solved using the SALQR and the simulator to solve the constraints (12.3.2)–(12.3.4). The overall optimization is attained when both the maximum bound violations of the constraints and the maximum differences of the control variables between iteration J and iteration $J - 1$ reach the prespecified small values. The relation of the optimizer and the simulator is shown conceptually in Figure 12.2, which illustrates that the control variables X and Z are determined by the optimizer (SALQR) and the state variable ω, and variables Y and F are attained by the simulator (the modified HYDRUS).

The modified HYDRUS model is based on a partial differential equation (one-dimensional Richard's equation). The control variables (the ap-

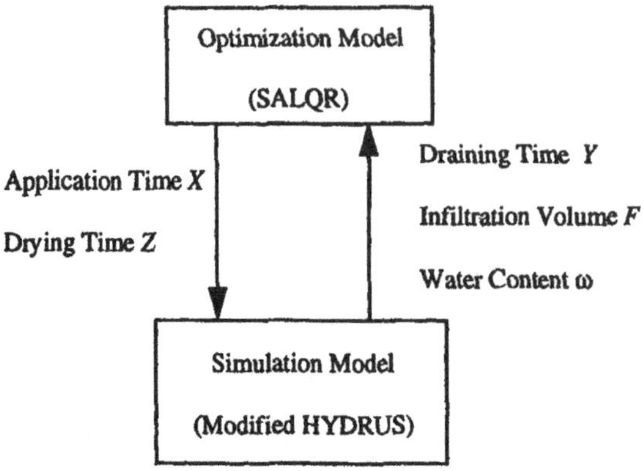

Figure 12.2 Relation of optimizer and simulator.

plication time X and the drying time Z) are boundary conditions. The derivatives of the transition equation with respect to the control variables and the state variable, and the derivatives of the infiltration volume and the draining time with respect to the control variables and state variable cannot be calculated analytically. The finite-difference approximation method was used to calculate these derivatives. More specifically, these derivatives are computed by a perturbation of X, Z, and ω and running the modified HYDRUS repeatedly. The calculations of these derivatives are listed in Appendix 12.A. The overall solution procedure is further illustrated through the flowchart in Figure 12.3. There are two loops in this procedure, with the outer loop determining the penalty parameters. The inner loop solves the reduced problem using the optimizer (SALQR), with the penalty parameters fixed at the values determined by the outer loop. Once an inner loop is completed, the convergence criteria check the maximum bound violations and the maximum difference of control variable between iteration J and iteration $J - 1$. If they are small enough, the procedure terminates; otherwise, the procedure returns to the outer loop and updates the penalty parameter, and then goes to the inner loop and solves the new problem with the updated penalty parameters from the outer loop. This procedure continues until an optimal solution is found. Obviously, there is no guarantee of convergence to a global optimal solution. However, with the use of initial trial solutions and engineering judgment, an optimal solution can be found.

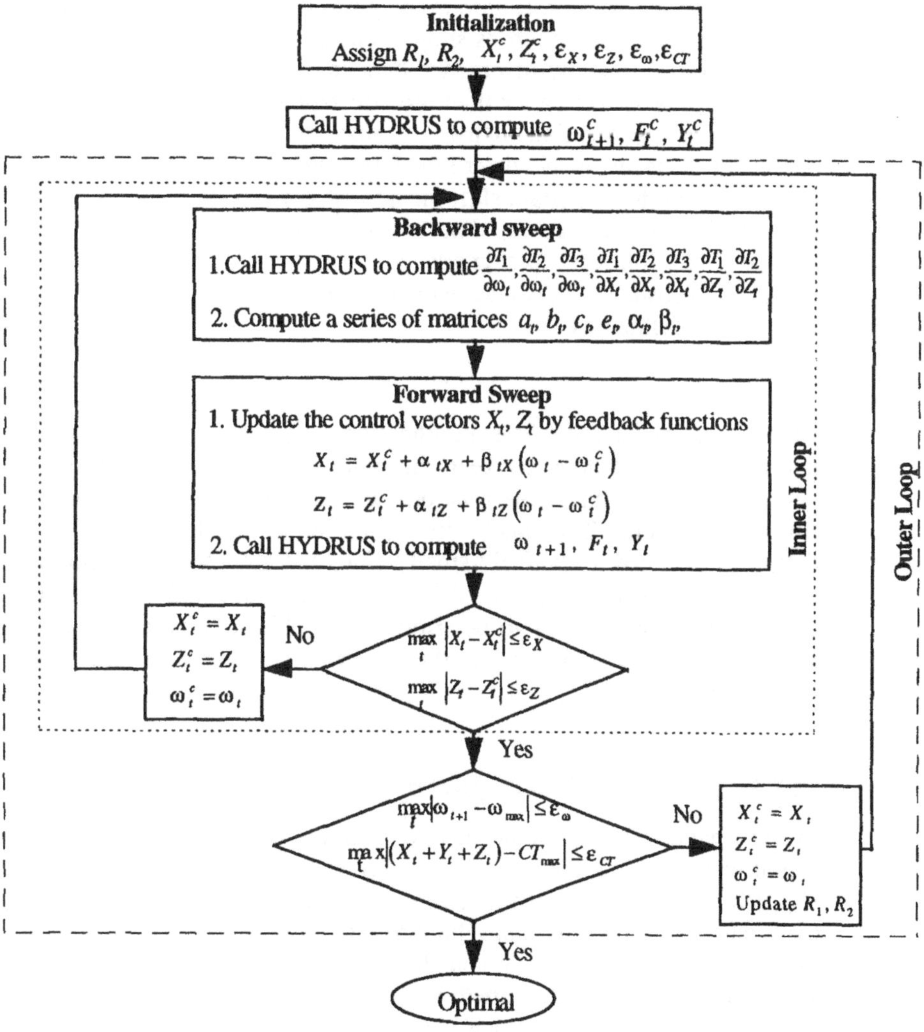

Figure 12.3 Flowchart of the SALQR interfaced with HYDRUS.

12.3.2 Convergence Procedure

The SALQR method requires the Hessian matrices C_t calculated in the algorithm to be positive definite, otherwise the SALQR method may not be convergent. Yakowitz and Rutherford (1984) suggested using a constant shift procedure to make C_t positive definite. Liao and Shoemaker (1991) used an adaptive shift procedure in the groundwater management problem. The constant shift procedure is simple, but the convergence speed of using this method is slow and divergence happens in some cases. The adaptive shift procedure works better than the constant shift procedure, but the eigenvalues of the Hessian matrices c_t need to be calculated, and a new parameter δ needs to be determined by numerical experiments. In this chapter, a modified constant shift method is proposed to obtain the positive definite Hessian matrices c_t. The modified constant shift method is described in the following paragraph.

Pick a constant λ (λ is independent of t) such that c_v is positive definite where c_v is defined as

$$c_v = c_t + \lambda I_m \tag{12.3.10}$$

where I_m is the identity. The modification to the constant shift method is that the constant λ does not keep the same value during the iteration process. The λ value has some relations with the step size calculated in the SALQR algorithm. The smaller the λ, the larger the step size. Therefore, first, λ is assigned a relatively small value, and after some iterations, λ is increased to a relatively large value. A series of numerical experiments are conducted in this chapter to determine a suitable λ value and to analyze the effect of the λ value on the convergence speed of the SALQR procedure.

12.3.3 Operation of SAT Systems

In this chapter, the operation of SAT systems is described by an optimization method. First, the pond is filled by water with an initial loading rate Q_0. When the ponding depth reaches the maximum ponding depth $D_{\max}$, the loading rate is set to equal the infiltration rate to maintain the constant maximum ponding depth. When the application time equals the optimal application time X determined by the optimizer, water application is stopped. The water in the pond is drained into the soil. The draining time Y is determined by the simulator. After Y days, all the water has infiltrated into the soil. The optimal drying time Z is determined by the optimizer. After Z days of drying, the bottom of the pond reaches an aerobic state and then the next cycle begins.

A computer model SATOM have been developed to interface the SALQR procedure with the modified HYDRUS to solve the SAT problem. The SATOM computer model is a very useful tool for engineers and agency to perform preliminary analysis on SAT systems to determine the optimal operations (loading schedule).

12.4 APPLICATION EXAMPLES

The SATOM model application focuses on three key issues: (1) to test different shift parameters λ; (2) to reduce the number of calls to the modified HYDRUS; and (3) to explore the global optimum issue. All of the application runs of the model were performed on an IBM RS/6000 computer. The hydraulic and numerical parameters for all of the test cases are listed in Table 12.1.

The total infiltration and the operation schedule for the example problem are listed in Table 12.2. The results indicate that the SALQR method converges very fast, with only seven iterations necessary for the example problem. The average infiltration rate is 3.84 cm/day. When the application time equals 2.5 days, the ponding depth reaches the maximum ponding depth D_{max} and then the influent water is maintained at a rate equal to the infiltration rate to keep a full pond. When the application time is about 6 days, the operation is stopped to allow the water to infiltrate the ground. After about 6 days of draining, the pond is left to dry for about 13 days. From cycle 2 to cycle 10, the operation schedules are the same. This indicates that the hydraulic condition is stable after the second cycle. The

Table 12.1 Hydraulic and Numerical Parameters Used in the Example Problem

Parameter	Symbol	Quantity	Unit
Initial loading rate	Q_0	30	cm/day
Initial water content	ω_0	0.1	—
Upper bound of water content	ω_{max}	0.2	—
Upper bound of cycle time	CT_{max}	25	days
Maximum ponding depth	D_{max}	30	cm
Saturated conductivity	K_s	25	cm/day
Soil column depth	D_{col}	1.85	m
Coefficient on clogging substance	α_c	0.001	—
Concentration of suspended solids	c_s	20	mg/L
Concentration of algae	c_a	20	mg/L
Percent of suspended solids being intercepted	δ_s	0.9	—
Percent of algae being intercepted	δ_a	0.99	—
Number of cycles	—	10	—
Criteria for checking water content and cycle	ε_ω	0.001	cm
time bound violation	ε_{CT}	0.01	days
Criteria for checking inner optimal	ε_X	0.001	days
	ε_Z	0.001	days

Table 12.2 Optimal Solutions for Example Problem

Cycle	Application time X (days)	Draining time Y (days)	Drying time Z (days)	Cycle time CT (days)	Water content ω
1	5.87	6.47	12.65	24.99	0.2001
2	5.86	6.44	12.68	24.99	0.2000
.	.	.	.	.	.
.	.	.	.	.	.
.	.	.	.	.	.
10	5.86	6.44	12.68	24.99	0.2000

Application time for constant loading = 2.45 (days).
Total infiltration volume = 959 (cm^3/cm^2).
Average infiltration rate = 3.84 (cm/day).
Total operation time = 250 (days).
Number of iterations = 7.
Number of calls to HYDRUS = 126.
CPU time = 654 (seconds).

changes of the loading rate, the infiltration rate, and the ponding depth with respect to the operation schedule for one operation cycle are presented in Figure 12.4.

12.4.1 Effects of Shift Parameters

Different λ values were employed to analyze the effects of λ on the convergence speed of the optimizer. The results for different λ are summarized in Table 12.3. As indicated in Table 12.3, the optimizer does not converge with $\lambda = 0.1$ (after five iterations, $\lambda = 1$). When $\lambda = 1000$ (after five iterations, $\lambda = 10{,}000$), the optimizer converges, but the convergence speed is very slow. After $\lambda \geq 10$, as λ increases, both the number of iterations and the CPU time increase, which indicates that the convergence speed of the optimizer decreases as the λ value increases. Therefore, λ is a very important parameter needed to be determined by numerical experiments. For the example problem, $\lambda = 10$ (after five iterations, $\lambda = 100$) is used and the number of iterations is only seven. This indicates that the modified constant shift procedure used in conjunction with the SALQR is very efficient.

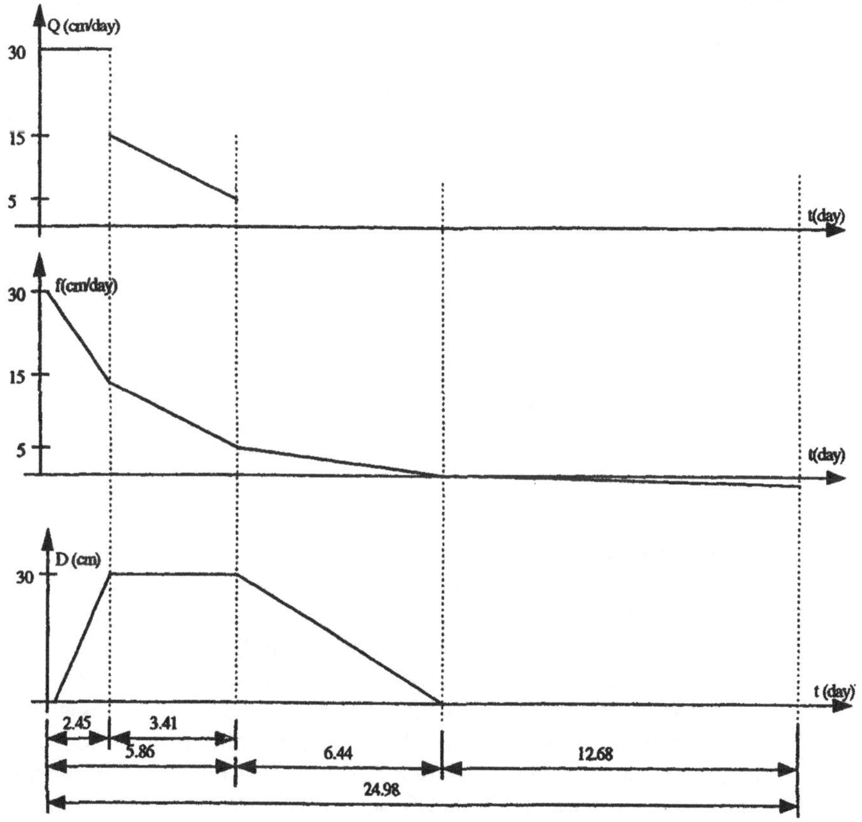

Figure 12.4 Loading rate, infiltration rate, and ponding depth versus operation schedule.

12.4.2 Method to Reduce the Number of Calls to Simulator

To reduce the number of calls to the modified HYDRUS and save CPU time, the transition derivatives and the derivatives of the infiltration volume, and the draining time with respect to the control variables and the state variable are calculated according to a simplified method. The main idea of this simplified method is that if the maximum differences of the control variables between iteration J and iteration $J - 1$ are very small, it is not necessary to call the modified HYDRUS to calculate the derivative terms. Therefore, after the first iteration, the maximum differences of the control variables in the iteration J and iteration $J - 1$ are checked. If the differ-

Table 12.3 Effect of Shift Parameter on the Convergence Speed of the SALQR Procedure

Shift parameters	No. of iterations	CPU (sec)	Convergent?	Optimal?
$\lambda = 0.1$, after 5 iterations	50			
$\lambda = 1$	(user stopped)	5557	No	No
$\lambda = 1$, after 5 iterations		2125	Yes	Yes
$\lambda = 10$	19			
$\lambda = 10$, after 5 iterations		654	Yes	Yes
$\lambda = 100$	7			
$\lambda = 100$, after 5 iterations		753	Yes	Yes
$\lambda = 1000$	9			
$\lambda = 1000$, after 5 iterations	50			
$\lambda = 10000$	(user stopped)	3665	Yes	No

ences are less than the prespecified criteria, let the derivative terms in the iteration J equal the derivative terms in iteration $J - 1$. Otherwise, the modified HYDRUS is called to calculate these derivative terms.

The solution of using this simplified method and the results of calling the modified HYDRUS to calculate the derivative terms in each iteration are listed in Table 12.4. The differences of the optimal solutions are very small. The difference of the total infiltration volume is 0.4 cm and the difference of the average infiltration rate is 0.0031 cm/day. But the number of calls to the modified HYDRUS and CPU times decreases significantly when the simplified method is used. The number of calls to the modified HYDRUS is decreased from 400 to 126 and the CPU time is reduced from 2003 sec to 654 sec on the IBM RS/6000 computer.

12.4.3 Effects of Different Initial Values

Although the SALQR method cannot guarantee a global solution, a variety of starting points can help decide the local–global optimal issue. If all starting points yield approximately the same final solution, the solution is more likely global optimal (Ladson and Waren, 1989). Several numerical experiments have been conducted in this work to find the global optimal solution. The results are presented in Tables 12.5 and 12.6. With an initial Z value equal to 10 days, the difference of optimal solution caused by different initial X values are small. The biggest difference of the total infiltration is 3.48 cm and the biggest difference of the average infiltration rate is 0.006 cm/day. Varying initial Z values and keeping the initial X

Table 12.4 Comparison of Using Different Methods to Compute the Derivatives of Simulator

		Call HYDRUS to compute the derivatives	Using simplified method to compute derivatives
Total infiltration volume V (cm^3/cm^2)		959.1	959.5
Total operation time (days)		249.97	249.86
Average infiltration rate (cm/day)		3.84	3.84
Operation schedule	X (days)	5.85	5.86
	Y (days)	6.44	6.44
	Z (days)	12.70	12.68
	CT (days)	24.99	24.98
	ω	0.2000	0.2000
Number of calls to HYDRUS		400	126
CPU (sec)		2003	654

value as a constant still leads to a minor difference in optimal solutions. Therefore, it is concluded that the local optimal solution obtained in the example problem is more likely a global optimal or at least near global optimal.

12.4.4 Selection of Bound Values and Other Parameters

The upper bound values ω_{max} and CT_{max} of the constraints (12.2.5) and (12.2.6), the initial loading rate Q_0, and the maximum ponding depth D_{max} are given by decision-makers. A series of numerical experiments were performed to analyze the change of infiltration with respect to ω_{max}, CT_{max}, Q_0, and D_{max}. The method used is to vary one parameter in each run and to keep the other parameters constant. The results for different ω_{max}, CT_{max}, Q_0, and D_{max} are presented in Tables 12.7–12.10, respectively. Table 12.7 indicates that when $\omega_{max} \leq 0.25$, the total infiltration volume and the average infiltration rate increase in proportional to ω_{max}. When $\omega_{max} > 0.25$, both the total infiltration volume and the average infiltration rate decrease with respect to ω_{max}. $\omega_{max} = 0.25$ is the most efficient with respect to the average infiltration rate and the maximum infiltration volume. ω_{max} should also be considered from the viewpoint of easy operation. If

Table 12.5 Optimal Solutions for Different Initial Application Time X_{int} (for Initial Drying Time Z_{int} = 10 days)

X_{int} (days)	$\sum_{t=1}^{10} F_t$ (cm^3/cm^2)	$\sum_{t=1}^{10} (CT)_t$ (days)	$\bar{f} = \frac{\sum F}{\sum CT}$ (cm/day)	X (days)	Y (days)	Z (days)	CT (days)	ω
4	959.45	249.98	3.8381	5.85	6.45	12.70	25.00	0.2000
6	960.05	250.27	3.8360	5.87	6.45	12.70	25.02	0.2000
8	962.84	251.14	3.8339	5.94	6.51	12.70	25.14	0.2000
10	959.87	249.87	3.8399	5.86	6.44	12.68	24.99	0.2000
12	959.35	250.05	3.8367	5.86	6.45	12.70	25.01	0.2000
14	960.65	250.44	3.8358	5.88	6.47	12.70	25.05	0.2000

Note: Parameters used in the model are D_{max} = 30 cm, Q_0 = 30 cm/day, ω_{max} = 0.2, CT_{max} = 25 days.

Table 12.6 Optimal Solutions for Different Initial Drying Time Z_{int} (for Different Initial Application Time X_{int} = 10 days)

Z_{int} (days)	$\sum_{t=1}^{10} F_t$ (cm^3/cm^2)	$\sum_{t=1}^{10} (CT)_t$ (days)	$\bar{f} = \frac{\sum F}{\sum CT}$ (cm/day)	X (days)	Y (days)	Z (days)	CT (days)	ω
4	959.19	250.00	3.8367	5.85	6.44	12.71	25.00	0.2000
6	959.86	250.36	3.8389	5.87	6.45	12.68	25.00	0.2001
8	960.62	250.48	3.8352	5.88	6.47	12.71	25.06	0.2000
10	959.47	249.87	3.8399	5.86	6.44	12.68	24.99	0.2000
12	959.09	249.95	3.8371	5.85	6.44	12.70	24.99	0.2000
14	959.76	250.07	3.8379	5.86	6.44	12.69	24.99	0.2001

Note: Parameters used in the model are D_{max} = 30 cm, Q_0 = 30 cm/day, ω_{max} = 0.2, CT_{max} = 25 days.

Table 12.7 Optimal Solutions for Different Maximum Water Content ω_{max}

ω_{max}	$\sum_{t=1}^{10} F_t$ (cm³/cm²)	$\sum_{t=1}^{10} (CT)_t$ (days)	$\bar{f} = \frac{\sum F}{\sum CT}$ (cm/day)	X (days)	Y (days)	Z (days)	CT (days)	ω
0.200	959.47	249.87	3.84	5.86	6.44	12.68	24.99	0.2000
0.225	1115.26	250.97	4.44	10.83	9.12	5.15	25.10	0.2250
0.250	1139.90	250.03	4.56	13.57	9.76	1.67	25.00	0.2498
0.275	1139.48	250.03	4.56	13.95	9.78	1.27	25.00	0.2547
0.275	1138.06	250.05	4.55	14.28	9.79	0.94	25.01	0.2598

Note: Parameters used in the model are CT_{max} = 25 days, D_{max} = 30 cm, Q_0 = 30 cm/day.

ω_{max} is large, the pond may not reach an aerobic state. As shown in Table 12.8, when $CT_{max} \leq 25$ days, the average infiltration rate increases, and when $CT_{max} > 25$ days, the infiltration value decreases as a result of increase in CT_{max}. $CT_{max} = 25$ day is a suitable upper-bound value of the cycle time for the example problem. Table 12.9 indicates that as the initial loading rate Q_0 increases, both the total infiltration volume and the average infiltration rate increase. But the rate of increase is small after $Q_0 > 30$ cm/day. Therefore, $Q_0 = 30$ cm/day was used in the example problem. Table 12.10 indicates that the average infiltration rates increase when $D_{max} \leq 45$ cm. After $D_{max} > 45$ cm, the average infiltration rate decrease. This is because that as the ponding depth increases, the clogging of the upper soil layer has an impact on the infiltration rate. $D_{max} = 45$ cm is the most efficient with respect to the average infiltration rate for the example problem. In general, D_{max} equals to 10–30 cm may be the most desirable for maximizing the infiltration rate and ease of maintenance (Bouwer and Lance, 1989).

12.4.5 Conclusions

The results of the numerical application of the new methodology indicate that the SATOM model is improved over the approach by Mushtaq et al. (1994). Briefly, the new methodology has the capability of incorporating the clogging effect and has the potential to incorporate the water quality aspect. The SATOM computer model is a powerful tool for the decision-maker to perform preliminary analysis on the SAT system to determine operations (loading schedule).

The following conclusions are drawn from the model development and case application:

1. SATOM converges within a few iterations, which shows that the modified constant shift method proposed in this chapter is very efficient.
2. The number of calls to HYDRUS and the CPU time are reduced significantly by using the simplified method to calculate the transition derivatives and the derivatives of the infiltration volume, and the draining time with respect to the control variables and the state variable.
3. By trying different initial X and Z values, assurance of at least a near-global optimal solution is obtained.

Table 12.8 Optimal Solutions for Different Maximum Cycle Time CT_{max}

CT_{max} (days)	$\sum_{t=1}^{10} F_t$ (cm^3/cm^2)	$\sum_{t=1}^{10} (CT)_t$ (days)	$\bar{f} = \frac{\sum F}{\sum CT}$ (cm/day)	X (days)	Y (days)	Z (days)	CT (days)	ω
15	283.33	150.03	1.89	1.02	0.91	13.07	15.00	0.2000
20	746.21	200.67	3.72	3.00	4.08	12.98	20.06	0.2000
25	959.47	249.87	3.84	5.86	6.44	12.68	24.99	0.2000
30	1087.72	299.96	3.63	8.90	8.38	12.71	29.99	0.2000
35	1180.89	349.91	3.37	12.14	10.04	12.80	34.99	0.2000
40	1255.83	399.94	3.14	15.67	11.52	12.80	39.99	0.2000
45	1316.89	449.90	2.93	19.31	12.79	12.88	44.99	0.2001

Note: Parameters used in the model are $D_{max} = 30$ cm, $Q_0 = 30$ cm/day, $\omega_{max} = 0.2$.

Table 12.9 Optimal Solutions for Different Initial Loading Rate Q_0

Q_0 (cm/day)	$\sum_{t=1}^{10} F_t$ (cm³/cm²)	$\sum_{t=1}^{10} (\mathrm{CT})_t$ (days)	$\bar{F} = \frac{\sum F}{\sum \mathrm{CT}}$ (cm/day)	X (days)	Y (days)	Z (days)	CT (days)	ω
10	837.03	249.92	3.35	8.88	3.00	13.11	24.99	0.2000
15	887.83	249.66	3.56	6.66	5.36	12.94	24.96	0.2000
20	926.43	249.99	3.71	6.14	6.04	12.82	25.00	0.2000
25	947.96	250.28	3.79	5.98	6.30	12.74	25.03	0.2000
30	959.47	249.87	3.84	5.86	6.44	12.68	24.99	0.2000
35	967.50	250.16	3.87	5.78	6.55	12.67	25.00	0.2000
40	973.61	250.10	3.89	5.74	6.63	12.66	25.03	0.2000
45	978.77	250.36	3.91	5.72	6.70	12.64	25.06	0.2000
50	979.98	250.11	3.92	5.66	6.72	12.63	25.01	0.2000
55	981.55	250.25	3.92	5.64	6.71	12.64	24.99	0.2000
60	982.22	250.20	3.93	5.67	6.75	12.64	25.06	0.2000

Note: Parameters used in the model are D_{max} = 30 cm, CT_{max} = 25 days, ω_{max} = 0.2.

Table 12.10 Optimal Solutions for Different Maximum Ponding Depth D_{max}

D_{max} (cm)	$\sum_{t=1}^{10} F_t$ (cm^3/cm^2)	$\sum_{t=1}^{10} (CT)_t$ (days)	$\bar{F} = \frac{\sum F}{\sum CT}$ (cm/day)	X (days)	Y (days)	Z (days)	CT (days)	ω
10	885.79	250.07	3.54	9.74	2.33	12.95	25.02	0.2000
15	912.17	250.01	3.65	8.68	3.47	12.86	25.01	0.2000
20	932.56	250.04	3.73	7.66	4.55	12.79	25.00	0.2000
25	948.00	249.98	3.79	6.71	5.55	12.73	24.99	0.2000
30	959.47	249.87	3.84	5.86	6.44	12.68	24.99	0.2000
35	967.43	250.02	3.87	5.10	7.23	12.67	25.00	0.2000
40	972.64	250.06	3.89	4.48	7.88	12.65	25.01	0.2000
45	974.36	249.90	3.90	3.95	8.38	12.66	24.99	0.2000
50	975.65	250.29	3.90	3.66	8.71	12.67	25.04	0.2000
60	971.21	250.45	3.87	3.64	8.57	12.83	25.04	0.2000

Note: Parameters used in the model are CT_{max} = 30 days, Q_0 = 30 cm/day, ω_{max} = 0.2.

APPENDIX: DERIVATIVE CALCULATIONS

The transition derivatives and the derivatives of the infiltration volume and the draining time with respect to the state variable are

$$\frac{\partial T_1}{\partial \omega_t} = \frac{T_1(\omega_t + \Delta\omega, X_t, Z_t, t) - T_1(\omega_t, X_t, Z_t, t)}{\Delta\omega} \tag{12.A.1}$$

$$\frac{\partial T_2}{\partial \omega_t} = \frac{T_2(\omega_t + \Delta\omega, X_t, Z_t, t) - T_2(\omega_t, X_t, Z_t, t)}{\Delta\omega} \tag{12.A.2}$$

$$\frac{\partial T_3}{\partial \omega_t} = \frac{T_3(\omega_t + \Delta\omega, X_t, t) - T_3(\omega_t, X_t, t)}{\Delta\omega} \tag{12.A.3}$$

The transition derivatives and the derivatives of infiltration volume and the draining time with respect to the control variables are

$$\frac{\partial T_1}{\partial X_t} = \frac{T_1(\omega_t, X_t + \Delta X, Z_t, t) - T_1(\omega_t, X_t, Z_t, t)}{\Delta X} \tag{12.A.4}$$

$$\frac{\partial T_2}{\partial X_t} = \frac{T_2(\omega_t, X_t + \Delta X, Z_t, t) - T_2(\omega_t, X_t, Z_t, t)}{\Delta X} \tag{12.A.5}$$

$$\frac{\partial T_3}{\partial X_t} = \frac{T_3(\omega_t, X_t + \Delta X, t) - T_3(\omega_t, X_t, t)}{\Delta X} \tag{12.A.6}$$

$$\frac{\partial T_1}{\partial Z_t} = \frac{T_1(\omega_t, X_t, Z_t + \Delta Z, t) - T_1(\omega_t, X_t, Z_t, t)}{\Delta Z} \tag{12.A.7}$$

$$\frac{\partial T_2}{\partial Z_t} = \frac{T_2(\omega_t, X_t, Z_t + \Delta Z, t) - T_2(\omega_t, X_t, Z_t, t)}{\Delta Z} \tag{12.A.8}$$

The objective function derivatives with respect to the state variable are

$$\frac{\partial G}{\partial \omega_t} = \frac{\partial T_2}{\partial \omega_t} + 2R_1[\min(0, \omega_{\max} - \omega_{t+1})]\frac{\partial T_1}{\partial \omega_t} + 2R_2[\min(0, \mathrm{CT}_{\max} - (X_t + Y_t + Z_t))]\frac{\partial T_3}{\partial \omega_t} \tag{12.A.9}$$

$$\frac{\partial^2 G}{\partial \omega_t^2} = -2R_1\left(\frac{\partial T_1}{\partial \omega_t}\right)^2 - 2R_2\left(\frac{\partial T_3}{\partial \omega_t}\right)^2 \tag{12.A.10}$$

Objective function derivatives with respect to the control variables are

$$\frac{\partial G}{\partial X_t} = \frac{\partial T_2}{\partial X_t} + 2R_1[\min(0,\ \omega_{\max} - \omega_{t+1})]\ \frac{\partial T_1}{\partial X_t}$$

$$+\ 2R_2[\min(0,\ \mathrm{CT}_{\max} - (X_t + Y_t + Z_t))]\left(1 + \frac{\partial T_3}{\partial X_t}\right) \quad (12.A.11)$$

$$\frac{\partial^2 G}{\partial X_j^2} = -2R_1\left(\frac{\partial T_1}{\partial X_{i,t}}\right)^2 - 2R_2\left(1 + \frac{\partial T_3}{\partial X_t}\right)^2 \quad (12.A.12)$$

$$\frac{\partial G}{\partial Z_t} = \frac{\partial T_2}{\partial Z_t} + 2R_1[\min(0,\ \omega_{\max} - \omega_{t+1})]\ \frac{\partial T_1}{\partial Z_t}$$

$$+\ 2R_2[\min(0,\ \mathrm{CT}_{\max} - (X_t + Y_t + Z_t))] \quad (12.A.13)$$

$$\frac{\partial^2 G}{\partial Z_t^2} = -2R_1\left(\frac{\partial T_1}{\partial Z_t}\right)^2 - 2R_2 \quad (12.A.14)$$

$$\frac{\partial^2 G}{\partial \omega_t\, \partial X_t} = -2R_1\ \frac{\partial T_1}{\partial X_t}\,\frac{\partial T_1}{\partial \omega_t} - 2R_2\left(1 + \frac{\partial T_3}{\partial X_t}\right)\frac{\partial T_3}{\partial \omega_t} \quad (12.A.15)$$

$$\frac{\partial^2 G}{\partial \omega_t\, \partial Z_t} = -2R_1\ \frac{\partial T_1}{\partial Z_t}\,\frac{\partial T_1}{\partial \omega_t} - 2R_2\ \frac{\partial T_3}{\partial \omega_t} \quad (12.A.16)$$

REFERENCES

Bouwer, H. and Lance, R. C., Effect of Water Depth in Groundwater Recharge Basins on Infiltration Rate, *Journal of Irrigation and Drainage Engineering, ASCE* Vol. 115, No. 4, pp. 556–568, 1989.

Chang, L.-C., Shoemaker, C. A., and Liu, P. L.-F., Optimal Time Varying Pumping Rates for Groundwater Remediation: Application of Constrained Optimal Control Algorithm, *Water Resources Research*, Vol. 28, No. 12, pp. 3157–3173, 1992.

Cournoyer, L. F. and Kriege, D., Operation and Maintenance of Recharge Facilities, Artificial Recharge of Groundwater, Proceedings of the International Symposium, 1988, pp. 487–494.

Jacobson, D. H. and Mayne, D. Q., *Differential Dynamic Programming*, Elsevier Scientific, New York, 1970.

Kool J. B. and van Genuchten, M. Th., *HYDRUS: One-Dimensional Variably Saturated Flow and Transport Model, Including Hysteresis and Root Uptake, v3.31*, U.S. Salinity Laboratory, U.S. Dept. of Agriculture, Agriculture Service, Pineside, CA, 1991.

Lasdon, L. S. and Waren, A. D., *GRG2 User's Guide*, Department of General Business, University of Texas at Austin, 1989.

Liao, L. and Shoemaker, C. A., Convergence in Unconstrained Discrete-Time Differential Dynamic Programming, *IEEE Transactions on Automatic Control*, Vol. AC-36, No. 6, pp. 692–706, 1991.
Li, G., Differential Dynamic Programming for Estuarine Management, Ph.D. Dissertation, Dept. of Civil Engineering, Arizona State University, Tempe, 1994.
Mushtaq, H., Mays, L. W., and Lansey, K. E., Optimal Operation of Recharge Basins, *Journal of Water Resources Planning and Management*, Vol. 120, No. 6, pp. 927–943, 1994.
Tang, Z., Li, G., Mays, L. W., and Fox, P., Optimal Operation of Soil Aquifer Treatment Systems: Its Hydraulic and Hydrologic Considerations. 7th Symposium on Artificial Recharge of Groundwater, 1995.
U.S. Environmental Protection Agency, *Manual: Guidelines for Water Reuse*, Report EPA-625/R-92-004, U.S. EPA, Washington, DC, 1992.
Yakowitz, S., Algorithm and Computational Techniques in Differential Dynamic Programming, *Control and Dynamic Systems*, Vol. 31, pp. 75–91, 1987.
Yakowitz, S. and Rutherford, B., Computational Aspects of Discrete-Time Optimal Control, *Applied Mathematics and Computation*, Vol. 15, pp. 29–49, 1984.

Chapter 13

Real-Time Optimal Control: Linear Quadratic Feedback for Lumped Systems

13.1 INTRODUCTION

As discussed in Chapter 1, a system is characterized by "(1) a system boundary which is a rule that determines whether an element is to be considered as a part of the system or of the environment; (2) statement of input and output interactions with the environment; and (3) statements of interrelationships between the system elements, inputs and outputs, called feedback" (Mays and Tung, 1992). The essential elements of the control problem are "(1) a mathematical model (system) to be controlled, (2) a desired output of the system, (3) a set of admissible inputs or controls, and (4) a performance index or cost functional which measures the effectiveness of a given control action" (Athans and Falb, 1966). The optimal control problem is to "steer" the system by controlling the inputs to generate the desired output and to optimize the chosen performance index.

Generally speaking, system control can be classified into open-loop control and closed-loop control (Brogan, 1974). If the control is determined as a function of the initial state and other given system parameters, the control is said to be open looped (Fig. 13.1). It can be seen from Figure 13.1 that the input is not influenced by the output of the system. If there are unexpected disturbances to an open-loop system, or if the behavior of the system is not completely known, then the output may not be what one expects. If the control is determined as a function of the current state, then it is a closed-loop or feedback control (Fig. 13.2). In the closed-loop system, the control $u(t)$ is modified by considering the system output. A feed-

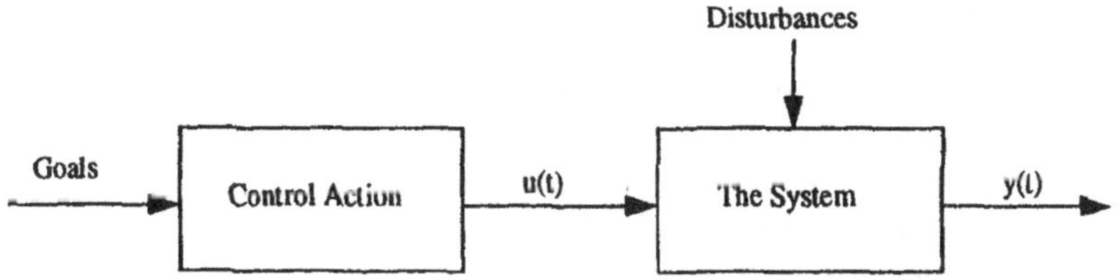

Figure 13.1 An open-loop control system. In an open-loop control system the control action is independent of the output. The control action is the quantity responsible for activating the system to produce the output. The ability of open-loop control system to perform accurately is determined by their calibration. Calibration is the establishment of the input–output relation to obtain a desired system accuracy.

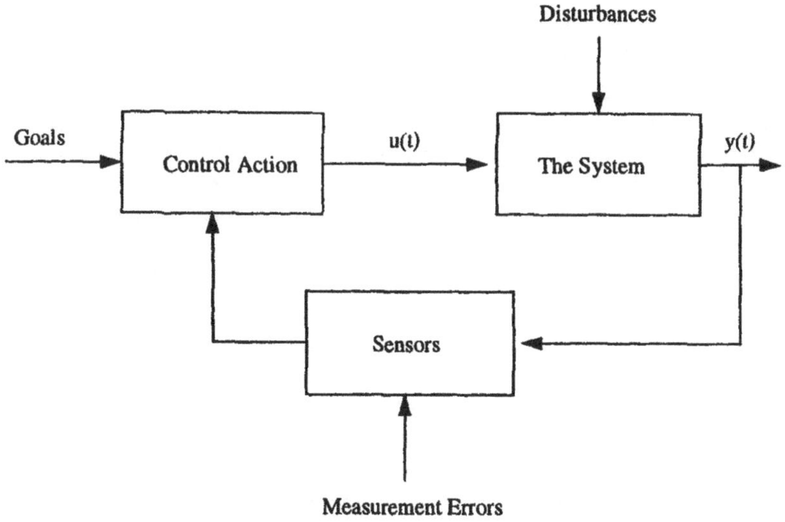

Figure 13.2 A closed-loop control system. A closed-loop control system is one in which the control action is dependent on the output. Closed-loop systems are more commonly referred to as feedback control systems. Feedback is the property which allows the output or some other controlled variable to be computed with the system input so that a control action can be formed. This control action may be formed as a function of the input and output.

back system is usually better than the open-loop system when the unexpected disturbances and uncertainties occur. However, when the measured output have large errors and when unexpected disturbances are relatively small, an open-loop control can be better than a closed-looped control.

Disturbances and measurement errors in Figure 13.2 are uncertainties. If the system operation also involves uncertainties, the optimal control problem may be classified as a stochastic optimal control problem. The stochastic optimal control reflects the real-world system more than does the deterministic optimal control, as everything in the world is subject to various kinds of uncertainties. Uncertainty could simply be defined as the occurrence of events that are beyond our control (Mays and Tung, 1992).

In general, uncertainties in water resource engineering projects can be divided into four basic categories: hydrologic, hydraulic, structural, and economic. The hydrologic uncertainty for any hydrosystem design problem can be further classified into three types: inherent, parameter, and model uncertainties. Those uncertainties are likely to cause inaccuracies when a numerical model is applied to a real-world system for the prediction of the behavior of the real-world system. Therefore, the resulting optimal solution to an optimization model may never be truly optimal for the real-world system, although it may be mathematically optimal for the assumed mathematical system. Thus, various uncertainties should be incorporated in optimization and optimal control models for systems.

Without exception, an estuarine system management model based on optimization and optimal control should involve the consideration of uncertainties. The chance-constraint stochastic programming technique has been successfully applied to the salinity and fish harvest regression equations for the optimal control framework and multiobjective analysis (Tung et al., 1990; Mao and Mays, 1994). However, there are many other inherent, parameter, and model uncertainties, some of which are often impossible to be statistically incorporated in the modeling. Even after many uncertainties are successfully incorporated in the modeling by various kinds of assumptions such as those for probability distributions, the solution to the optimization and optimal control model may often be intractable due to significantly large computation. In addition, the unavoidable trial and error for the initial point for most optimization techniques makes the solution more "uncertain." Therefore, a practical stochastic optimal control method for the estuarine system management is needed to deal with the uncertainties without involving too many assumptions and too much computation.

Feedback control is such a practical control method, which feeds back the new state observation to the determination of the control. The control is readjusted whenever a new observation is made. The motivations for feedback control are listed by Van De Vegte (1990) as (1) reducing the effects of parameter variations, (2) reducing the effects of disturbance inputs, (3) improving transient response characteristics, and (4) reducing steady-state errors. Because the feedback control considers the new observation, it can also be called a real-time control problem.

The linear quadratic feedback optimal control problem has been one of the most important feedback control problems in control engineering since the 1960s. An analytical feedback control law can be derived for a linear system with a quadratic performance index by dynamic programming or Pontryagin's minimum principle. Although there exists model uncertainty for linear systems due to the fact that real-world systems are not linear, the model uncertainty is reduced after each new observation is made and is fed back to the determination of control. In addition, the solution to a linear quadratic control problem requires little computation and is free of trial and error for the initial point. The beauty of the linear quadratic control is its simplicity and efficiency, as an analytical feedback control law can be derived (Jacobson et al. 1980; Bertsekas, 1987).

When there exists white Gaussian noise in the system equation and measurement equation, linear quadratic Gaussian (LQG) optimal control can be used (Maybeck, 1982; Bertsekas, 1987). "Whiteness" implies that the noise value is not correlated in time (Maybeck, 1979). The LQG is based on the *separation theorem for linear systems and quadratic cost*, which may be stated that the optimal controller can be designed independently as optimal solutions of an estimation and control problem. The procedure of LQG has two steps. First, Kalman filtering is used to estimate the optimal state vector. Second, the optimal control vector is computed by the feedback control law. However, if the measurement errors in the observed state vector are considered insignificant, then the feedback control law can be directly applied after the state vector is observed. Applications of LQG in reservoir system management can be found elsewhere (Wasimi and Kitanidis, 1983; Georgakakos and Marks, 1987).

If the parameters in the system equation are not exactly known, they are usually estimated by historical data and updated by new data. The feedback control law is then computed based on the system equation with the updated parameters. The recursive least squares technique is considered as an efficient method to update the parameters. A complete discussion on parameter estimation techniques can be found in Åström and Eykhoff (1970). Ljung (1987) gave a more updated discussion on parameter estimation techniques. It may be noted that the combination of the recursive parameter estimation and the stochastic linear quadratic control can be considered as one of the adaptive control methods in control engineering. However, if the parameters in the system equation vary slowly in a short time period, the parameters may not have to be updated every time new data are obtained.

In this chapter, a new type of estuarine system management model based on discrete-time, stochastic, linear, quadratic feedback, optimal control is proposed, which can feed back the new observations for the salinity or other nutrient levels for the determination of freshwater inflows. The precipitation, evaporation, and ungauged freshwater inflows are considered to be the elements of a random vector with its mean and covariance matrix. The feedback control law is analytically derived by using the dynamic programming principle. Decision-makers can use this model to determine the optimal amount of freshwater inflows into an estuary during a time interval, after the salinity and nutrient levels at the specified locations in the estuary are observed at the beginning of the time interval. The optimal freshwater inflows are determined so that the salinity and nutrient levels starting from the beginning of the time interval to the end of the terminal time interval are as close as possible to the desirable levels in the sense of statistical expectation. The freshwater inflows can be controlled by operating the upstream reservoir gates during this time interval.

The control problem in this chapter is concerned with the discrete-time, stochastic, linear, quadratic optimal control for a lumped-parameter system, whereas Chapter 14 considers a distributed-parameter system. The lumped-parameter and continuous-time system is described by ordinary differential equations, whereas the lumped-parameter and discrete-time system is described by difference equations. In many fields, a general, linear, stochastic system is described by an input–output difference equation which is sometimes called ARMAX (AutoRegressive, Moving Average, with eXogenous input). The estuarine system management in this phase is a type of ARMAX model. In this chapter, various linear, quadratic control models are discussed, which are followed by a discussion on a general, stochastic, discrete-time, linear, quadratic optimal control. This general, stochastic, discrete-time, linear, quadratic optimal control is presented in the framework of an estuarine system management model. The analytical feedback control law for this estuarine system management model is derived by using the dynamic programming principle.

13.2 DISCRETE-TIME, LINEAR, QUADRATIC REGULATOR (LQR)

13.2.1 LQR with a Deterministic System

The LQR with a deterministic system (LQRD) is used to "steer" a deterministic linear system during a fixed period of time $[t_0, t_f]$ such that the states of the system are as close as possible to a desirable constant state with the least control energy. Herein, the desirable constant state is the zero vector. The purpose is to find an optimal control law at a given initial state which "steers" a deterministic, discrete-time, linear system

$$\mathbf{s}_{t+1} = \mathbf{A}_t\mathbf{s}_t + \mathbf{B}_t\mathbf{u}_t \tag{13.2.1}$$

such that the following quadratic performance index is minimized:

$$J(\mathbf{s}_t) = \mathbf{s}_{t_f}^T\mathbf{Q}_{t_f}\mathbf{s}_{t_f} + \sum_{t}^{t_f-1} \{\mathbf{s}_t^T\mathbf{Q}_t\mathbf{s}_t + \mathbf{u}_t^T\mathbf{R}_t\mathbf{u}_t\} \tag{13.2.2}$$

where the superscript T denotes the transpose of a matrix, the vector $\mathbf{s}_t$ is the state vector at the beginning of time interval t, the vector $\mathbf{u}_t$ is the control vector applied during time interval t, the weighting matrix $\mathbf{Q}_t$ is symmetric positive semidefinite, the weighting matrix $\mathbf{R}_t$ is symmetric positive definite, the matrices $\mathbf{A}_t$ and $\mathbf{B}_t$ are the parameter matrices which describe the system, and t_f is the terminal time. $\mathbf{J}(\mathbf{s}_t)$ is often called the cost-to-go performance index. The term $\mathbf{u}_t^T\mathbf{R}_t\mathbf{u}_t$ is the measure of the energy for the control. For example, $\mathbf{u}_t^T\mathbf{R}_t\mathbf{u}_t$ measures the magnitude of the control vector when $\mathbf{R}_t$ is an identity matrix.

13.2.2 LQR with a Stochastic System

The LQR with a stochastic system (LQRS) is used to "steer" a stochastic linear system during a fixed period of time $[t_0, t_f]$ such that the states of the system are as close as possible to, in the sense of statistical expectation, a desirable constant state with the least control energy. Herein, the desirable constant state is the zero vector. The purpose is to find an optimal control law which "steers" a stochastic, discrete-time, linear system

$$\mathbf{s}_{t+1} = \mathbf{A}_t\mathbf{s}_t + \mathbf{B}_t\mathbf{u}_t + \mathbf{w}_t \tag{13.2.3}$$

such that the following quadratic performance index is minimized:

$$J(\mathbf{s}_t) = \mathrm{E}[\mathbf{s}_{t_f}^T\mathbf{Q}_{t_f}\mathbf{s}_{t_f} + \sum_{t}^{t_f-1} \{\mathbf{s}_t^T\mathbf{Q}_t\mathbf{s}_t + \mathbf{u}_t^T\mathbf{R}_t\mathbf{u}_t\}] \tag{13.2.4}$$

where E[ ] is the statistical expectation operator; the superscript T denotes the transpose of a matrix, the vector $\mathbf{s}_t$ is the state vector at the beginning of time interval t, the vector $\mathbf{u}_t$ is the control vector applied during time interval t, $\mathbf{w}_t$ is the random vector with a given mean and covariance matrix and is independent of $\mathbf{w}_{t-1}, \ldots, \mathbf{w}_0$, the weighting matrix $\mathbf{Q}_t$ is symmetric positive semidefinite, the weighting matrix $\mathbf{R}_t$ is symmetric positive definite, the matrices $\mathbf{A}_t$ and $\mathbf{B}_t$ are the parameter matrices which describe the system, and t_f is the terminal time.

13.3 DISCRETE-TIME, LINEAR, QUADRATIC TRACKER (LQT)

13.3.1 LQT with a Deterministic System

The LQT with a deterministic system (LQTD) is used to "steer" a deterministic linear system during a fixed period of $[t_0, t_f]$ such that the states of the system are as close as possible to the desirable states with the least control energy. The desirable state for a tracker is a function of time. The purpose is to find an optimal control law which "steers" a discrete-time, linear system

$$\mathbf{s}_{t+1} = \mathbf{A}_t\mathbf{s}_t + \mathbf{B}_t\mathbf{u}_t \tag{13.3.1}$$

such that the following quadratic performance index is minimized:

$$J(\mathbf{s}_t) = [\mathbf{s}_{t_f} - \mathbf{s}_{t_f}^D]^{\mathrm{T}}\mathbf{Q}_{t_f}[\mathbf{s}_{t_f} - \mathbf{s}_{t_f}^D] + \sum_{t}^{t_f-1} \{[\mathbf{s}_t - \mathbf{s}_t^D]^{\mathrm{T}}\mathbf{Q}_t[\mathbf{s}_t - \mathbf{s}_t^D] + \mathbf{u}_t^{\mathrm{T}}\mathbf{R}_t\mathbf{u}_t]\} \tag{13.3.2}$$

where the superscript T denotes the transpose of a matrix, the vector $\mathbf{s}_t$ is the state vector at the beginning of time interval t, the vector $\mathbf{s}_t^D$ is the desirable state vector at the beginning of time interval t, the vector $\mathbf{u}_t$ is the control vector applied during time interval t, the weighting matrix $\mathbf{Q}_t$ is symmetric positive semidefinite, the weighting matrix $\mathbf{R}_t$ is symmetric positive definite, the matrices $\mathbf{A}_t$ and $\mathbf{B}_t$ are the parameter matrices which describe the system, and t_f is the terminal time. The regulators are special cases of trackers.

13.3.2 LQT with a Stochastic System

The LQT with a stochastic system (LQTS) is used to "steer" a stochastic linear system during a fixed period of time $[t_0, t_f]$ such that the states of the system are as close as possible to the desirable states with the least control energy in the sense of statistical expectation. The purpose is to find an optimal control law which "steers" a discrete-time, linear system

$$\mathbf{s}_{t+1} = \mathbf{A}_t\mathbf{s}_t + \mathbf{B}_t\mathbf{u}_t + \mathbf{w}_t \tag{13.3.3}$$

such that the following quadratic performance index is minimized:

$$J(\mathbf{s}_t) = E[[\mathbf{s}_{t_f} - \mathbf{s}_{t_f}^D]^T \mathbf{Q}_{t_f}[\mathbf{s}_{t_f} - \mathbf{s}_{t_f}^D] + \sum_{t}^{t_f-1} \{[\mathbf{s}_t - \mathbf{s}_t^D]^T \mathbf{Q}_t[\mathbf{s}_t - \mathbf{s}_t^D] + \mathbf{u}_t^T \mathbf{R}_t \mathbf{u}_t\}] \tag{13.3.4}$$

where $E[\ ]$ is the statistical expectation operator; the superscript T is the transpose of a matrix, the vector $\mathbf{s}_t$ is the state vector at the beginning of time interval t, the vector $\mathbf{s}_t^D$ is the desirable state vector at the beginning of time interval t, $\mathbf{w}_t$ is the random vector with a given mean and covariance matrix and is independent of $\mathbf{w}_{t-1}, \ldots, \mathbf{w}_0$, the vector $\mathbf{u}_t$ is the control vector applied during time interval t, the weighting matrix $\mathbf{Q}_t$ is symmetric positive semidefinite, the weighting matrix $\mathbf{R}_t$ is symmetric positive definite, the matrices $\mathbf{A}_t$ and $\mathbf{B}_t$ are the parameter matrices which describe the system, and t_f is the terminal time.

13.4 GENERAL, DISCRETE-TIME, STOCHASTIC, LINEAR, QUADRATIC, OPTIMAL CONTROL FOR ESTUARINE MANAGEMENT

The aforementioned LQRD, LQRS, LQTD, and LQTS are special cases of a general, discrete-time, stochastic, linear, quadratic, optimal control which is presented herein in the framework of the estuarine system management (Zhao, 1994; Zhao and Mays, 1995). The quadratic performance index for the proposed estuarine management model is

$$J(\mathbf{s}_t) = \mathrm{E}\{[\mathbf{s}_{t_f} - \mathbf{s}_{t_f}^D]^T \mathbf{Q}_{t_f}[\mathbf{s}_{t_f} - \mathbf{s}_{t_f}^D] + \sum_{t}^{t_f-1} \{[\mathbf{s}_t - \mathbf{s}_t^D]^T \mathbf{Q}_t[\mathbf{s}_t - \mathbf{s}_t^D] + [\mathbf{q}_{c,t} - \mathbf{q}_{c,t}^D]^T \mathbf{R}_t[\mathbf{q}_{c,t} - \mathbf{q}_{c,t}^D]\}\} \tag{13.4.1}$$

in which $\mathrm{E}[\ ]$ is the expectation operator, the superscript T denotes the transpose of a matrix or a vector, the superscript D denotes the desirable values for the state vector or control vector, given by the decision-makers, t_f is the terminal time, $\mathbf{s}_t$ is the state vector for salinity and nutrient levels at specified locations in the estuarine system at the beginning of the time interval t, $t = 1, 2, \ldots, t_f$, $\mathbf{q}_{c,t}$ is the control vector for freshwater inflows into the estuary during the time interval t, the weighting matrix $\mathbf{Q}_t$ for the state vector is symmetric and positive semidefinite, and the weighting matrix $\mathbf{R}_t$ for the control vector is symmetric and positive definite.

The estuarine system is considered as a linear system which can be expressed as a type of ARMAX model:

$$\mathbf{s}_{t+1} = \mathbf{A}_t\mathbf{s}_t + \mathbf{B}_t\mathbf{q}_{c,t} + \mathbf{C}_t\mathbf{w}_t + \mathbf{D}_t\mathbf{v}_t \tag{13.4.2}$$

in which $\mathbf{s}_t$ and $\mathbf{q}_{c,t}$ have the same definitions as before. The random vector $\mathbf{w}_t$ with the given mean vector and covariance matrix consists of three stacked vectors: $\mathbf{q}_{c,t}$ is the vector of ungauged freshwater inflows; $\mathbf{p}_t$ is the vector of rainfall amounts in different regions in the estuarine system; and $\mathbf{e}_t$ is the vector of evaporation amounts in different regions in the estuarine system. $\mathbf{v}_t$ is a known deterministic vector which can be used to define the deterministic uncontrollable freshwater inflows during the time interval t, and $\mathbf{A}_t$, $\mathbf{B}_t$, $\mathbf{C}_t$, and $\mathbf{D}_t$ are the parameter matrices.

The purpose of the proposed estuarine system management model is to determine the freshwater inflows into the estuarine system such that the cost-to-go performance index at time t is minimized. The upstream reservoir releases during each month can then be calculated. When the freshwater inflows into the estuarine system must compete with the water demand from upstream water users, the desirable freshwater inflows may be chosen to be very small. In this case, the proposed discrete-time, stochastic, linear, quadratic, optimal control is to use the freshwater inflows to control the estuarine system such that the performance index is minimized with the least release from the reservoirs. When the marsh inundation needs for the flushing of nutrients from riverine marshes into the estuary are considered, the desirable freshwater inflows can be selected based on the water demand from marsh inundation.

The discrete-time, stochastic, linear, quadratic, optimal control model can be analytically derived by using dynamic programming, yielding the feedback control law which expresses the optimal amount of freshwater inflow from each river at the present time interval as a linear function of the state vector at the beginning of the present time. Once the present state vector is measured at the beginning of the present time interval, the optimal amount of freshwater inflows during the time interval can be directly computed.

13.5 ANALYTICAL FEEDBACK CONTROL LAW FOR GENERAL, DISCRETE-TIME, STOCHASTIC, LINEAR, QUADRATIC, OPTIMAL CONTROL FOR ESTUARINE MANAGEMENT

13.5.1 Feedback Control Law

The feedback control law for the optimal freshwater inflow can be expressed as

$$\mathbf{q}_{c,t}^{*} = \mathbf{M}_t[\mathbf{A}_t[\mathbf{s}_t - \mathbf{s}_t^D] + \mathbf{e}_t] + \mathbf{q}_{c,t}^D \tag{13.5.1}$$

in which

$$\mathbf{M}_t = -\mathbf{L}_t^{-1}\mathbf{B}_t^T\mathbf{K2}_{t+1} \tag{13.5.2}$$

$$\mathbf{L}_t = \mathbf{R}_t + \mathbf{B}_t^T\mathbf{K2}_{t+1}\mathbf{B}_t \tag{13.5.3}$$

$$\mathbf{K2}_t = \mathbf{Q}_t + \mathbf{A}_t^T\mathbf{K2}_{t+1}\mathbf{A}_t - \mathbf{A}_t^T\mathbf{K2}_{t+1}\mathbf{B}_t\mathbf{L}_t^{-1}\mathbf{B}_t^T\mathbf{K2}_{t+1}\mathbf{A}_t \tag{13.5.4}$$

$$\mathbf{e}_t = \mathbf{A}_t\mathbf{s}_t^D + \mathbf{B}_t\mathbf{q}_{c,t}^D + \mathbf{C}_t\overline{\mathbf{w}}_t + \mathbf{D}_t\mathbf{v}_t - \mathbf{s}_{t+1}^D \tag{13.5.5}$$

and

$$\mathbf{K2}_{t_f} = \mathbf{Q}_{t_f} \tag{13.5.6}$$

The minimized cost-to-go function is

$$J^*(\mathbf{s}_t) = [\mathbf{s}_t - \mathbf{s}_t^D]^T\mathbf{K2}_t[\mathbf{s}_t - \mathbf{s}_t^D] + 2[\mathbf{s}_t - \mathbf{s}_t^D]^T\mathbf{k1}_t + k0_t \tag{13.5.7}$$

in which

$$\mathbf{k1}_t = \mathbf{k1}_{t+1} + \mathbf{A}_t^T\mathbf{K2}_{t+1}\mathbf{e}_t - \mathbf{A}_t^T\mathbf{K2}_{t+1}\mathbf{B}_t\mathbf{L}_t^{-1}\mathbf{B}_t^T\mathbf{K2}_{t+1}\mathbf{e}_t \tag{13.5.8}$$

$$k0_t = k0_{t+1} + \mathbf{e}_t^T[\mathbf{K2}_{t+1} - \mathbf{K2}_{t+1}\mathbf{B}_t\mathbf{L}_t^{-1}\mathbf{B}_t^T\mathbf{K2}_{t+1}]\mathbf{e}_t + \text{Trace}\{\mathbf{C}_t^T\mathbf{K2}_{t+1}\mathbf{C}_t\ \text{cov}(\mathbf{w}_t)\} \tag{13.5.9}$$

and

$$\mathbf{k1}_{t_f} = \mathbf{0}, \qquad k0_{t_f} = 0 \tag{13.5.10}$$

The operator Trace for a matrix in Eq. (13.5.9) is defined as the sum of the diagonal elements of the matrix. If the state vector $\mathbf{s}_t$, which corresponds to the beginning of time interval t, can be measured or quantified with insignificant error, the control vector for the optimal amounts of freshwater inflows during the time interval t is computed by Eq. (13.5.1). Therefore, this control law can directly handle the real-time control problem. The diagram for this feedback control is illustrated in Figure 13.3. Figure 13.4 shows the flowchart for computing the control law for the lumped-parameter estuary system. The covariance matrix of $\mathbf{w}_t$, cov($\mathbf{w}_t$), only appears in $k0_t$ in Eq. (13.5.9). Because the cov($\mathbf{w}_t$) represents the uncertainty in the random vector $\mathbf{w}_t$, the feedback control law given by Eq. (13.5.1) is not affected by the uncertainty in $\mathbf{w}_t$. This may be referred to as the manifestation of the certainty equivalence principle. However, the uncertainty in $\mathbf{w}_t$ affects the minimized cost-to-go performance index given by Eq. (13.5.7).

13.5.2 Derivation of the Analytical Feedback Control Law by Dynamic Programming

Dynamic programming can be used to derive the feedback control law for the discrete-time, stochastic, linear, quadratic, optimal control estuarine

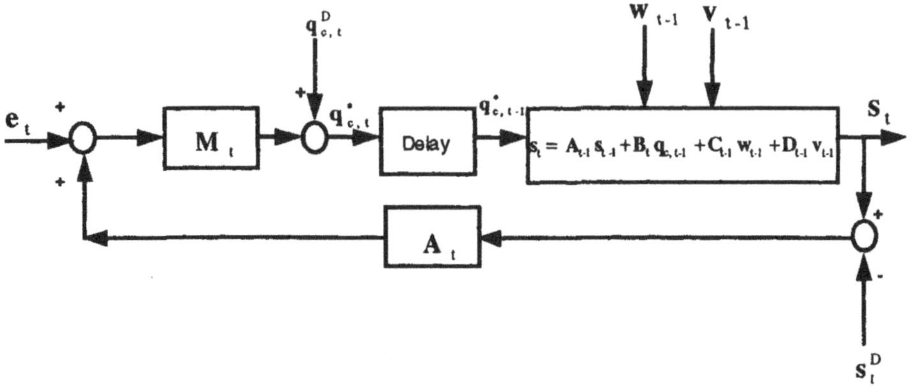

Figure 13.3 Stochastic feedback control system (Zhao, 1994).

management model. The basic idea of the derivation is to start from the terminal time t_f and proceed backward in time. For time $t = 1, 2, \ldots, t_f - 1$, the state vector from the linear system equation is substituted into the quadratic performance index $J(\mathbf{s}_t)$, which becomes a quadratic function of the control vector. The optimal control vector for time t is found by taking the derivative of $J(\mathbf{s}_t)$ and setting it to zero. This control vector optimizes the cost-to-go performance index $J(\mathbf{s}_t)$. Denote the optimal cost-to-go performance index by $J^*(\mathbf{s}_t)$. The next step is $t - 1$. Based on the dynamic programming principle, the optimal performance index starting from $t - 1$ to t_f also implies the optimal performance index starting from t to t_f. The objective is to minimize $J(\mathbf{s}_{t-1}) = E[g(\mathbf{s}_{t-1}, \mathbf{q}_{c,t-1}) + J^*(\mathbf{s}_t)]$ over $\mathbf{q}_{c,t-1}$ in which

$$g(\mathbf{s}_{t-1}, \mathbf{q}_{c,t-1}) = (\mathbf{s}_{t-1} - \mathbf{s}_{t-1}^D)^{\mathrm{T}}\mathbf{Q}_{t-1}(\mathbf{s}_{t-1} - \mathbf{s}_{t-1}^D) + (\mathbf{q}_{c,t-1} - \mathbf{q}_{c,t-1}^D)^{\mathrm{T}}\mathbf{R}_{t-1}(\mathbf{q}_{c,t-1} - \mathbf{q}_{c,t-1}^D) \quad (13.5.11)$$

Substituting the state vector into $J(\mathbf{s}_{t-1})$ yields the performance index $J(\mathbf{q}_{c,t-1})$, which is only a function of the control $\mathbf{q}_{c,t-1}$. Taking the derivative of the resulting $J(\mathbf{q}_{c,t-1})$ with respect to the control vector results in the optimal control $\mathbf{q}_{c,t-1}^*$. As an analogous example to the fastest route problem mentioned earlier, $J^*(\mathbf{s}_{t-1})$ and $J^*(\mathbf{s}_t)$ may be considered as the fastest route from Los Angeles to Boston and the fastest route from Chicago to Boston, respectively. An optimal control law for any time t can be generalized by proceeding backward for a few time steps. The following gives a more detailed derivation. Because the derivation involves tedious algebraic work, some algebraic details have to be omitted.

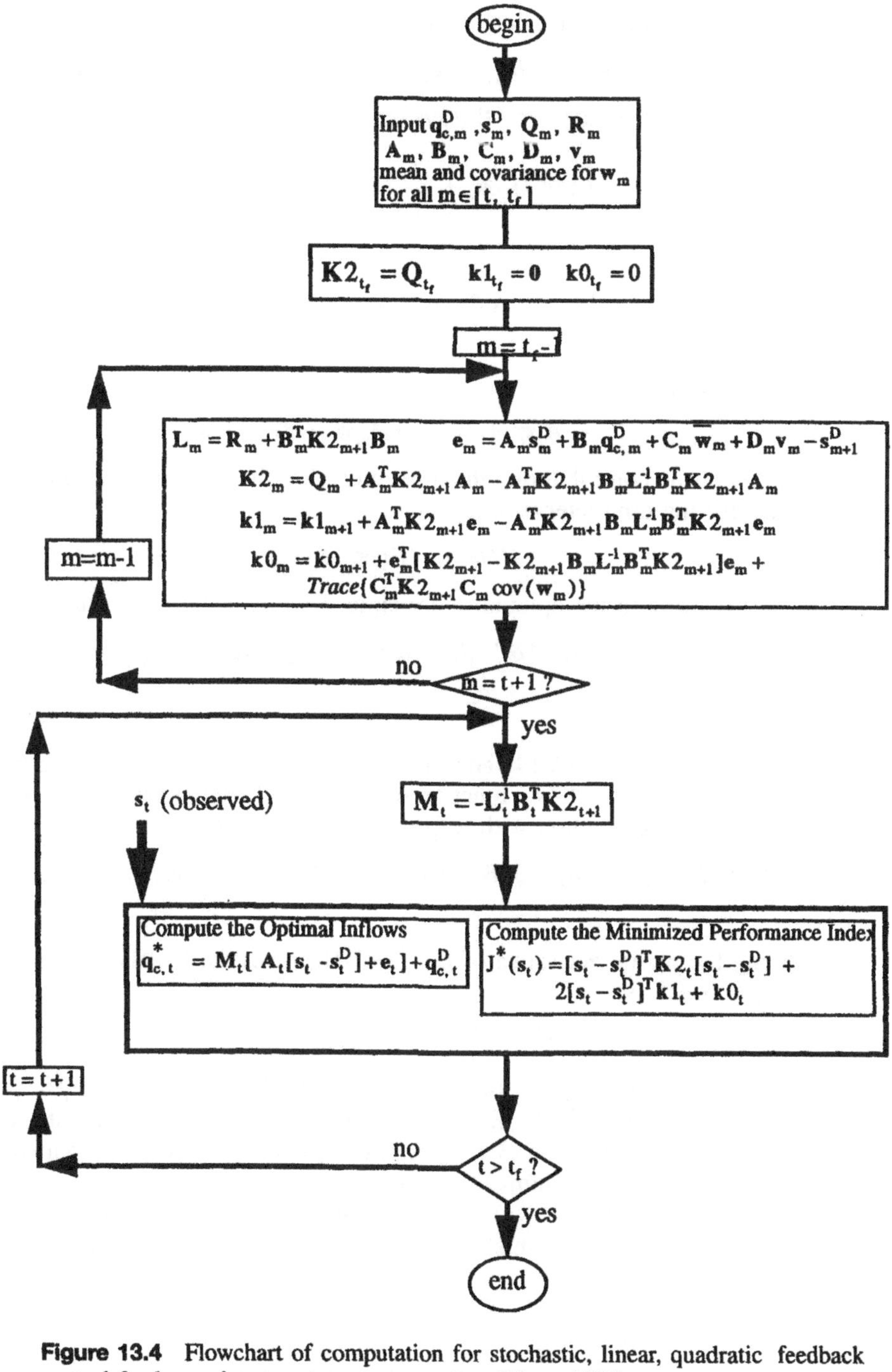

Figure 13.4 Flowchart of computation for stochastic, linear, quadratic feedback control for lumped-parameter estuary system (Zhao, 1994).

For the sake of simplicity, define

$$\delta \mathbf{s}_t = \mathbf{s}_t - \mathbf{s}_t^D, \qquad \delta \mathbf{q}_{c,t} = \mathbf{q}_{c,t} - \mathbf{q}_{c,t}^D, \qquad \delta \mathbf{w}_t = \mathbf{w}_t - \overline{\mathbf{w}}_t \quad (13.5.12)$$

The performance index and the system equation become,

$$J(\delta \mathbf{s}_t) = E\{\delta \mathbf{s}_{t_f}^{\mathrm{T}} \mathbf{Q}_{t_f} \delta \mathbf{s}_{t_f} + \sum_{t}^{t_f} \{\delta \mathbf{s}_t^{\mathrm{T}} Q_t \delta \mathbf{s}_t + \delta \mathbf{q}_{c,t}^{\mathrm{T}} \mathbf{R}_t \delta \mathbf{q}_{c,t}\}\} \quad (13.5.13)$$

and

$$\delta \mathbf{s}_{t+1} = \mathbf{A}_t \delta \mathbf{s}_t + \mathbf{B}_t \delta \mathbf{q}_{c,t} + \mathbf{C}_t \delta \mathbf{w}_t + \mathbf{e}_t \quad (13.5.14)$$

respectively, in which

$$\mathbf{e}_t = \mathbf{A}_t \mathbf{s}_t^D + \mathbf{B}_t \mathbf{q}_{c,t}^D + \mathbf{C}_t \overline{\mathbf{w}}_t + \mathbf{D}_t \mathbf{v}_t - \mathbf{s}_{t+1}^D \quad (13.5.15)$$

As there is no control during the time interval t_f, the minimum performance index for the time interval t_f is

$$J^*(\delta \mathbf{s}_{t_f}) = E\{\delta \mathbf{s}_{t_f}^{\mathrm{T}} \mathbf{Q}_{t_f} \delta \mathbf{s}_{t_f}\} \quad (13.5.16)$$

According to the optimality principle from dynamic programming,

$$J^*(\delta \mathbf{s}_{t_f-1}) = \min_{\delta \mathbf{q}_{c,t_f-1}} \underset{\delta \mathbf{w}_{t_f-1}}{E} \{\delta \mathbf{s}_{t_f-1}^{\mathrm{T}} \mathbf{Q}_{t_f-1} \delta \mathbf{s}_{t_f-1}^{\mathrm{T}} + \delta \mathbf{q}_{c,t_f-1}^{\mathrm{T}} \mathbf{R}_{t_f-1} \delta \mathbf{q}_{c,t_f-1} + J^*(\delta \mathbf{s}_{t_f})\} \quad (13.5.17)$$

Substituting Eqs. (13.5.14) and (13.5.16) into Eq. (13.5.17) gives a performance index which is a function of $\delta \mathbf{q}_{c,t_f-1}$. Taking the partial derivative of the resulting function with respect to $\delta \mathbf{q}_{c,t_f-1}$ results in

$$\mathbf{G} = 2([\mathbf{R}_{t_f-1} + \mathbf{B}_{t_f-1}^{\mathrm{T}} \mathbf{Q}_{t_f} \mathbf{B}_{t_f-1}] \delta \mathbf{q}_{c,t_f-1} + \mathbf{B}_{t_f-1}^{\mathrm{T}} \mathbf{Q}_{t_f} [\mathbf{A}_{t_f-1} \delta \mathbf{s}_{t_f-1} + \mathbf{e}_{t_f-1}]) \quad (13.5.18)$$

Taking the partial derivative of Eq. (13.5.18) gives the Hessian matrix for $\delta \mathbf{q}_{c,t_f-1}$, which is

$$\mathbf{H} = 2[\mathbf{R}_{t_f-1} + \mathbf{B}_{t_f-1}^{\mathrm{T}} \mathbf{Q}_{t_f} \mathbf{B}_{t_f-1}] \quad (13.5.19)$$

Because the matrix $\mathbf{Q}_t$ is positive semidefinite, the matrix $\mathbf{B}_{t_f-1}^{\mathrm{T}} \mathbf{Q}_{t_f} \mathbf{B}_{t_f-1}$ is also positive semidefinite. Because the matrix $\mathbf{R}_t$ is positive definite, the Hessian matrix equation (13.5.19) is positive definite; hence, the optimality problem becomes a minimization problem.

Setting Eq. (13.5.18) to a zero vector yields the necessary condition for the minimization, which can be arranged as (algebraic details are omitted)

$$\delta \mathbf{q}^*_{c,t_f-1} = -\mathbf{L}^{-1}_{t_f-1}\mathbf{B}^{\mathrm{T}}_{t_f-1}\mathbf{Q}_{t_f}[\mathbf{A}_{t_f-1}\delta \mathbf{s}_{t_f-1} + \mathbf{e}_{t_f-1}] \tag{13.5.20}$$

in which

$$\mathbf{L}_{t_f-1} = \mathbf{R}_{t_f-1} + \mathbf{B}^{\mathrm{T}}_{t_f-1}\mathbf{Q}_{t_f}\mathbf{B}_{t_f-1} \tag{13.5.21}$$

The optimal performance index becomes

$$J^*(\delta \mathbf{s}_{t_f-1}) = \delta \mathbf{s}^{\mathrm{T}}_{t_f-1}\mathbf{K2}_{t_f-1}\delta \mathbf{s}_{t_f-1} + 2\delta \mathbf{s}^{\mathrm{T}}_{t_f-1}\mathbf{k1}_{t_f-1} + k0_{t_f-1} \tag{13.5.22}$$

in which

$$\mathbf{K2}_{t_f-1} = \mathbf{Q}_{t_f-1} + \mathbf{A}^{\mathrm{T}}_{t_f-1}\mathbf{Q}_{t_f}\mathbf{A}_{t_f-1} - \mathbf{A}^{\mathrm{T}}_{t_f-1}\mathbf{Q}_{t_f}\mathbf{B}_{t_f-1}\mathbf{L}^{-1}_{t_f-1}\mathbf{B}^{\mathrm{T}}_{t_f-1}\mathbf{Q}_{t_f}\mathbf{A}_{t_f-1} \tag{13.5.23}$$

$$\mathbf{k1}_{t_f-1} = \mathbf{A}^{\mathrm{T}}_{t_f-1}\mathbf{Q}_{t_f}\mathbf{e}_{t_f-1} - \mathbf{A}^{\mathrm{T}}_{t_f-1}\mathbf{Q}_{t_f}\mathbf{B}_{t_f-1}\mathbf{L}^{-1}_{t_f-1}\mathbf{B}^{\mathrm{T}}_{t_f-1}\mathbf{Q}_{t_f}\mathbf{e}_{t_f-1} \tag{13.5.24}$$

$$\begin{aligned} k0_{t_f-1} &= \mathbf{e}^{\mathrm{T}}_{t_f-1}[\mathbf{Q}_{t_f} - \mathbf{Q}_{t_f}\mathbf{B}_{t_f-1}\mathbf{L}^{-1}_{t_f-1}\mathbf{B}^{\mathrm{T}}_{t_f-1}\mathbf{Q}_{t_f}]\mathbf{e}_{t_f-1} \\ &\quad + E\{\delta \mathbf{w}^{\mathrm{T}}_{t_f-1}\mathbf{C}^{\mathrm{T}}_{t_f-1}\mathbf{Q}_{t_f}\mathbf{C}_{t_f-1}\delta \mathbf{w}_{t_f-1}\} \end{aligned} \tag{13.5.25}$$

The above equation can be simplified to

$$\begin{aligned} k0_{t_f-1} &= \mathbf{e}^{\mathrm{T}}_{t_f-1}[\mathbf{Q}_{t_f} - \mathbf{Q}_{t_f}\mathbf{B}_{t_f-1}\mathbf{L}^{-1}_{t_f-1}\mathbf{B}^{\mathrm{T}}_{t_f-1}\mathbf{Q}_{t_f}]\mathbf{e}_{t_f-1} \\ &\quad + \mathrm{Trace}\{\mathbf{C}^{\mathrm{T}}_{t_f-1}\mathbf{Q}_{t_f}\mathbf{C}_{t_f-1}\ \mathrm{cov}(\mathbf{w}_{t_f-1})\} \end{aligned} \tag{13.5.26}$$

in which the operator Trace for a matrix is the sum of the diagonal elements of the matrix. This simplification is based on the following:

$$\begin{aligned} E[\mathbf{x}^{\mathrm{T}}\mathbf{A}\mathbf{x}] &= \sum_{i=1}\sum_{j=1} (a_{ij}\ E[x_i x_j]) \\ &= \sum_{i=1}\sum_{j=1} (a_{ij}[\mathrm{cov}(x_i x_j) + \bar{x}_i\bar{x}_j]) \\ &= \mathrm{Trace}[\mathbf{A}\ \mathrm{cov}(\mathbf{x}) + \bar{\mathbf{x}}^{\mathrm{T}}\mathbf{A}\bar{\mathbf{x}}] \end{aligned} \tag{13.5.27}$$

in which $\mathbf{A}$ is a deterministic matrix and $\mathbf{x}$ is a random column vector with mean $\bar{\mathbf{x}}$ and covariance matrix cov($\mathbf{x}$). The last step in Eq. (13.5.27) is based on $\mathbf{x}^{\mathrm{T}}\mathbf{A}\mathbf{x} = \mathrm{Trace}(\mathbf{A}\mathbf{x}\mathbf{x}^{\mathrm{T}})$ which can be found in Graybill (1983). It should be noted that $E[\delta \mathbf{w}] = \mathbf{0}$ and $\mathrm{cov}(\mathbf{w}) = \mathrm{cov}(\delta \mathbf{w})$.

According to the optimality principle from dynamic programming, the optimal performance index at time $t_f - 2$ is

$$J^*(\delta \mathbf{s}_{t_f-2}) = \min_{\delta \mathbf{q}_{c,t_f-2}}\ \mathop{\mathrm{E}}_{\delta \mathbf{w}_{t_f-2}}\ \{\delta \mathbf{s}^{\mathrm{T}}_{t_f-2}\mathbf{Q}_{t_f-2}\delta \mathbf{s}_{t_f-2} + \delta \mathbf{q}^{\mathrm{T}}_{t_f-2}\mathbf{R}_{t_f-2}\delta \mathbf{q}_{c,t_f-2} + J^*(\delta \mathbf{s}_{t_f-1})\} \tag{13.5.28}$$

By following the same procedure for time $t_f - 1$, the optimal control vector during time $t_f - 2$ can be derived as (algebraic details are omitted)

$$\delta\mathbf{q}^*_{c,t_f-2} = -\mathbf{L}^{-1}_{t_f-2}\mathbf{B}^{\mathrm{T}}_{t_f-2}\mathbf{K2}_{t_f-1}[\mathbf{A}_{t_f-2}\delta\mathbf{s}_{t_f-2} + \mathbf{e}_{t_f-2}] \quad (13.5.29)$$

in which

$$\mathbf{L}_{t_f-2} = \mathbf{R}_{t_f-2} + \mathbf{B}^{\mathrm{T}}_{t_f-2}\mathbf{K2}_{t_f-1}\mathbf{B}_{t_f-2} \quad (13.5.30)$$

The minimized cost-to-go function at time $t_f - 2$ is

$$J^*(\delta\mathbf{s}_{t_f-2}) = \delta\mathbf{s}^{\mathrm{T}}_{t_f-2}\mathbf{K2}_{t_f-2}\delta\mathbf{s}_{t_f-2} + 2\delta\mathbf{s}^{\mathrm{T}}_{t_f-2}\mathbf{k1}_{t_f-2} + k0_{t_f-2} \quad (13.5.31)$$

in which

$$\begin{aligned}\mathbf{K2}_{t_f-2} &= \mathbf{Q}_{t_f-2} + \mathbf{A}^{\mathrm{T}}_{t_f-2}\mathbf{K2}_{t_f-1}\mathbf{A}_{t_f-2} \\ &\quad - \mathbf{A}^{\mathrm{T}}_{t_f-2}\mathbf{K2}_{t_f-1}\mathbf{B}_{t_f-2}\mathbf{L}^{-1}_{t_f-2}\mathbf{B}^{\mathrm{T}}_{t_f-2}\mathbf{K2}_{t_f-1}\mathbf{A}_{t_f-2} \quad (13.5.32)\end{aligned}$$

$$\begin{aligned}\mathbf{k1}_{t_f-2} &= \mathbf{k1}_{t_f-1} + \mathbf{A}^{\mathrm{T}}_{t_f-2}\mathbf{K2}_{t_f-1}\mathbf{e}_{t_f-2} \\ &\quad - \mathbf{A}^{\mathrm{T}}_{t_f-2}\mathbf{K2}_{t_f-1}\mathbf{B}_{t_f-2}\mathbf{L}^{-1}_{t_f-2}\mathbf{B}^{\mathrm{T}}_{t_f-2}\mathbf{K2}_{t_f-1}\mathbf{e}_{t_f-2} \quad (13.5.33)\end{aligned}$$

$$\begin{aligned}k0_{t_f-2} &= k0_{t_f-1} + \mathbf{e}^{\mathrm{T}}_{t_f-2}[\mathbf{K2}_{t_f-1} - \mathbf{K2}_{t_f-1}\mathbf{B}_{t_f-2}\mathbf{L}^{-1}_{t_f-2}\mathbf{B}^{\mathrm{T}}_{t_f-2}\mathbf{K2}_{t_f-1}]\mathbf{e}_{t_f-2} \\ &\quad + E\{\delta\mathbf{w}^{\mathrm{T}}_{t_f-2}\mathbf{C}^{\mathrm{T}}_{t_f-2}\mathbf{K2}_{t_f-1}\mathbf{C}_{t_f-2}\delta\mathbf{w}_{t_f-2}\} \quad (13.5.34)\end{aligned}$$

Consequently, the results for any time t can be generalized to be Eqs. (13.5.1)–(13.5.10).

13.6 APPLICATION OF STOCHASTIC, LINEAR, QUADRATIC FEEDBACK OPTIMAL CONTROL TO WATER RESOURCES ENGINEERING

13.6.1 Reservoir System Management

One of the reservoir system management problems is to find the optimal releases during each time interval which "steers" the reservoir system from a given initial storage at time t_0 to time t_f such that the storage and control release for all times are as close as possible to the desirable storage and release in the sense of statistical expectation. Mathematically, the control problem is to find the optimal release u_t during time interval t, which "steers" the reservoir system governed by

$$s_{t+1} = s_t + I_t + e_t + p_t - u_t \quad (13.6.1)$$

$$s_{t_0} = s_0 \quad (13.6.2)$$

such that the following quadratic performance index is minimized:

$$J(s_t) = E\{[s_{t_f} - s^{\mathrm{D}}_{t_f}]^2 Q_{t_f} + \sum_{t}^{t_f-1} \{[s_t - s^{\mathrm{D}}_t]^2 Q_t + [s_t - s^{\mathrm{D}}_t]^2 R_t\}\} \quad (13.6.3)$$

in which s_t is the storage at the beginning of time interval t; I_t, e_t, and p_t are the inflow to the reservoir, evaporation, and precipitation, respectively, during the time interval; Q_t is the non-negative number; and R_t is the positive number. It should be noted that I_t, e_t, and p_t are considered to be uncertain, and their means and variances are given.

Because this reservoir system management problem is a special case of the proposed general, stochastic, discrete-time, linear, quadratic, feedback optimal control, the optimal control release law follows Eq. (13.5.1). Therefore, after the storage at the beginning of time t is observed without significant error, the optimal release during time interval t can be directly computed by using Eq. (13.5.1). It may be pointed out that this real-time, optimal control problem can be applied in different time scales such as minutes, hours, or months.

13.6.2 Hydrologic Routing

One of the river system management problems is to find the optimal inflow to the river system during each time interval which "steers" the river system from a given initial storage at time t_0 to time t_f such that the storage and inflow for all times are as close as possible to the desirable storage and inflow in the sense of statistical expectation. Mathematically, the control problem is to find the optimal inflow I_t during time interval t which "steers" the river system governed by

$$s_{t+1} = \left(1 - \frac{1}{K(1 - X)}\right) s_t + \frac{1}{1 - X} I_t + I_{u,t} \tag{13.6.4}$$

$$s_{t_0} = s_0 \tag{13.6.5}$$

such that the following quadratic performance index is minimized:

$$J(s_t) = E\{[s_{t_f} - s_{t_f}^D]^2 Q_{t_f} + \sum_{t}^{t_f - 1} \{[s_t - s_t^D]^2 Q_t + [s_t - s_t^D]^2 R_t\}\} \tag{13.6.6}$$

in which s_t is the storage for the river system at the beginning of time interval t; I_t and $I_{u,t}$ are respectively the controllable inflow and uncontrollable inflow to the river system during the time interval; Q_t is the non-negative number, and R_t is the positive number; and K and X are the coefficients for the Muskingum routing model. It should be noted that $I_{u,t}$ is considered to be uncertain, and its means and variances are given.

Because this river system management problem is a special case of the proposed general, stochastic, discrete-time, linear, quadratic, feedback optimal control, the optimal controllable inflow law follows Eq. (13.5.1). Therefore, after the storage at the beginning of time t is observed without significant error, the optimal inflow (or release from the reservoir) during time interval t can be directly computed by using Eq. (13.5.1). It may be pointed out that this real-time, optimal control problem can be applied in different time scales such as minutes, hours, or months.

13.7 PARAMETER ESTIMATION FOR LUMPED-PARAMETER SYSTEM

13.7.1 Introduction

In Sec. 13.4, the parameters in the estuarine system equation are assumed to be known. Actually, the parameters in the system equation are estimated by using historical data. When new data are obtained, the parameters should be updated. The feedback control law is then computed based on the system equation with the updated parameters.

Parameter estimation is the major part of system identification. As defined by Hsia (1977), the problem of system identification is generally referred to as the determination of a mathematical model for a system or a process by observing its input–output relationships. Åström and Eykhoff (1970) gave an excellent survey on system identification. Ljung (1987) gave a more updated discussion on parameter estimation techniques. The methods for the system identification have been well summarized by Sinha and Kuszta (1983) as follows:

I. Classical Methods
   (a) Frequency response identification
   (b) Impulse response identification by deconvolution
   (c) Step response identification
   (d) Identification from correlation functions
II. Equation-Error Approach
   (a) Least squares
   (b) Generalized least squares
   (c) Maximum likelihood
   (d) Minimum variance
   (e) Gradient methods
III. Model Adjustment Techniques (On-Line)
   (a) Recursive least squares
   (b) Recursive generalized least squares
   (c) Instrumental variables
   (d) Bootstrap
   (e) Recursive maximum likelihood
   (f) Recursive correlation
   (g) Stochastic approximation

Least squares was first proposed by Karl Gauss for carrying out his work on orbit prediction of planets (Hsia, 1977). Due to its simplicity and statistical properties that the least squares estimate is consistent, unbiased, and efficient, the least squares method has been one of the most useful system identification techniques. The least squares method essentially uses the historical and experimental data to estimate the parameters in the system equation in such a way that an error criterion is minimized. When new data are obtained, the parameters can be estimated through the least squares method by including the new data into all previous data. As more and more data are available as time goes, the computation will be tremendous because all data have to be used in the estimation. Instead of using all data, the recursive least squares method can be used to update the previously estimated parameters based on the new data to obtain the current parameters.

The recursive least squares (RLS) technique is considered as an efficient method to update the parameters. It can be used to update the parameter matrices in the estuarine system equation. The advantages of the recursive least squares over the ordinary least squares are less computation time, less data storage, and numerical stability. It may be noted that the combination of the recursive parameter estimation and the stochastic linear quadratic control can be considered as one of the adaptive control methods in control engineering. The theory of adaptive control has been extensively studied in the control engineering (Kumar and Varaiya, 1986; Bitmead et al., 1990; Gerenéser and Caines, 1991; Kokotovic, 1991). Most results from the theoretical study require some restrictive assumptions. A survey of stochastic adaptive control was given by Kumar (1985). He classified the stochastic adaptive control into Bayesian adaptive control problems and non-Bayesian adaptive control problems. For the Bayesian adaptive control, a probability distribution is given for the value of the unknown parameter, whereas no probability distribution is given for the non-Bayesian adaptive control. Applications of adaptive control in water resources can be found in Ko et al. (1982) and Jones (1992). However, if the parameters in the system equation vary slowly in a short time period, the parameters may not have to be updated every time new data are obtained. In this section, the recursive least squares is derived for the estuarine system equation expressed in the form of vectors and matrices.

13.7.2 Recursive Least Squares for Estuarine Management

Consider a linear estuarine system equation,

$$\mathbf{s}_{t+1} = \mathbf{A}_t\mathbf{s}_t + \mathbf{B}_t\mathbf{q}_{c,t} + \mathbf{C}_t\mathbf{w}_t + \mathbf{D}_t\mathbf{v}_t \qquad (13.7.1)$$

in which the parameter matrices $\mathbf{A}_t$, $\mathbf{B}_t$, $\mathbf{C}_t$, and $\mathbf{D}_t$ can be estimated from historical data by using the parameter estimation technique. When new data are available, the parameter matrices can be updated by recursive parameter estimation techniques such as the recursive least squares. In this section, the ordinary least squares (OLS) is briefly reviewed and followed by a discussion and derivation of the recursive least squares (RLS).

Suppose there are t time-series observations for state vectors, $\{\mathbf{s}_i, i = 1, 2, \ldots, t\}$ and $t - 1$ observations for $\{\mathbf{q}_{c,i}, \mathbf{w}_i, \mathbf{v}_i, i = 1, 2, \ldots, t - 1\}$. Consider a time-invariant case where the parameter matrices do not change with time (i.e., $\mathbf{A}_t = \mathbf{A}$, $\mathbf{B}_t = \mathbf{B}$, $\mathbf{C}_t = \mathbf{C}$, and $\mathbf{D}_t = \mathbf{D}$). The stochastic linear estuarine system equation can be written in the standard linear regression format as

$$\mathbf{X}_t\boldsymbol{\Theta}_t = \mathbf{Y}_t \qquad (13.7.2)$$

in which

$$\mathbf{X}_t = \begin{bmatrix} \mathbf{s}_1^{\mathrm{T}} & \mathbf{q}_{c,1}^{\mathrm{T}} & \mathbf{w}_1^{\mathrm{T}} & \mathbf{v}_1^{\mathrm{T}} \\ \mathbf{s}_2^{\mathrm{T}} & \mathbf{q}_{c,2}^{\mathrm{T}} & \mathbf{w}_2^{\mathrm{T}} & \mathbf{v}_2^{\mathrm{T}} \\ \vdots & \vdots & \vdots & \vdots \\ \mathbf{s}_{t-1}^{\mathrm{T}} & \mathbf{q}_{c,t-1}^{\mathrm{T}} & \mathbf{w}_{t-1}^{\mathrm{T}} & \mathbf{v}_{t-1}^{\mathrm{T}} \end{bmatrix} \qquad (13.7.3)$$

$$\boldsymbol{\Theta}_t = \begin{bmatrix} \mathbf{A}_t^{\mathrm{T}} \\ \mathbf{B}_t^{\mathrm{T}} \\ \mathbf{C}_t^{\mathrm{T}} \\ \mathbf{D}_t^{\mathrm{T}} \end{bmatrix} \qquad (13.7.4)$$

and

$$\mathbf{Y}_t = \begin{bmatrix} \mathbf{s}_2^{\mathrm{T}} \\ \mathbf{s}_3^{\mathrm{T}} \\ \vdots \\ \mathbf{s}_t^{\mathrm{T}} \end{bmatrix} \qquad (13.7.5)$$

The ordinary least squares (OLS) estimate of the parameter matrix can be derived as

$$\hat{\Theta}_t = [X_t^T X_t]^{-1} X_t^T Y_t \qquad (13.7.6)$$

Suppose that a new data point $\{s_{t+1}, x_{t+1}\}$ is observed, in which $x_{t+1} = [s_t^T\ q_{c,t}^T\ w_t^T\ v_t^T]$. Although the two matrices [Eqs. (13.7.3) and (13.7.5)] can be appended by the new data point and the new parameter matrix estimate can be computed by Eq. (13.7.6), it involves much more computation and may have numerical instability due to the inversion of the matrix which contains all previous data and the new data point. Instead of inverting the matrix associated with all data, one can find the parameter matrix estimate at time $t + 1$ by updating the parameter matrix estimate at time t through

$$\hat{\Theta}_{t+1} = \hat{\Theta}_t + \gamma_{t+1} P_t x_{t+1}^T [y_{t+1} - x_{t+1} \hat{\Theta}_t] \qquad (13.7.7)$$

in which

$$\gamma_{t+1} = \frac{1.0}{1.0 + x_{t+1} P_t x_{t+1}^T} \qquad (13.7.8)$$

$$P_{t+1} = P_t - \gamma_{t+1} P_t x_{t+1}^T x_{t+1} P_t \qquad (13.7.9)$$

$$x_{t+1} = [s_t^T\ q_{c,t}^T\ w_t^T\ v_t^T] \qquad (13.7.10)$$

$$y_{t+1} = s_{t+1}^T \qquad (13.7.11)$$

and the initial matrix for P_t is $P_0 = [X_0^T X_0]^{-1}$.

It can be seen that once the initial $P_0 = [X_0^T X_0]^{-1}$ is computed, the parameter matrix is updated from the previous parameter matrix by using the new observation data. Thus, the inversion of matrix $X_t^T X_t$ for each time is not necessary for the RLS. It may be noted that the matrix P_t must be stored in the computer in order to update P_{t+1} and $\hat{\Theta}_{t+1}$. However, the storage size for P_t is much smaller than that for all original data which are needed for OLS. For example, consider the case that Eq. (13.7.3) has dimensions of $(m \times n)$, Eq. (13.7.4) has dimensions of $(n \times k)$, and Eq. (13.7.5) has dimensions of $(m \times k)$, and the matrix P_t has dimension of $(n \times n)$. Because the size for the historical data is often much larger than that for the parameter matrix or $m \gg n$ and $m \gg k$, the size for storing P_t is much smaller than that for the OLS. In addition, the OLS requires frequent matrix inversion, which may involve numerical instability when the matrix $X_t^T X_t$ is ill-conditioned. Furthermore, the computation time for the OLS is larger than that for the RLS due to the inversion of a large matrix. Therefore, the RLS is better than the OLS for parameter estimation.

The derivation of Eqs. (13.7.7)–(13.7.11) is based on the matrix inversion relationship

$$(A + BCD)^{-1} = A^{-1} - A^{-1}B(C^{-1} + DA^{-1}B)^{-1}DA^{-1} \qquad (13.7.12)$$

where the matrices A, C, and A + BCD are nonsingular square matrices.

A derivation for a single-variable linear system can be found in Hsia (1977). The following is a brief derivation for the RLS for a system expressed in terms of matrices and vectors. Appending the new point to Eq. (13.7.2),

$$\mathbf{X}_{t+1}\mathbf{\Theta}_{t+1} = \mathbf{Y}_{t+1} \tag{13.7.13}$$

in which

$$\mathbf{X}_{t+1} = \begin{bmatrix} \mathbf{X}_t \\ \\ \mathbf{x}_{t+1} \end{bmatrix} \tag{13.7.14}$$

$$\mathbf{Y}_{t+1} = \begin{bmatrix} \mathbf{Y}_t \\ \\ \mathbf{y}_{t+1} \end{bmatrix}$$

$$= \begin{bmatrix} \mathbf{Y}_t \\ \\ \mathbf{s}_{t+1}^{\mathrm{T}} \end{bmatrix} \tag{13.7.15}$$

then,

$$\hat{\mathbf{\Theta}}_{t+1} = [\mathbf{X}_{t+1}^{\mathrm{T}}\mathbf{X}_{t+1}]^{-1}\mathbf{X}_{t+1}^{\mathrm{T}}\mathbf{Y}_{t+1} \tag{13.7.16}$$

Define

$$\mathbf{P}_t = [\mathbf{X}_t^{\mathrm{T}}\mathbf{X}_t]^{-1}, \qquad \mathbf{P}_{t+1} = [\mathbf{X}_{t+1}^{\mathrm{T}}\mathbf{X}_{t+1}]^{-1} \tag{13.7.17}$$

then,

$$\hat{\mathbf{\Theta}}_{t+1} = \mathbf{P}_{t+1}\mathbf{X}_{t+1}^{\mathrm{T}}\mathbf{Y}_{t+1} \tag{13.7.18}$$

$$\begin{aligned} \mathbf{P}_{t+1} &= [\mathbf{X}_{t+1}^{\mathrm{T}}\mathbf{X}_{t+1}]^{-1} \\ &= [\mathbf{X}_{t+1}^{\mathrm{T}}\mathbf{X}_t + \mathbf{x}_{t+1}^{\mathrm{T}}\mathbf{x}_{t+1}]^{-1} \end{aligned} \tag{13.7.19}$$

By using Eq. (13.7.12) and defining

$$\mathbf{A} = \mathbf{X}_t^{\mathrm{T}}\mathbf{X}_t, \qquad \mathbf{B} = \mathbf{x}_{t+1}^{\mathrm{T}}, \qquad \mathbf{C} = 1.0, \qquad \mathbf{D} = \mathbf{B}^{\mathrm{T}} \tag{13.7.20}$$

one obtains

$$\mathbf{P}_{t+1} = \mathbf{P}_t - \mathbf{P}_t\mathbf{x}_{t+1}^{\mathrm{T}}[1.0 + \mathbf{x}_{t+1}\mathbf{P}_t\mathbf{x}_{t+1}^{\mathrm{T}}]^{-1}\mathbf{x}_{t+1}\mathbf{P}_t \tag{13.7.21}$$

Substituting Eq. (13.7.19) into Eq. (13.7.18) results in

$$\begin{aligned} \hat{\mathbf{\Theta}}_{t+1} &= \{\mathbf{P}_t - \mathbf{P}_t\mathbf{x}_{t+1}^{\mathrm{T}}[1.0 + \mathbf{x}_{t+1}\mathbf{P}_t\mathbf{x}_{t+1}^{\mathrm{T}}]^{-1}\mathbf{x}_{t+1}\mathbf{P}_t\}\mathbf{X}_{t+1}^{\mathrm{T}}\mathbf{Y}_{t+1} \\ &= \{\mathbf{P}_t - \mathbf{P}_t\mathbf{x}_{t+1}^{\mathrm{T}}[1.0 + \mathbf{x}_{t+1}\mathbf{P}_t\mathbf{x}_{t+1}^{\mathrm{T}}]^{-1}\mathbf{x}_{t+1}\mathbf{P}_t\} \\ &\quad \cdot [\mathbf{X}_t^{\mathrm{T}}\mathbf{Y}_t + \mathbf{x}_{t+1}^{\mathrm{T}}\mathbf{y}_{t+1}] \end{aligned} \tag{13.7.22}$$

By defining $\gamma_{t+1} = 1.0/(1.0 + \mathbf{x}_{t+1}\mathbf{P}_t\mathbf{x}_{t+1}^T)$ and using $\hat{\mathbf{\Theta}}_t = \mathbf{P}_t\mathbf{X}_t^T\mathbf{Y}_t$, Eq. (13.7.22) can be arranged as follows:

$$
\begin{aligned}
\hat{\mathbf{\Theta}}_{t+1} &= \hat{\mathbf{\Theta}}_t + \mathbf{P}_t\mathbf{x}_{t+1}^T\mathbf{y}_{t+1} - \mathbf{P}_t\mathbf{x}_{t+1}^T\gamma_{t+1}\mathbf{x}_{t+1}\hat{\mathbf{\Theta}}_t - \mathbf{P}_t\mathbf{x}_{t+1}^T\gamma_{t+1}\mathbf{x}_{t+1}\mathbf{P}_t\mathbf{x}_{t+1}^T\mathbf{y}_{t+1} \\
&= \hat{\mathbf{\Theta}}_t - \mathbf{P}_t\mathbf{x}_{t+1}^T\gamma_{t+1}\mathbf{x}_{t+1}\hat{\mathbf{\Theta}}_t + \mathbf{P}_t\mathbf{x}_{t+1}^T[1.0 - \gamma_{t+1}\mathbf{x}_{t+1}\mathbf{P}_t\mathbf{x}_{t+1}^T]\mathbf{y}_{t+1} \\
&= \hat{\mathbf{\Theta}}_t - \mathbf{P}_t\mathbf{x}_{t+1}^T\gamma_{t+1}\mathbf{x}_{t+1}\hat{\mathbf{\Theta}}_t + \mathbf{P}_t\mathbf{x}_{t+1}^T \\
&\quad \cdot \left(1.0 - \frac{1.0}{1.0 + \mathbf{x}_{t+1}\mathbf{P}_t\mathbf{x}_{t+1}^T}\mathbf{x}_{t+1}\mathbf{P}_t\mathbf{x}_{t+1}^T\right)\mathbf{y}_{t+1} \\
&= \hat{\mathbf{\Theta}}_t - \mathbf{P}_t\mathbf{x}_{t+1}^T\gamma_{t+1}\mathbf{x}_{t+1}\hat{\mathbf{\Theta}}_t + \mathbf{P}_t\mathbf{x}_{t+1}^T\left(\frac{1.0}{1.0 + \mathbf{x}_{t+1}\mathbf{P}_t\mathbf{x}_{t+1}^T}\right)\mathbf{y}_{t+1} \\
&= \hat{\mathbf{\Theta}}_t - \mathbf{P}_t\mathbf{x}_{t+1}^T\gamma_{t+1}\mathbf{x}_{t+1}\hat{\mathbf{\Theta}}_t + \mathbf{P}_t\mathbf{x}_{t+1}^T\gamma_{t+1}\mathbf{y}_{t+1} \\
&= \hat{\mathbf{\Theta}}_t + \gamma_{t+1}\mathbf{P}_t\mathbf{x}_{t+1}^T[\mathbf{y}_{t+1} - \mathbf{x}_{t+1}\hat{\mathbf{\Theta}}_t]
\end{aligned}
$$

which yields Eq. (13.7.7).

13.8 APPLICATIONS TO ESTUARY MANAGEMENT

13.8.1 Introduction

Recently, estuary management problems based on hydraulic control for the Lavaca-Tres Palacios Estuary in Texas have been studied (see Chapters 7 and 10). The estuary management for this estuary is defined as maintaining the ecologically sound estuarine condition by controlling the amount of freshwater inflow to the estuary (Bao and Mays, 1994a). In order to maintain an ecologically sound condition for the Lavaca-Tres Palacios Estuary, Martin (1987) modeled an estuary management problem for this estuary by using a linear programming framework. To deal with the nonlinear problems in modeling the estuaries, Tung et al. (1990) developed a nonlinear programming estuarine model with the consideration of uncertainty in the regression equation and solved the model using GRG2 (Lasdon and Waren, 1989). Mao and Mays (1994) developed two multiobjective models based on goal programming procedures and solved these by the GAMS/MINOS computer programs (Brooke et al., 1988) and GRG2. As described in Chapter 7, Bao and Mays (1994a, 1994b) formulated the management problem as a discrete-time, optimal control problem and developed an optimization approach which interfaced GRG2 with HYD-SAL, a two-dimensional hydrodynamic salinity transport model.

As previously discussed, uncertainties often cause enormous inaccuracies when a numerical model is applied to a real-world system for the prediction of the behavior of the real-world system. Therefore, the resulting optimal solution to an optimization model may never be truly optimal for the real-world system, although it may be mathematically optimal for the assumed mathematical system. Thus, various uncertainties should be incorporated in optimization and optimal control models for the estuarine system management modeling.

The chance-constraint stochastic programming technique has been used to deal with the uncertainties in the regression equations for the estuary management, which requires probabilistic assumptions (Chapter 7; Tung et al., 1990; Mao and Mays, 1994). However, there are many inherent, parameter, and model uncertainties, some of which are often impossible to be statistically incorporated in the modeling. Therefore, the feedback control method is needed in order to reduce various kinds of uncertainties by feeding back the new state observation to the control. The proposed feedback control model for the estuarine system management is the discrete-time, stochastic, linear, quadratic, optimal control.

A new type of estuary management model based on discrete-time, stochastic, linear, quadratic feedback, optimal control has been presented in Sec. 13.5. It is a feedback control model which enables the decision-makers to determine the upstream reservoir releases during a time interval after the salinity and nutrients levels are observed at specified locations in the estuary at the beginning of the time interval. The optimal upstream reservoir releases are determined so that the salinity and nutrient level at these locations are as close as possible to the desirable levels for the following time intervals in the sense of statistical expectation. The ungauged inflows, precipitation, and evaporation are incorporated into the model as random variables. The control vector for the estuarine system consists of the freshwater inflows into the estuary, whereas the state vector contains the salinity and nutrient levels at specified locations for measurement in the estuary. The parameter matrices in the system equation can be recursively updated by the recursive least squares (RLS).

In this section, the proposed discrete-time, stochastic, linear, quadratic, optimal control and the RLS technique are applied to the Lavaca-Tres Palacios Estuary in Texas. The numerical examples for parameter estimation show that the RLS is more efficient than the OLS. The numerical examples for the discrete-time, stochastic, linear, quadratic, optimal control are classified into three cases due to many unpredictable uncertainties. These three cases are as follows: (1) the observed state vector at the beginning of each month is the desirable state vector; (2) the observed state vector at the beginning of the month increases and deviates from the desirable state vector as time goes from January to November; and (3) the observed state vector at the beginning of the month decreases and approaches to the desirable state vector from January to November. For each case, different weights are given to the desirable state vector for different time periods and bays. Numerical examples show that the optimal monthly inflows decrease (increase) as the observed state vector is approaching to (deviating from) the desirable state vector.

13.8.2 Application of Discrete-Time, Stochastic, Linear, Quadratic Feedback Optimal Control to Estuary Management

The proposed discrete-time, stochastic, linear, quadratic, optimal control estuarine management model is applied to the Lavaca-Tres Palacios estuarine system. The freshwater inflows into the estuary from the Lavaca-Navidad and Colorado Rivers during each month are considered as the components of the control vector in the optimal control framework, whereas the salinity and other nutrients in the estuary are considered as the components of the state vector. Based on the historical data, the percentages for the ungauged rivers, precipitation, and ungauged return flows are 25%, 23%, and 2%, respectively. Therefore, the ungauged rivers, precipitation, and evaporation should be incorporated into the estuary model. The ungauged river inflows, precipitation, and evaporation should be considered as random variables whose statistical information may be obtained from the rainfall-runoff model.

The optimal freshwater inflows during each month are determined so that the salinity and nutrient levels are as close as possible to the desired levels in the sense of statistical expectation for the rest of year. The desired or prescribed salinity and nutrient levels can be obtained from biological and environmental studies and historical data. For example, the desirable salinity and nutrient levels can be selected to be those corresponding to the largest amount of fish harvest in a particular historical year.

Based on the previous statistical analysis by the Texas Department of Water Resources (1980), a highly linear relationship exists between $\ln(\mathbf{s}_t)$ and $\ln(\mathbf{q}_{c,t})$ in which ln denotes the natural logarithm. In order to reduce the model error, it is reasonable to model the optimal control framework in the logarithmic space (using natural logs) as

$$\begin{aligned} J(\ln(\mathbf{s}_t)) = E\{[\ln(\mathbf{s}_{t_f}) - \ln(\mathbf{s}_{t_f}^D)]^{\mathrm{T}}\mathbf{Q}_{t_f}[\ln(\mathbf{s}_{t_f}) - \ln(\mathbf{s}_{t_f}^D)] \\ + \sum_{t}^{t_f-1} \{[\ln(\mathbf{s}_t) - \ln(\mathbf{s}_t^D)]^{\mathrm{T}}\mathbf{Q}_t[\ln(\mathbf{s}_t) - \ln(\mathbf{s}_t^D)] \\ + [\ln(\mathbf{q}_{c,t}) - \ln(\mathbf{q}_{c,t}^D)]^{\mathrm{T}}\mathbf{R}_t[\ln(\mathbf{q}_{c,t}) - \ln(\mathbf{q}_{c,t}^D)]\}\} \end{aligned} \quad (13.8.1)$$

$$\ln(\mathbf{s}_{t+1}) = \mathbf{A}_t \ln(\mathbf{s}_t) + \mathbf{B}_t \ln(\mathbf{q}_{c,t}) + \mathbf{C}_t \ln(\mathbf{w}_t) + \mathbf{D}_t \ln(\mathbf{v}_t) \quad (13.8.2)$$

in which

$$\mathbf{s}_t = \begin{bmatrix} \text{salinity in the Lavaca Bay} \\ \text{nitrogen in the Lavaca Bay} \\ \text{phosphorus in the Lavaca Bay} \\ \text{salinity in the Matagorda Bay} \\ \text{nitrogen in the Matagorda Bay} \\ \text{phosphorus in the Matagoba Bay} \\ \text{organic carbon in the Matagorda Bay} \end{bmatrix}$$

$$\mathbf{q}_{c,t} = \begin{bmatrix} \text{monthly inflow from the Lavaca} - \text{Navidada River} \\ \text{monthly inflow from the Colorado River} \end{bmatrix}$$

$$\mathbf{w}_t = \begin{bmatrix} \text{monthly precipitation for the Lavaca} - \text{Tres Palacios Estuary} \\ \text{monthly evaporation for the Lavaca} - \text{Tres Palacios Estuary} \\ \text{monthly ungauged freshwater inflow for the Lavaca} \\ \quad - \text{Tres Palacios Estuary} \end{bmatrix}$$

The optimal control law becomes

$$\mathbf{q}^*_{c,t} = \exp(\mathbf{M}_t[\mathbf{A}_t[\ln(\mathbf{s}_t) - \ln(\mathbf{s}^D_t)] + \mathbf{e}_t] + \ln(\mathbf{q}^D_{c,t})) \tag{13.8.3}$$

in which

$$\mathbf{M}_t = -\mathbf{L}_t^{-1}\mathbf{B}_t^{\mathrm{T}}\mathbf{K2}_{t+1} \tag{13.8.4}$$

$$\mathbf{L}_t = \mathbf{R}_t + \mathbf{B}_t^{\mathrm{T}}\mathbf{K2}_{t+1}\mathbf{B}_t \tag{13.8.5}$$

$$\mathbf{K2}_t = \mathbf{Q}_t + \mathbf{A}_t^{\mathrm{T}}\mathbf{K2}_{t+1}\mathbf{A}_t - \mathbf{A}_t^{\mathrm{T}}\mathbf{K2}_{t+1}\mathbf{B}_t\mathbf{L}_t^{-1}\mathbf{B}_t^{\mathrm{T}}\mathbf{K2}_{t+1}\mathbf{A}_t \tag{13.8.6}$$

$$\mathbf{e}_t = \mathbf{A}_t \ln(\mathbf{s}^D_t) + \mathbf{B}_t \ln(\mathbf{q}^D_{c,t}) + \mathbf{C}_t \ln(\overline{\mathbf{w}}_t) + \mathbf{D}_t \ln(\mathbf{v}_t) - \ln(\mathbf{s}^D_{t+1}) \tag{13.8.7}$$

and

$$\mathbf{K2}_T = \mathbf{Q}_T \tag{13.8.8}$$

The minimized cost-to-go performance index is

$$\begin{aligned} J^*(\mathbf{s}_t) = {} & [\ln(\mathbf{s}_t) - \ln(\mathbf{s}^D_t)]^{\mathrm{T}}\mathbf{K2}_t[\ln(\mathbf{s}_t) - \ln(\mathbf{s}^D_t)] \\ & + 2[\ln(\mathbf{s}_t) - \ln(\mathbf{s}^D_t)]^{\mathrm{T}}\mathbf{k1}_t + k0_t \end{aligned} \tag{13.8.9}$$

in which

$$\mathbf{k1}_t = \mathbf{k1}_{t+1} + \mathbf{A}_t^{\mathrm{T}}\mathbf{K2}_{t+1}\mathbf{e}_t - \mathbf{A}_t^{\mathrm{T}}\mathbf{K2}_{t+1}\mathbf{B}_t\mathbf{L}_t^{-1}\mathbf{B}_t^{\mathrm{T}}\mathbf{K2}_{t+1}\mathbf{e}_t \tag{13.8.10}$$

$$\begin{aligned} k0_t = {} & k0_{t+1} + \mathbf{e}_t^{\mathrm{T}}[\mathbf{K2}_{t+1} - \mathbf{K2}_{t+1}\mathbf{B}_t\mathbf{L}_t^{-1}\mathbf{B}_t^{\mathrm{T}}\mathbf{K2}_{t+1}]\mathbf{e}_t \\ & + \mathrm{Trace}\{\mathbf{C}_t^{\mathrm{T}}\mathbf{K2}_{t+1}\mathbf{C}_t \operatorname{cov}(\ln(\mathbf{w}_t))\} \end{aligned} \tag{13.8.11}$$

and

$$\mathbf{k1}_T = \mathbf{0}, \qquad k0_T = 0 \tag{13.8.12}$$

A flowchart for computing the freshwater inflows with parameter estimation is shown in Figure 13.5.

13.8.3 Numerical Examples for Parameter Estimation

The parameter matrices $\mathbf{A}_t$, $\mathbf{B}_t$, $\mathbf{C}_t$, and $\mathbf{D}_t$ in Eq. (13.4.2) must be known before Eq. (13.5.1) is solved. The data in Texas Department of Water Resources (1980) can be used to estimate those parameter matrices. Numerical examples are performed for parameter estimation by the OLS and RLS (Zhao, 1994). In the numerical examples, the deterministic vector $\ln(\mathbf{v}_t)$ is a zero vector and the matrix $\mathbf{D}_t$ is, thus, unnecessary. A computer program is developed by using MATLAB language (MathWorks, 1989) which is a high-level computer code and is very efficient in matrix operation and computation.

The developed MATLAB program is run on a PC-486-33MHz. The estimated parameter matrices estimated by the OLS and RLS are found to be the same for the example presented in Zhao (1994). It should be pointed out that the RLS is better than the OLS from the viewpoint of saving computation time and numerical stability. The computation time by the OLS is much larger than that by the RLS. The comparison in computation time between the OLS and RLS is made by assuming 2000 new observations. For each new observation, the OLS and RLS are applied separately and the computation time for 2000 computations is compared. It has been found from the numerical examples that the computation time for the OLS is much larger than that for the RLS, which confirms the theory.

13.8.4 Numerical Examples of Discrete-Time, Stochastic, Linear, Quadratic Feedback Optimal Control

Numerical examples are used to illustrate the proposed discrete-time, stochastic, linear, quadratic, optimal control estuarine management model. A number of figures are plotted in Zhao (1994) to summarize the computed optimal monthly inflows. Because a real-world system is often unpredictable due to many unknown uncertainties, it is necessary to study the feedback control law under different observed state vectors. Three general cases are considered. First, the observed state vector at the beginning of each month is assumed to be the desirable state vector over the time horizon. Second, the observed state vector at the beginning of each month is assumed to increase and gradually deviate from the desirable state vector as time goes from January to November. Third, the observed state vector at the beginning of each month is assumed to decrease and gradually approach the desirable state vector as time goes from January to November.

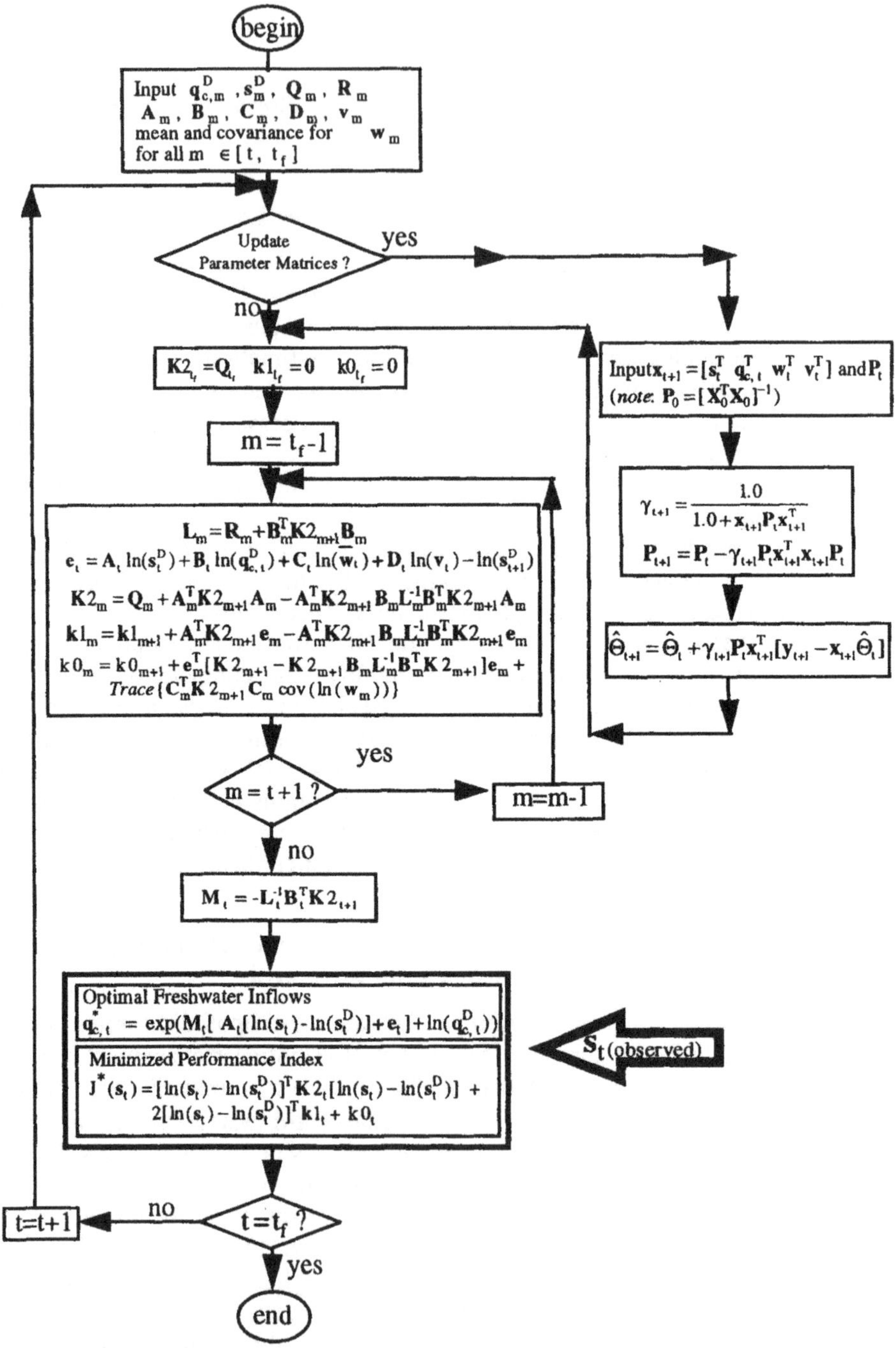

Figure 13.5 Flowchart of computation for stochastic, linear, quadratic feedback control with recursive parameter estimation for lumped-parameter estuary system (Zhao, 1994).

Computer programs are developed by using MATLAB language to compute the optimal monthly inflows. The computation for the monthly inflows for 11 months reqeires 2.69 sec CPU time and 474646 floating-point operations on a PC-486-33MHz. The numerical results show that the optimal monthly inflows from the Lavaca-Navidad River and the Colorado River reasonably respond to different kinds of patterns of the observed salinity values.

REFERENCES

Åström, K. J. and Eykhoff, P., System Identification—A Survey, *Automatica*, Vol. 7, pp. 123–162, 1970.

Athans, M. and Falb, P. L., *Optimal Control*, McGraw-Hill, New York, 1966.

Bao, Y. X., and Mays, L. W., Optimization of Freshwater Inflows to the Lavaca-Tres Palacios Estuary, *Journal of Water Resources Planning and Management*, Vol. 120, No. 2, pp. 199–217, 1994a.

Bao, Y. X., and Mays, L. W., New Methodology for Optimization of Freshwater Inflows to Estuaries, *Journal of Water Resources Planning and Management*, Vol. 120, No. 2, pp. 218–236.

Bertsekas, D. P., *Dynamic Programming: Deterministic and Stochastic Models*, Prentice-Hall, Englewood Cliffs, NJ 1987.

Bitmead, R. R., Bevers, M., and Wertz, V., *Adaptive Optimal Control*, Prentice-Hall, Englewood Cliffs, NJ, 1990.

Brogan, W.L., *Modern Control Theory*, Quantum Publishers, New York, 1974.

Brooke, A., Kendrick, D., and Meeraus, A., *GAMS, a User's Guide*, The Scientific Press, South San Francisco, CA, 1988.

Georgakakos, A. P. and Marks, D. H., A New Method for the Real-Time Operation of Reservoir Systems, *Water Resources Research*, Vol. 23, No. 7, pp. 1376–1390, 1987.

Gerenéser, L. and Caines, P. E., *Topics in Stochastic Systems: Modelling, Estimation and Adaptive Control*, Springer-Verlag, Berlin, 1991.

Graybill, F. A., *Matrices with Applications in Statistics*, Wadsworth, Inc., Belmont, CA, 1983.

Hsia, T. C., System Identification, Lexington Books, Lexington, MA, 1977.

Jacobson, D. H., Martin, D. H., Pachter, M., and Geveci, T., Extensions of Linear-Quadratic Control Theory, in *Lecture Notes in Control and Information Sciences*, A. V., Balakrishnan and M. Thoman (eds.), Springer-Verlag, Berlin, 1980.

Jones, L., Adaptive Control of Ground-Water Hydraulics, *Journal of Water Resources Planning and Magagement*, Vol. 118, No. 1, pp. 1–17, 1992.

Ko, K., McInnis, B., and Goodwin, G., Adaptive Control and Identification of the Dissolved Oxygen Process, *Automatica*, Vol. 18, No. 6, pp. 727–730, 1982.

Kokotovic, P. V., *Foundations of Adaptive Control*, M. Thoma and A. Wyner (eds.), Springer-Verlag, Berlin, 1991.

Kumar, P. R., A Survey of Some Results in Stochastic Adaptive Control, *SIAM Journal of Control and Optimization*, Vol. 23, No. 3, pp. 329–380, 1985.

Kumar, P. R. and Varaiya, P., *Stochastic Systems: Estimation, Identification, and Adaptive Control*, Prentice-Hall, Englewood Cliffs, NJ, 1986.

Lasdon, L. S. and Waren, A. D., *GRG2 User's Guide*, Department of General Business, University of Texas, Austin, 1989.

Ljung, L., *System Identification: Theory for the User*, Prentice-Hall, Englewood Cliffs, NJ, 1987.

Mao, N. and Mays, L. W., Goal Programming Models for Determining Freshwater Inflows to Bays and Estuaries. *Journal of Water Resources Planning and Management*, Vol. 120, No. 3, pp. 316–329, 1994.

Martin, Q. W., Estimating Freshwater Inflow Needs for Texas Estuaries by Mathematical Programming, *Water Resources Research*, Vol. 23, No. 2, pp. 230–238, 1987.

MathWorks, *MATLAB User's Guide*, The MathWorks, Inc., South Natick, MA, 1989.

Maybeck, P. S., *Stochastic Models, Estimation, and Control, Vol. 1*, Academic Press, New York, 1979.

Maybeck, P. S., *Stochastic Models, Estimation, and Control, Vol. 3*, Academic Press, New York, 1982.

Mays, L. W. and Tung, Y. K., *Hydrosystems Engineering and Management*, McGraw-Hill, New York, 1992.

Sinha, N. K. and Kuszta, B., *Modeling and Identification of Dynamic Systems*, Van Nostrand Reinhold Publishing, New York, 1983.

Texas Department of Water Resources, Lavaca-Tres Palacios Estuary: A study of the Influence of Freshwater Flows, Technical Report LP-106, 1980.

Tung, Y. K., Bao, Y. X., Mays, L. W., and Ward, W. H., Optimization of Freshwater Inflow to Estuaries, *Journal of Water Resources Planning and Management*, Vol. 116, No. 4, pp. 567–584, 1990.

Van de Vegte, J., *Feedback Control Systems*, Prentice-Hall, Englewood Cliffs, NJ, 1990.

Wasimi, S. A. and Kitanidis, P. K., Real-Time Forecasting and Daily Operation of a Multireservoir System During Floods by Linear Quadratic Gaussian Control, *Water Resources Research*, Vol. 19, No. 6, pp. 1511–1522, 1983.

Zhao, B., Stochastic Optimal Control and Parameter Estimation for an Estuary System, Ph.D. Dissertation, Department of Civil Engineering, Arizona State University, Tempe, 1994.

Zhao, B. and Mays, L. W., Estuary Management by Discrete-Time Stochastic Linear Quadratic Optimal Control, *Journal of Water Resources Planning and Management*, Vol. 121, No. 5, pp. 382–391, 1995.

Chapter 14
Stochastic Optimal Control for Estuarine Management

14.1 INTRODUCTION

In Chapter 13, parameter estimation and discrete-time, stochastic, linear, quadratic feedback, optimal control for a lumped estuarine system were discussed and applied to the Lavaca-Tres Palacios Estuary in Texas. The parameter estimation techniques for a lumped estuarine system were the ordinary least squares (OLS) and the recursive least squares (RLS) methods. In Chapter 13, the RLS method was shown to be better than the OLS. The estuary management model based on the discrete-time, stochastic, optimal control for a lumped-parameter system is essentially a feedback control which can feed back the new observations of the state vector in the determination of the control vector. The feedback control model can implicitly reduce various kinds of uncertainties such as inherent, parameter, and model uncertainties without involving too much computation and too many probability distribution assumptions. The proposed estuary management model enables the decision-makers to determine the optimal monthly inflows during a month after the salinity or nutrient levels are observed at the beginning of the month.

In this chapter, optimal control and parameter estimation for a distributed-parameter system are discussed and applied to the Lavaca-Tres Palacios Estuary. The optimal control for a distributed-parameter system is discussed and applied. As defined in Chapter 1, a distributed-parameter system is governed by a set of partial differential equations (PDEs) which involve spatial parameters. Since the 1960s, a great deal of research work has been done on the theory and computation analysis for distributed-parameter optimal control, including deterministic and stochastic control. It should be noted that optimal control for a distributed-parameter system is much more difficult than that for a lumped-parameter system. Very little work has been done in the application of the distributed-parameter system control theory to water resources engineering.

Robinson (1971) gave a complete survey of optimal control of distributed-parameter systems with more than 250 references discussed. The results surveyed include (1) methods of reduction from partial differential equations to a set of ordinary differential or difference equations, (2) methods for deriving necessary conditions and sufficient conditions, (3) existence of solutions, (4) controllability and observability, (5) determination of optimal open-loop control, (5) feedback control, and (6) computation methods. Stochastic problems and identification of distributed-parameter systems were also discussed in the survey. The survey summarized the applications of the distributed-parameter optimal control to (1) heating, cooling, and drying problems, (2) hydrodynamics and magneto-hydrodynamics, (3) altitude control of vehicles with flexible structure and/or fluid components, (4) the circling line, and (5) melting, freezing, and crystal growth.

The methods of reduction from distributed-parameter system optimal control to lumped-parameter system (governed by the ordinary differential equation) optimal control include eigenfunction expansions (Kim and Erzberger, 1967), space quantization (Dodd, 1968), time and space quantization (Grahamna D'Souza, 1969), and transfer function approximations. The necessary conditions for the distributed-parameter system control can be developed by using variational methods (Butkovsky, 1961), moment method (Butkovsky, 1964), dynamic programming (Wang, 1964), and function space methods (Butkovsky, 1963). It should be pointed out that the computation problem for distributed-parameter system optimal control is much more complex than that for lumped-parameter system optimal control. Many results of distributed-parameter optimal control are restricted to particular distributed-parameter systems.

Another optimal control problem associated with the distributed-parameter system is boundary control where the control appears in the boundary conditions. Fattorini (1967) gave discussions on the boundary control for first-order and second-order systems. Many boundary control problems can be directly solved by using the methods in the regular distributed-parameter system optimal control.

Since the 1960s, application of the distributed-parameter system optimal control has been a very active field. Applications of distributed-parameter system optimal control to chemical engineering have been reported by Ray (1977). Computation methods for optimizing distributed-parameter system were summarized by Teo and Wu (1984). The distributed-parameter system they considered were first and second boundary-value problems of a linear, second parabolic, partial differential equation.

Stochastic optimal control of distributed-parameter systems has been extensively discussed by Omatu and Seinfeld (1989). The distributed-parameter system herein is stochastic partial differential equation which may be coupled with measurement equations. They used dynamic programming to derive the Hamilton–Jacobi differential equation for a stochastic, optimal control problem of a linear, quadratic, distributed-parameter system with measurement equations. Optimal sensor and actuator locations problems have also been discussed by Omatu and Seinfeld (1989).

Recently, some optimization methods have been applied to optimal control problems such as groundwater management problems (Wanakule, 1984; Wanakule et al., 1985, 1986; Chang et al., 1992; Culver and Shoemaker, 1992, 1993; Whiffen and Shoemaker, 1993), reservoir operation (Murray and Yakowitz, 1979; Unver, 1987; Unver et al., 1987; Carriaga and Mays, 1994), water distribution systems design and operation (Lansey, 1987; Lansey and Mays, 1989; Brion, 1990; Brion and Mays, 1989), and estuarine system management (Bao, 1992; Bao and Mays, 1994a, 1994b; Li and Mays, 1995). This approach is more general compared with the classic approach, in that it considers the distributed-parameter system as a simulator or a function. The simulator is usually a numerical model for solving the partial differential equations. Because the distributed-parameter system is considered as a function, optimization methods can be used directly. Generally speaking, this approach consists of two different methods in that the optimization methods are different. One method uses the GRG2 optimization code (Lasdon and Waren, 1989) to "communicate" with a simulator (or a numerical model for solving the partial differential equation). Another method is based on the differential dynamic programming (Jacobson and Mayne, 1970; Yakowitz and Brian Rutherford, 1984). The first method is more general than the second one because the differential dynamic programming requires that the performance index be a summation of functions over the time.

The advantage of this new approach over the classic deterministic distributed-parameter system control is that this approach can be applied to many deterministic distributed-parameter systems without "knowing" the detailed structure of the partial differential equation as long as there is a numerical model for solving the partial differential equation. The advantage is based on the fact that the optimizor considers the simulator as a function. It is the fact that also leads to a disadvantage that some important information in the structure of the partial differential equation is ignored, which may result in excessive computational effort and the convergence problems. Because the detailed structure of the partial differential equation is ignored, some important theoretical results, such as the necessary conditions that the classic approach offers and the stochastic consideration in the partial differential equations, become very difficult to obtain.

In this chapter, the proposed estuary management model for a distributed-parameter system is based on discrete-time, stochastic, linear, quadratic feedback, optimal control. The distributed-parameter system is governed by the two-dimensional continuity, momentum, and salinity transport partial differential equations (PDEs), which are approximated by a linear lumped-parameter system through numerical linearization of the PDEs about the operating points (see Chapter 7). The PDEs are converted into the ordinary difference equations describing the lumped-parameter system. The results of feedback control from Chapter 13 can be used directly. The operating points are usually predetermined from the previous deterministic model. For example, the approach developed in Chapter 7 for the deterministic distributed-parameter system optimal control can be used to find the operating points which are desirable in the sense of optimization. As there are many unknown uncertainties, the predetermined optimal monthly inflows may be adjusted after the salinity values are observed. The operating points chosen in this study are the desirable state, desirable control, and mean of the random vector. It is assumed that the estuary system has been stable at the operating points. Once the lumped-parameter system is established through numerically solving the numerical model for the PDEs, the feedback control law discussed in Chapter 13 can be directly applied to update the predetermined control vector after the state observations are made. The measurement error in the state vector is considered insignificant. The proposed estuarine management models are applied to the Lavaca-Tres Palacios Estuary. The simulation model, HYD-SAL (Masch and Associates, 1971; Bao and Mays, 1994b; Chapter 7), is incorporated in the proposed estuary management control.

The parameters in the PDEs are usually unknown and need to be determined before the PDEs are used in simulation and model control. The determination of the parameters in the PDEs by using the observed values for the dependent variable in the spatial domain is called the inverse problem. More theoretical discussions on parameter estimation for the general distributed-parameter systems were given by Omatu and Seinfeld (1989). They classified the parameter estimation methods into stochastic approximation methods, least squares identification, and discrete regularization. Application of parameter estimation for a distributed-parameter estuary system can be found in Ten Brummelhuis (1989, 1990).

A complete review of the inverse problem in groundwater hydrology was given by Yeh (1986). He classified the parameter estimation techniques for the groundwater distributed-parameter system into the equation-error criterion (direct method) method and output-error criterion method (indirect method). The generalized matrix method based on equation-error criterion method employed in the groundwater inverse problem requires the substitution of the head observations into the finite-difference scheme. An error term is added to the finite-difference equation. Then an explicit formulation of the unknown parameters can be obtained from which the unknown parameters can be solved. The Gauss–Newton minimization method based on the output-error criterion seeks the optimal parameters such that the difference between the observation heads and the calculated heads is minimized. This method considers the groundwater numerical model as a black box. If there is a numerical model for solving a distributed-parameter system, this method is very easy to use. Yeh (1986) also presented several methods for determining the sensitivity coefficients which are needed in the output-error criterion method. Because many cells in the discretized domain may have the same values for the parameters, the reduction of parameter dimension can be achieved by the zonation method or interpolation method (Yeh, 1986). The Gauss–Newton minimization technique is an effective method for solving an inverse problem associated with a numerical model, as it only requires the evaluation of the Jacobian matrices rather than the Hessian matrices. The computation of the Hessian matrices requires a rather large amount of CPU time when the inverse problem involves a numerical model.

Because PDEs involve many uncertainties, such as the uncertainties in the boundary and initial conditions, the estimated parameter vector by any parameter estimation technique such as the Gauss–Newton minimization technique can be considered as a random variable. Because the mean of a parameter is a more representative value for the true parameter, the estimation of the mean of the parameter is needed. The uncertainty analysis method in conjunction with the Gauss–Newton minimization technique can be used to find the mean of the parameter. Three uncertainty analysis methods are discussed in this chapter, which are the First-Order Second Moment (F.O.S.M.), Rosenblueth's point estimate, and Harr's point estimate methods. After the mean of the parameter is found, the PDEs use the mean as the true value for the parameter. Numerical simulation and system control can then be performed.

The F.O.S.M. method is essentially a first-order uncertainty analysis method by which the PDEs are numerically approximated by the first-order Taylor series expanded about the mean of the random vector consisting of boundary conditions. This method is efficient, but the results may be inaccurate when the functions describing the boundary conditions and the parameter vector are highly nonlinear. In addition, the F.O.S.M method only uses the first moment information of the random variable. Therefore, Rosenblueth's point estimate (Rosenblueth, 1975, 1981; Tung and Yen, 1993) should be used when the highly nonlinear problems exist and the second moment or higher is significant.

For a univariate case (one boundary condition), Rosenblueth's point estimate method approximates the probability distribution of this boundary condition by assuming that the entire probability mass of the boundary condition is "concentrated" at two points. After the two points with their probability masses are computed, the mean of the parameter can then be found. For a multivariate case with N boundary conditions, 2^N points and their probability masses need to be computed. With Rosenblueth's point estimate method, the mean, covariance matrix, and skews for the boundary conditions can be included in the determination of the mean of the parameter vector. It should be noted that the computation time may be intractable when N is fairly large. When N is very large and the third moment for the boundary conditions are not large, Harr's point estimate method (Harr, 1987, 1989) can be used, as it only requires $2N$ points. Harr's point estimate uncertainty analysis method is based on eigenvalue–eigenvector decomposition. The application of F.O.S.M., Rosenblueth's point estimate, and Harr's point estimate uncertainty analysis methods can be found in Mays et al. (1994).

In this chapter, the inverse problem based on the Gauss–Newton minimization algorithm is coupled with the uncertainty analysis methods to find the mean of the parameter in the PDEs. The PDEs are approximated by a numerical model which may use finite differences or finite elements as the solution technique. It should be noted that the inverse problem with the uncertainty analysis methods may need repeated solutions of the numerical model.

14.2 STOCHASTIC OPTIMAL CONTROL FOR A DISTRIBUTED-PARAMETER SYSTEM

14.2.1 Perturbation System Equation

An estuary management model based on the discrete-time, stochastic, optimal control for a discrete-time system is proposed. A lumped-parameter perturbation system equation is used to approximate the two-dimensional PDEs by numerical linearization about the operating points. The operating points herein are chosen to be the desirable states, desirable controls, and mean of the random vector. It is assumed that the estuary system has been stable at operating points. Because the PDEs are reduced to a lumped-parameter system, the feedback control law discussed in Chapter 13 can be applied directly. The decision-makers can determine the optimal control vector after the state vector is observed. By using the feedback control law, the decision-makers can determine the optimal monthly inflows during a month after the salinity values are measured at the test sites in the estuary at the beginning of the month.

The perturbation system equation for a distributed-parameter system can be derived as

$$\delta \mathbf{s}_{t+1} = \mathbf{A}_t \delta \mathbf{s}_t + \mathbf{B}_t \delta \mathbf{q}_{c,t} + \mathbf{C}_t \delta \mathbf{w}_t + \boldsymbol{\varepsilon}_t \qquad (14.2.1)$$

in which

$$\delta \mathbf{s}_t = \mathbf{s}_t - \mathbf{s}_t^D \qquad (14.2.2)$$

$$\delta \mathbf{q}_{c,t} = \mathbf{q}_{c,t} - \mathbf{q}_{c,t}^D \qquad (14.2.3)$$

$$\delta \mathbf{w}_t = \mathbf{w}_t - \overline{\mathbf{w}}_t \qquad (14.2.4)$$

$$\mathbf{A}_t = \left. \frac{\partial \mathbf{s}_{t+1}}{\partial \mathbf{s}} \right|_{\mathbf{s}=\mathbf{s}_t^D} \qquad (14.2.5)$$

$$\mathbf{B}_t = \left. \frac{\partial \mathbf{s}_{t+1}}{\partial \mathbf{q}} \right|_{\mathbf{q}=\mathbf{q}_{c,t}^D} \qquad (14.2.6)$$

$$\mathbf{C}_t = \left. \frac{\partial \mathbf{s}_{t+1}}{\partial \mathbf{w}} \right|_{\mathbf{w}=\overline{\mathbf{w}}_t} \qquad (14.2.7)$$

$$\boldsymbol{\varepsilon}_t = \mathbf{f}(\mathbf{s}_t^D, \mathbf{q}_{c,t}^D, \overline{\mathbf{w}}_t) - \mathbf{s}_{t+1}^D \qquad (14.2.8)$$

in which the superscript D denotes the desirable values for the state vector or control vector, given by the decision-makers, $\mathbf{s}_t$ is the state vector at the beginning of the time interval t, the elements of the state vector $\mathbf{s}_t$ correspond to the state values at different measurement locations in the distributed-parameter system, $\mathbf{q}_{c,t}$ is the control vector during the time interval t, $\mathbf{w}_t$ is the random vector during the time interval t with mean vector $\overline{\mathbf{w}}_t$ and covariance matrix $\text{cov}(\mathbf{w}_t)$, and the function $\mathbf{f}$ describes the discrete-time dynamic system $\mathbf{s}_{t+1} = \mathbf{f}(\mathbf{s}_t, \mathbf{q}_t, \mathbf{w}_t)$. It should be mentioned that $\mathbf{s}_{t+1} = \mathbf{f}(\mathbf{s}_t, \mathbf{q}_t, \mathbf{w}_t)$ and the matrices $\mathbf{A}_t$, $\mathbf{B}_t$, and $\mathbf{C}_t$ can be evaluated by solving the numerical model for the distributed-parameter system.

The quadratic performance index to be minimized is

$$J(\delta \mathbf{s}_t) = E\left[\left[\delta \mathbf{s}_{t_f}^T \mathbf{Q}_{t_f} \delta \mathbf{s}_{t_f}\right] + \sum_{t}^{t_f-1} \left[\delta \mathbf{s}_t^T \mathbf{Q}_t \delta \mathbf{s}_t + \delta \mathbf{q}_{c,t}^T \mathbf{R}_t \delta \mathbf{q}_{c,t}\right]\right] \qquad (14.2.9)$$

in which E denotes the statistical expectation operator, the superscript T denotes the transpose of a matrix or a vector, t_f is the terminal time, the weighting matrix $\mathbf{Q}_t$ for the state vector is symmetric and positive semi-definite, and the weighting matrix $\mathbf{R}_t$ for the control vector is symmetric and positive definite. The purpose of the stochastic optimal control is to "steer" the distributed-parameter system by controlling $\mathbf{q}_{c,t}$ such that Eq. (14.2.9) is minimized.

14.2.2 Derivation of the Perturbation System Equation

Consider a discrete-time dynamic system defined by

$$\mathbf{s}_{t+1} = \mathbf{f}(\mathbf{s}_t, \mathbf{q}_t, \mathbf{w}_t) \tag{14.2.10}$$

in which $\mathbf{s}_t$ is the state vector at the beginning of the time interval t, the elements of the state vector $\mathbf{s}_t$ correspond to the state values at different measurement locations in the distributed-parameter system, $\mathbf{q}_{c,t}$ is the control vector during the time interval t, and $\mathbf{w}_t$ is the random vector during the time interval t with mean vector $\overline{\mathbf{w}}_t$ and covariance matrix $\text{cov}(\mathbf{w}_t)$.

Consider a nominal system equation

$$\mathbf{s}_{n,t+1} = \mathbf{f}(\mathbf{s}_{n,t}, \mathbf{q}_{n,t}, \mathbf{w}_{n,t}) \tag{14.2.11}$$

in which $\mathbf{s}_{n,t}$ is the nominal state vector at the beginning of time interval t, $\mathbf{q}_{n,t}$ is the control vector during time interval t, and $\mathbf{w}_{n,t}$ is the random vector during time interval t.

Define the desirable state vector, desirable control vector, mean of the random vector as the nominal state, control, and random vectors, respectively. Because the desirable state and control vectors and mean of the random vector may not satisfy the nominal system equation (14.2.10), there may be an error. Thus,

$$\mathbf{s}_{t+1}^D = \mathbf{f}(\mathbf{s}_t^D, \mathbf{q}_{c,t}^D, \overline{\mathbf{w}}_t) - \boldsymbol{\varepsilon}_t \tag{14.2.12}$$

in which $\boldsymbol{\varepsilon}_t$ is the error vector.

Performing Taylor's expansion about the "nominal" state, control, and random vectors for Eq. (14.2.10) results in

$$\begin{aligned}\mathbf{s}_{t+1} &= \mathbf{f}(\mathbf{s}_t^D + \delta\mathbf{s}_t, \mathbf{q}_{c,t}^D + \delta\mathbf{q}_{c,t}, \overline{\mathbf{w}}_t + \delta\mathbf{w}_t) \\ &= \mathbf{f}(\mathbf{s}_t^D, \mathbf{q}_{c,t}^D, \overline{\mathbf{w}}_t) + \left.\frac{\partial \mathbf{f}}{\partial \mathbf{s}}\right|_{\mathbf{s}=\mathbf{s}_t^D} \delta\mathbf{s}_t + \left.\frac{\partial \mathbf{f}}{\partial \mathbf{q}}\right|_{\mathbf{q}=\mathbf{q}_{c,t}^D} \delta\mathbf{q}_{c,t} + \left.\frac{\partial \mathbf{f}}{\partial \mathbf{w}}\right|_{\mathbf{w}=\overline{\mathbf{w}}} \delta\mathbf{w}_t \\ &\quad + \boldsymbol{\alpha}_{\text{nominal}}(\delta\mathbf{s}_{t+1}, \delta\mathbf{q}_{c,t}, \delta\mathbf{w}_t)\end{aligned} \tag{14.2.13}$$

in which $\boldsymbol{\alpha}$ is the higher-order error vector.

Subtracting Eq. (14.2.12) from Eq. (14.2.13) yields

$$\begin{aligned}\mathbf{s}_{t+1} - \mathbf{s}_{t+1}^D = {} & \left.\frac{\partial \mathbf{f}}{\partial \mathbf{s}}\right|_{\mathbf{s}=\mathbf{s}_t^D} \delta\mathbf{s}_t + \left.\frac{\partial \mathbf{f}}{\partial \mathbf{q}}\right|_{\mathbf{q}=\mathbf{q}_{c,t}^D} \delta\mathbf{q}_{c,t} + \left.\frac{\partial \mathbf{f}}{\partial \mathbf{w}}\right|_{\mathbf{w}=\bar{\mathbf{w}}} \delta\mathbf{w}_t + \boldsymbol{\varepsilon}_t \\ & + \boldsymbol{\alpha}_{\text{nominal}}(\delta\mathbf{s}_{t+1}, \delta\mathbf{q}_{c,t}, \delta\mathbf{w}_t)\end{aligned} \tag{14.2.14}$$

Thus, neglecting the higher-order error vector $\boldsymbol{\alpha}$ gives the linear perturbation model shown as Eq. (14.2.1).

14.2.3 Analytic, Linear, Quadratic Feedback Control Law

The analytic feedback control law for the lumped-parameter system can be modified and applied to the distributed-parameter system optimal control, because the distributed-parameter system optimal control has been reduced to the lumped-parameter, discrete-time, stochastic, linear, quadratic, optimal control. The analytic feedback control law is then expressed as

$$\mathbf{q}_{c,t}^* = \mathbf{M}_t[\mathbf{A}_t[\mathbf{s}_t - \mathbf{s}_t^D] + \boldsymbol{\varepsilon}_t] + \mathbf{q}_{c,t}^D \tag{14.2.15}$$

in which

$$\mathbf{M}_t = -\mathbf{L}_t^{-1}\mathbf{B}_t^T\mathbf{K2}_{t+1} \tag{14.2.16}$$

$$\mathbf{L}_t = \mathbf{R}_t + \mathbf{B}_t^T\mathbf{K2}_{t+1}\mathbf{B}_t \tag{14.2.17}$$

$$\mathbf{K2}_t = \mathbf{Q}_t + \mathbf{A}_t^T\mathbf{K2}_{t+1}\mathbf{A}_t - \mathbf{A}_t^T\mathbf{K2}_{t+1}\mathbf{B}_t\mathbf{L}_t^{-1}\mathbf{B}_t^T\mathbf{K2}_{t+1}\mathbf{A}_t \tag{14.2.18}$$

and

$$\mathbf{K2}_{t_f} = \mathbf{Q}_{t_f} \tag{14.2.19}$$

The minimized cost-to-go function is

$$J^*(\mathbf{s}_t) = [\mathbf{s}_t - \mathbf{s}_t^D]^T\mathbf{K2}_t[\mathbf{s}_t - \mathbf{s}_t^D] + 2[\mathbf{s}_t - \mathbf{s}_t^D]^T\mathbf{k1}_t + k0_t \tag{14.2.20}$$

in which

$$\mathbf{k1}_t = \mathbf{k1}_{t+1} + \mathbf{A}_t^T\mathbf{K2}_{t+1}\boldsymbol{\varepsilon}_t - \mathbf{A}_t^T\mathbf{K2}_{t+1}\mathbf{B}_t\mathbf{L}_t^{-1}\mathbf{B}_t^T\mathbf{K2}_{t+1}\boldsymbol{\varepsilon}_t \tag{14.2.21}$$

$$\begin{aligned}k0_t = {} & k0_{t+1} + \boldsymbol{\varepsilon}_t^T[\mathbf{K2}_{t+1} - \mathbf{K2}_{t+1}\mathbf{B}_t\mathbf{L}_t^{-1}\mathbf{B}_t^T\mathbf{K2}_{t+1}]\boldsymbol{\varepsilon}_t \\ & + \text{Trace}\{\mathbf{C}_t^T\mathbf{K2}_{t+1}\mathbf{C}_t\ \text{cov}(\mathbf{w}_t)\}\end{aligned} \tag{14.2.22}$$

and

$$\mathbf{k1}_{t_f} = \mathbf{0}, \qquad k0_{t_f} = 0 \tag{14.2.23}$$

A flowchart for computing the real-time, stochastic, feedback, optimal control for a distributed-parameter estuary system is shown in Figure 14.1.

14.3 PARAMETER ESTIMATION WITH UNCERTAINTY ANALYSIS

14.3.1 Parameter Estimation by Gauss–Newton Minimization Method for PDEs

Suppose there are n test sites where the states can be measured. The output error criterion to be minimized is defined as

$$J = [\mathbf{s} - \mathbf{s}^*]^T[\mathbf{s} - \mathbf{s}^*] \tag{14.3.1}$$

where $\mathbf{s}^*$ is the measured state vector of $(n \times 1)$ and $\mathbf{s}$ is the computed state vector of $(n \times 1)$ from a numerical model such as the HYD-SAL for an estuary problem. Let $\mathbf{p}$ be the parameter vector of $(m \times 1)$ to be estimated. The following parameter vector sequences will converge to the optimal parameter vector with the given starting point:

$$\mathbf{p}^{k+1} = \mathbf{p}^k - \rho^k\mathbf{d}^k \tag{14.3.2}$$

in which

$$\begin{aligned} \mathbf{A}^k\mathbf{d}^k &= \mathbf{g}^k \\ \mathbf{A}^k &= [\mathbf{J}_D(\mathbf{p}^k)]^T[\mathbf{J}_D(\mathbf{p}^k)] \\ \mathbf{g}^k &= [\mathbf{J}_D(\mathbf{p}^k)]^T[\mathbf{s}(\mathbf{p}^k) - \mathbf{s}^*] \end{aligned} \tag{14.3.3}$$

in which $\mathbf{J}_D(\mathbf{p}^k)$ is the Jacobian matrix

$$\mathbf{J}_D(\mathbf{p}^k) = \begin{bmatrix} \dfrac{\partial s_1}{\partial p_1} & \dfrac{\partial s_1}{\partial p_2} & \cdots & \dfrac{\partial s_1}{\partial p_m} \\ \dfrac{\partial s_2}{\partial p_1} & \dfrac{\partial s_2}{\partial p_2} & \cdots & \dfrac{\partial s_2}{\partial p_m} \\ \vdots & \vdots & & \vdots \\ \dfrac{\partial s_n}{\partial p_1} & \dfrac{\partial s_n}{\partial p_2} & \cdots & \dfrac{\partial s_n}{\partial p_2} \end{bmatrix}_{(n \times m)} \tag{14.3.4}$$

and it is evaluated by solving the numerical model for the partial differential equations. The step size ρ^k can be determined by a quadratic interpolation scheme such that $J(\mathbf{p}^{k+1}) < J(\mathbf{p}^k)$, or simply by a trial-and-error scheme (Yeh, 1986).

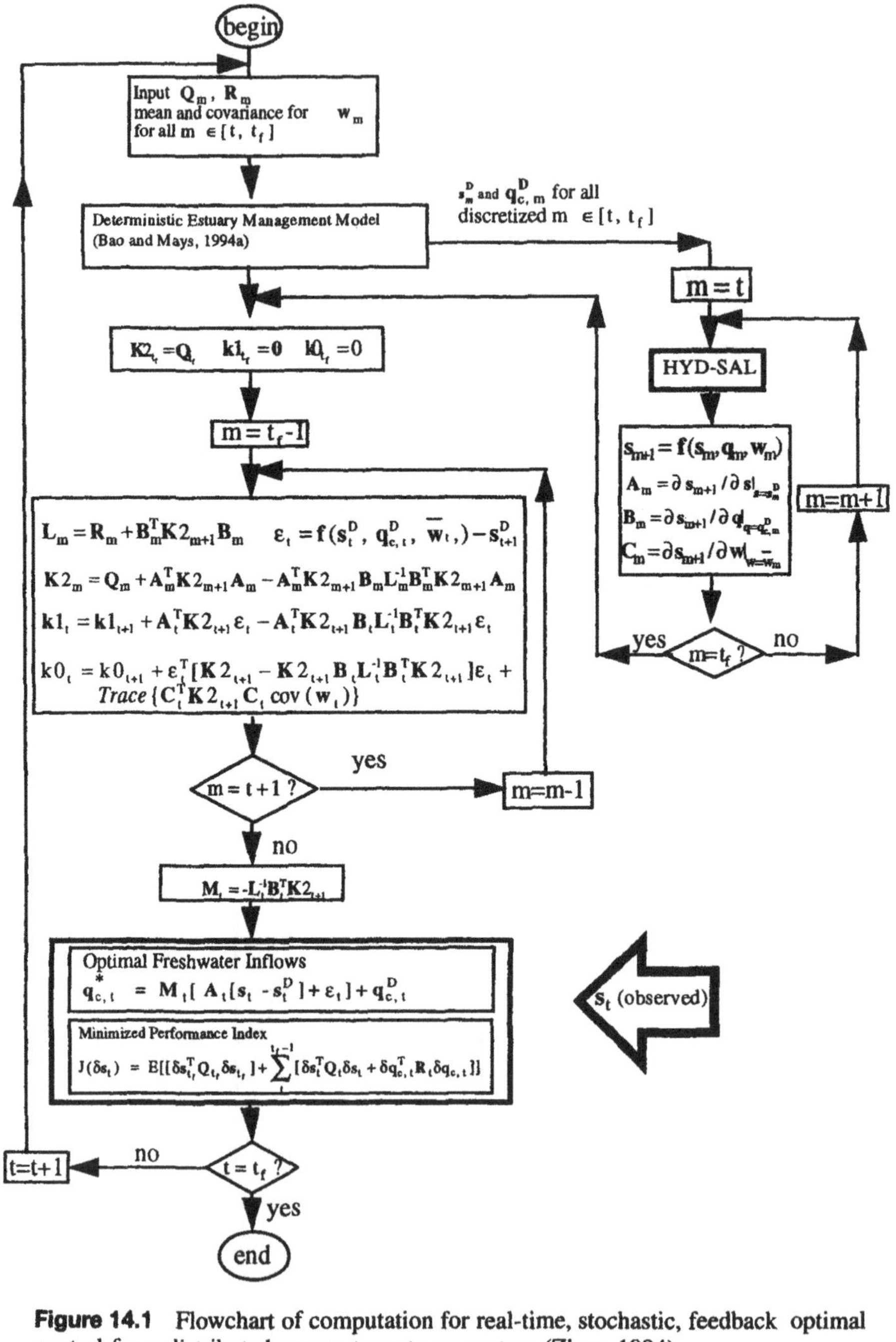

Figure 14.1 Flowchart of computation for real-time, stochastic, feedback optimal control for a distributed-parameter estuary system (Zhao, 1994).

A distributed-parameter system usually involves many uncertainties which may be due to various kinds of uncertain boundary conditions and other variables. For example, tidal flows, ungauged inflows, rainfall, and evaporation can be considered as random variables in the numerical model HYD-SAL in which the parameter vector **p** may consist of the spatial Manning's roughness coefficient and dispersion coefficient. Therefore, the parameter vector **p** estimated by the aforementioned optimization technique is also random. The following uncertainty analysis methods should be incorporated in the parameter estimation to find the mean of the parameter **p** which is considered as a more representative value for the true parameter.

14.3.2 Uncertainty Analysis Methods with Parameter Estimation for PDEs

14.3.2.1 First-Order Second Moment (F.O.S.M.) with Parameter Estimation for PDEs

The F.O.S.M. method discussed herein is essentially a first-order analysis method based on Taylor series expansion about the mean value. It enables the estimation of the mean and variance of a random variable which is functionally related to several random variables. Consider a continuous random variable $Y = g(\mathbf{X})$ in which $\mathbf{X} = (X_1, X_2, \ldots, X_k)$ is a random vector with a given mean and covariance matrix. Through a Taylor's expansion about the means of k random variables, the first-order approximation of the random variable Y can be expressed as

$$Y = g(\bar{\mathbf{x}}) + \sum_{i=1}^{k} \left[\frac{\partial g(\mathbf{X})}{\partial X_i}\right]_{\mathbf{X}=\bar{\mathbf{x}}} (X_i - \bar{x}_i) + \sum_{i=1}^{k}\sum_{j=1}^{k} \left[\frac{\partial^2 g(\mathbf{X})}{\partial X_i\, \partial X_j}\right]_{\mathbf{X}=\bar{\mathbf{x}}} (X_i - \bar{x}_i)(X_j - \bar{x}_j) + \cdots \quad (14.3.5)$$

in which $\bar{\mathbf{x}} = (\bar{x}_1, \bar{x}_2, \ldots, \bar{x}_k)$ is a vector containing the means of k random variables.

Neglecting the higher-order terms yields the first-order approximation as

$$Y \approx g(\bar{\mathbf{x}}) + \sum_{i=1}^{k} \left[\frac{\partial g(\mathbf{X})}{\partial X_i}\right]_{\mathbf{X}=\bar{\mathbf{x}}} (X_i - \bar{x}_i) \quad (14.3.6)$$

Applying the expectation and variance operators to the both sides of Eq. (14.3.6), one obtains the mean and variance of random variable Y as (Mays and Tung, 1992)

$$E[Y] \approx g(\bar{\mathbf{x}}) \tag{14.3.7}$$

$$\mathrm{Var}[Y] \approx \sum_{i=1}^{k} a_i^2\sigma_i^2 + 2\sum_{i=1}^{k}\sum_{j=1}^{k} a_i a_j \operatorname{cov}[X_i, X_j] \tag{14.3.8}$$

in which

$$a_i = \left.\frac{\partial g}{\partial x_i}\right|_{\bar{\mathbf{x}}} \tag{14.3.9}$$

$$\sigma_i^2 = \mathrm{Var}[X_i] \tag{14.3.10}$$

and $\operatorname{cov}[X_i, X_j]$ is the covariance between variables X_i and X_j. The coefficients a_i are called the sensitivity coefficients.

Because partial differential equations are essentially functions describing the relationships among the variables, the F.O.S.M. can be directly applied to the partial differential equations. Suppose that the boundary conditions are random and form a random vector **X** with a given mean and covariance matrix. Then the optimal parameter **p** is also random. The F.O.S.M. can be used to estimate the mean and variance for the random parameter **p**.

Based on F.O.S.M., the average values of the boundary conditions are used in the PDEs when the PDEs are solved for parameter estimation. The resulting optimal parameter is then the mean of the random parameter [see Eq. (14.3.7)]. In Eq. (14.3.7), the random vector **X** may be considered as the boundary conditions and $E[Y]$ is the mean of the parameter. The function g in Eq. (14.3.7) may be considered as the parameter estimation algorithm involving optimization techniques such as the Gauss–Newton algorithm in conjunction with the numerical model for PDEs such as the HYD-SAL. If the variance of the parameter is to be estimated, the sensitivity coefficients of the parameter with respect to the average boundary conditions are evaluated by numerically solving the PDEs such as the HYD-SAL for an estuary problem. Figure 14.2 shows a flowchart for the parameter estimation in conjunction with the F.O.S.M. uncertainty analysis method.

Because the F.O.S.M. method is based on first-order analysis, the results by the F.O.S.M. method may not be accurate when the functions are highly nonlinear. It also should be pointed out that the F.O.S.M. method cannot use the third moment information such as the skew vector for the random vector **X**. It may be noted that even the covariance matrix for the random vector **X** is not included in determining the mean vector of the parameter vector **p**. Therefore, the estimated mean for the parameter Y is not accurate when the skew for each element of the random vector **X** is large.

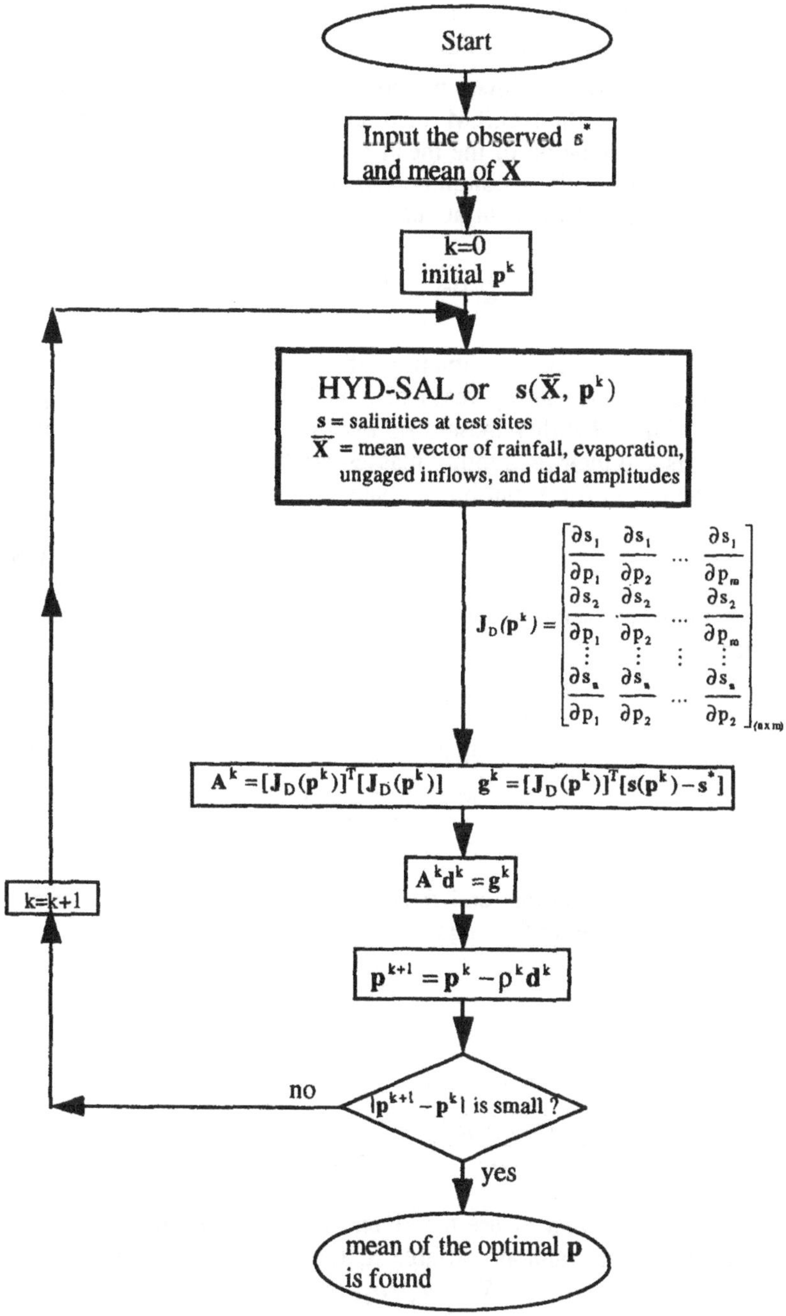

Figure 14.2 Flowchart of parameter estimation in conjunction with F.O.S.M. uncertainty analysis method for a distributed-parameter estuary system (Zhao, 1994).

14.3.2.2 Rosenblueth's Multivariate Point Estimate Method with Parameter Estimation for PDEs

Rosenblueth's point estimate method is a more accurate uncertainty analysis method than the F.O.S.M. method, as it can incorporate the skews for **X** in the determination of the mean for Y. Consider a univariate case, $Y = g(X)$, in which X is a continuous random variable with PDF $f_X(x)$. By definition, the mth moment about the origin for Y is

$$E(Y^m) = \int g^m(x) f_X(x)\, dx \qquad (14.3.11)$$

Rosenblueth's point estimate method approximates $f_X(x)$ by assuming that the entire probability mass of X is "concentrated" at two points x_- and x_+ with probability masses p_- and p_+, respectively. Therefore, if the two points and their probability masses can be found, then the mth moment about the origin for Y can be estimated by

$$E(Y^m) \approx g^m(x_-)p_- + g^m(x_+)p_+ \qquad (14.3.12)$$

Thus, the mean, standard deviation, and skew coefficient of Y can be computed by

$$\mu_Y = E(Y) \approx g(x_-)p_- + g(x_+)p_+ \qquad (14.3.13)$$

$$\sigma_Y^2 = E[Y^2] - \mu_Y^2 \qquad (14.3.14)$$

$$\gamma_Y = \frac{E[Y^3] - 3\mu_Y E[Y^2] + 2\mu_Y^3}{\sigma_Y^3} \qquad (14.3.15)$$

in which the two points and their probability masses are determined by solving the following four equations:

$$p_- + p_+ = 1.0 \qquad (14.3.16)$$

$$p_- x_- + p_+ x_+ = \mu_X \qquad (14.3.17)$$

$$p_-(x_- - \mu_X)^2 + p_+(x_+ - \mu_X) = \sigma_X^2 \qquad (14.3.18)$$

$$p_- \left(\frac{x_- - \mu_X}{\sigma_X}\right)^3 + p_+ \left(\frac{x_+ - \mu_X}{\sigma_X}\right)^3 = \gamma \qquad (14.3.19)$$

in which μ_X, σ_X, and γ_X are the given mean, standard deviation, and skew coefficient of the random variable X, respectively. Standardizing the random variable X by $Z = (X - \mu_X)/\sigma_X$, one obtains

$$p_- + p_+ = 1.0 \tag{14.3.20}$$

$$p_- z_- + p_+ z_+ = 0.0 \tag{14.3.21}$$

$$p_- z_-^2 + p_+ z_+^2 = 1.0 \tag{14.3.22}$$

$$p_- z_-^3 + p_+ z_+^3 = \gamma_Z = \gamma_X \tag{14.3.23}$$

Thus, the standardized two points and their probabilities can be derived as

$$z_- = \tfrac{1}{2}\gamma_X - \sqrt{1 + (\tfrac{1}{2}\gamma_X)^3} \tag{14.3.24}$$

$$z_+ = \gamma_X - z_- \tag{14.3.25}$$

$$p_- = \frac{z_+}{z_+ - z_-} \tag{14.3.26}$$

$$p_+ = 1.0 - p_- \tag{14.3.27}$$

The original two points can then be found by

$$x_- = \mu_X + z_- \sigma_X \tag{14.3.28}$$

$$x_+ = \mu_X + z_+ \sigma_X \tag{14.3.29}$$

Consider a multivariate case $Y = g(\mathbf{X})$ in which $\mathbf{X} = (X_1, X_2, \ldots, X_N)$ is the random vector with a given mean vector, covariance matrix, and skew vector. Each X_i has two mass points X_{i+} and X_{i-}, with probability masses p_{i+} and p_{i-}, respectively. By using the sign indicator for sake of simplicity, X_i takes two mass points X_{i,δ_i} with p_{i,δ_i}, where δ_i can be either $-$ or $+$. The probability p_{i,δ_i} for the ith variable is determined as in the univariate case. Therefore, the random vector $\mathbf{X}$ can take 2^N possible values, each of which has the probability mass $p(\delta_1, \delta_2, \ldots, \delta_N)$. This probability mass can be computed by (Tung and Yen, 1993)

$$p(\delta_1, \delta_2, \ldots, \delta_N) = \prod_{i=1}^{N} p_{i,\delta_i} + \prod_{i=1}^{N-1} \left[\sum_{j=i+1}^{N} \delta_i \delta_j a_{ij} \right] \tag{14.3.30}$$

in which

$$a_{ij} = \frac{\rho_{ij}/2^N}{\sqrt{\prod_{i=1}^{N} [1 + (\tfrac{1}{2}\gamma_i)^2]}} \tag{14.3.31}$$

in which ρ_{ij} is the correlation coefficient between X_i and X_j.

Suppose the boundary conditions for a distributed-parameter system are random and they form a random vector $\mathbf{X}$ with a given mean vector, skew coefficient vector, and covariance matrix, then Rosenblueth's point estimate method can be used to find the 2^N "concentrated" points with their probability masses. For each of the 2^N "concentrated" points, the inverse problem involving an optimization technique in conjunction with a numerical model for PDEs is solved to yield one value of the parameter Y. Then the mth moment about the origin for Y can be found by summing 2^N terms, each of which is the multiplication of the estimated parameter to its mth power and its probability mass. Figure 14.3 shows the flowchart for the parameter estimation in conjunction with Rosenblueth's point estimate for a distributed-parameter estuary system.

It should be pointed out that because Rosenblueth's point-estimate method requires 2^N "concentrated" points for N random variables, the computation time may be significantly large when N is very large. If the computation becomes intractable and the third or higher moments for the random vector $\mathbf{X}$ are fairly small, one may use Harr's multivariate point estimate method which only requires $2N$ "concentrated" points.

14.3.2.3 Harr's Multivariate Point Estimate Method with Parameter Estimation for PDEs

Consider $Y = g(\mathbf{X})$ in which $\mathbf{X}$ is the random vector of dimension $(N \times 1)$ with mean and covariance matrix. The $2N$ points for Harr's method are computed by

$$\mathbf{x}_{i\pm} = \mu \pm \sqrt{N}\, \boldsymbol{\Sigma}^{1/2}\mathbf{v}_i \tag{14.3.32}$$

in which μ is the mean vector of $\mathbf{x}$, $\boldsymbol{\Sigma}$ is a diagonal matrix of variances of $\mathbf{x}$, and $\mathbf{v}_i$ is the ith eigenvector of the correlation matrix of $\mathbf{x}$, $i = 1, 2, \ldots, N$. The eigenvalue–eigenvector decomposition for the correlation matrix is

$$\mathrm{Cor}(\mathbf{x}) = \mathbf{V}\boldsymbol{\Lambda}\mathbf{V}^t \tag{14.3.33}$$

in which each column of the eigenvector matrix is $\mathbf{V}_{(N\times N)}$ is $\mathbf{v}_i$ whose corresponding eigenvalue λ_i is the diagonal elements of the diagonal matrix Λ. Then, the mth moment of Y is

$$E[Y^m] = \frac{\sum_{i=1}^{N} \lambda_i \bar{y}_i^m}{N} \tag{14.3.34}$$

in which λ_i is the eigenvalue of the correlation matrix of $\mathbf{x}$, and

$$\bar{y}_i^m = \frac{g^m(\mathbf{x}_{i+}) + g^m(\mathbf{x}_{i-})}{2} \tag{14.3.35}$$

Consider the numerical model HYD-SAL. Suppose the rainfall, evap-

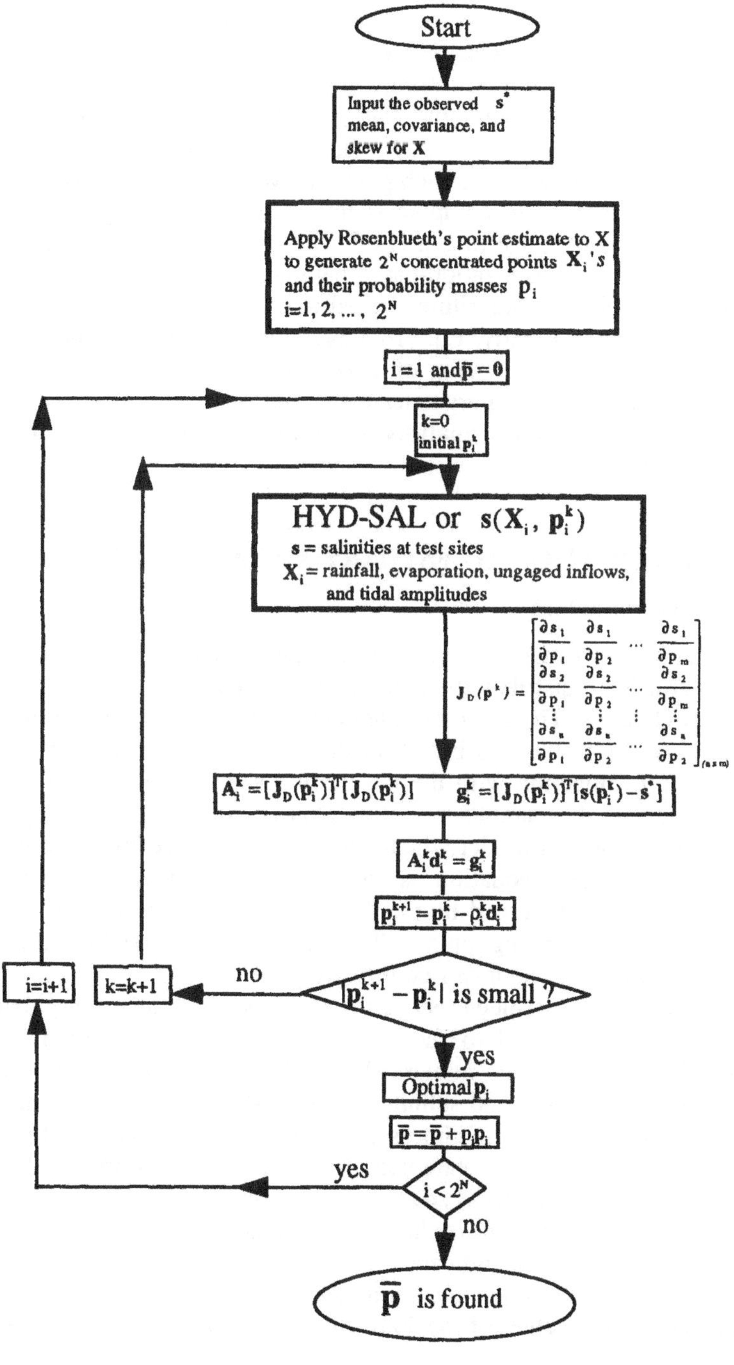

Figure 14.3 Flowchart of parameter estimation with Rosenblueth's point estimate uncertainty analysis method for a distributed-parameter estuary system (Zhao, 1994).

oration, ungauged inflows, and tide flows are random and they form the random vector **X** of dimension ($N \times 1$) with the known mean vector μ and correlation matrix. Performing eigenvector–eigenvalue decomposition [Eq. 14.3.33] on the correlation matrix for **X** yields N eigenvectors $\mathbf{v}_i$ and N eigenvalues λ_i, $i = 1, 2, \ldots, N$. Then, N pairs of points are computed by using Eq. (14.3.32). Note each point is a vector. For each pair of points, Eq. (14.3.35) is used to compute N intermediate values $\bar{y}_i^m$. It should be noted that the function g in Eq. (14.3.35) involves the inverse problem in which a particular optimization-based parameter estimation technique such as the Gauss–Newton algorithm is used in conjunction with the numerical model HYD-SAL. Finally, Eq. (14.3.34) is used to find the mth moment $E[Y^m]$. In particular, the mean of Y is found when $m = 1$. Figure 14.4 shows the flowchart for parameter estimation in conjunction with Harr's point estimate uncertainty analysis method for a distributed-parameter estuary system.

14.4 APPLICATION TO ESTUARY MANAGEMENT

14.4.1 Introduction

The parameter estimation technique and the discrete-time stochastic optimal control for a distributed-parameter system have been discussed in the previous chapters. The proposed parameter estimation technique for solving the inverse problem is based on the Gauss–Newton minimization algorithm in conjunction with the uncertainty analysis methods to find the mean vector of the parameter vector. This procedure involves repeatedly calling the numerical model which solves the PDEs. A lumped-parameter perturbation system equation is used to approximate the distributed-parameter system governed by the PDEs through the numerical linearization about the operating points. Then the feedback control law for a lumped-parameter system can be directly used.

The proposed parameter estimation technique and the estuary management model are applied to the Lavaca-Tres Palacios Estuary in Texas. HYD-SAL is used as the numerical model to solve the two-dimensional continuity, momentum, and salinity transport PDEs. For the parameter estimation, the Manning's roughness coefficient and dispersion coefficient in the spatial domain form a parameter vector **p** to be determined. Because many cells in the discretized space have the same values of the Manning's roughness coefficient and the dispersion coefficient, the dimension of the parameter vector **p** is significantly reduced. The output-error criterion required for the Gauss–Newton minimization algorithm is associated with the spatial salinity values of the 30 test sites. The random vector **X** required

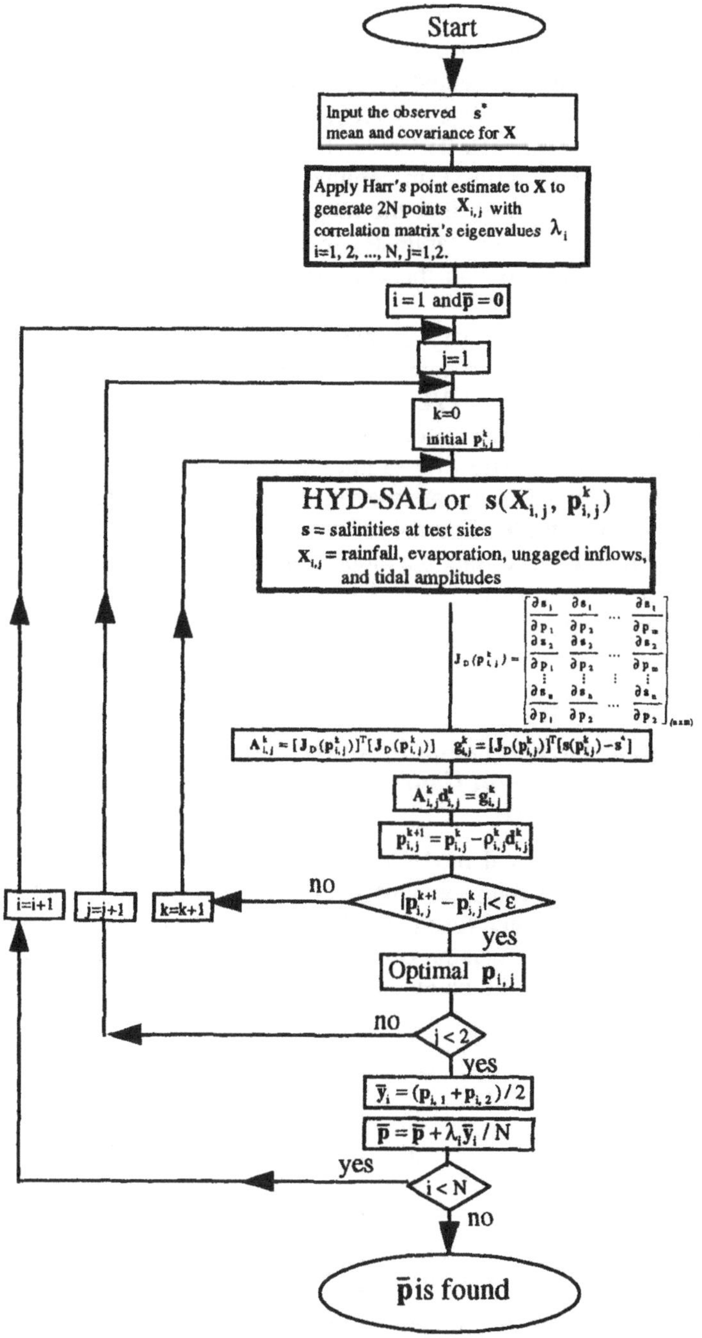

Figure 14.4 Flowchart of parameter estimation with Harr's point estimate uncertainty analysis method for a distributed-parameter estuary system (Zhao, 1994).

for the uncertainty analysis in HYD-SAL consists of the rainfall, evaporation, sum of the ungauged inflows, and sum of the tide amplitudes over two cycles. The eight ungauged creeks are the Carancahua Creek, Choc Creek, Garcitas Creek, Huisache Creek, Keller Creek, Powderhorn Creek, Tres Palacios Creek, and Turtle Creek. Three uncertainty analysis methods previously discussed are used, in conjunction with the Gauss–Newton algorithm and HYD-SAL, to find the mean vector of the parameter vector **p**. After the mean of each parameter in HYD-SAL is found, HYD-SAL with the new parameters can then be used in the simulation and system control.

For the proposed estuary management control model, a lumped-parameter system is used to approximate the PDEs at the operating points by numerically solving the HYD-SAL. The operating points chosen in this study are the desirable state vector, desirable control vector, and mean of the random vector **X**. After the salinity values at the 30 test sites in the Lavaca-Tres Palacios Estuary are measured without significant errors at the beginning of a month, the optimal inflows from the Lavaca-Navidad River and the Colorado River during the month can be computed by the analytical feedback control law. In the numerical examples, three different kinds of observed salinity patterns are assumed to study the response of the feedback control law.

14.4.2 Parameter Estimation by Uncertainty Analysis Methods in Conjunction with Gauss–Newton Minimization Method with HYD-SAL

HYD-SAL is a numerical model for solving the two-dimensional hydrodynamic and salinity transport partial differential equations (PDEs). The parameters to be estimated for the PDEs in this study are Manning's roughness coefficients and dispersion coefficients for each cell in the discretized spatial domain. Define the parameter vector **p** as

$$\mathbf{p} = \begin{bmatrix} n_1 \\ n_2 \\ \vdots \\ n_M \\ d_1 \\ d_2 \\ \vdots \\ d_M \end{bmatrix} \tag{14.4.1}$$

in which M is the number of cells in the discretized space, n_i is the Manning's roughness coefficient for the ith cell, $i = 1, 2, \ldots, M$, and d_i is the dispersion coefficient for the ith cell. However, as many cells have the same values for Manning's roughness coefficient, the number of different roughness coefficients to be estimated can be significantly reduced. This also is the case for the dispersion coefficient. Therefore, redefine the parameter vector as

$$\mathbf{p} = \begin{bmatrix} n_1 \\ n_2 \\ \vdots \\ n_N \\ d_1 \\ d_2 \\ \vdots \\ d_L \end{bmatrix} \tag{14.4.2}$$

or

$$\mathbf{p} = \begin{bmatrix} p_1 \\ p_2 \\ \vdots \\ p_{N+L} \end{bmatrix}_{(N+L \times 1)} \tag{14.4.3}$$

in which $N \ll M$ and $L \ll M$.

Because there are 30 test sites for salinity measurement in the Lavaca-Tres Palacios estuary, the dimension of the state vector is (30×1). The optimal parameter vector $\mathbf{p}$ minimizes the output error

$$J = [\mathbf{s} - \mathbf{s}^*]^T[\mathbf{s} - \mathbf{s}^*] \tag{14.4.4}$$

where $\mathbf{s}^*$ is the measured state vector of (30×1) consisting of the salinity values at 30 test sites, and $\mathbf{s}$ is the vector consisting of 30 salinity values computed from the HYD-SAL.

Define a vector $\mathbf{X}$ as

$$\mathbf{X} = \begin{bmatrix} \text{rainfall rate} \\ \text{evaporation rate} \\ \sum_{k=1}^{K_{\text{ungauge}}} Q_{\text{ungauge},k} \\ \sum_{k=1}^{K_{\text{tide}}} T_{\text{tide},k} \end{bmatrix}_{(4 \times 1)} \tag{14.4.5}$$

in which the unit for the rainfall rate is inch/year, the unit for the evaporation rate is inch/year, $Q_{\text{ungauge},k}$ is the discharge from the kth ungauged river (cfs), $T_{\text{tide},k}$ is the hourly gulf tidal amplitude above the mean sea level (feet), K_{ungauge} is the number of the ungauged rivers, and K_{tide} is the number of the ordinates for a tide. For the Lavaca-Tres Palacios Estuary, there are eight ungauged creeks or $K_{\text{ungauge}} = 8$. They are the Carancahua Creek, Choc Creek, Garcitas Creek, Huisache Creek, Keller Creek, Powderhorn Creek, Tres Palacios Creek, and Turtle Creek. The tidal excitation data are the two tidal cycle data. Each tidal cycle is 25 hr; thus, $K_{\text{tide}} = 50$.

If rainfall, evaporation, ungauged inflows, and tide flows for the Lavaca-Tres Palacios Estuary are considered to be random variables, then the vector $\mathbf{X}$ is a random vector of dimension (4×1). Thus, the estimated optimal parameter vector $\mathbf{p}$ is also a random vector. With the belief that the mean of the parameter vector $\mathbf{p}$ is considered to be a more representative value for the true parameter, it is necessary to use an uncertainty analysis method to estimate the mean of the random parameter vector $\mathbf{p}$. Therefore, the uncertainty analysis methods and the Gauss–Newton minimization algorithm are combined to find the mean of the parameter vector $\mathbf{p}$.

In Sec. 14.3, the F.O.S.M. method, Rosenblueth's point estimate method, and Harr's point estimate method in conjunction with the Gauss–Newton minimization and the numerical model HYD-SAL have been discussed. When the skew vector for the random vector $\mathbf{X}$ is a zero vector, the F.O.S.M method in conjunction with the Gauss–Newton minimization algorithm and HYD-SAL can be used to find the mean of the parameter vector $\mathbf{p}$. This is achieved by substituting the mean vector for $\mathbf{X}$ into the PDEs. Performing the Gauss–Newton minimization algorithm on the output-error criterion in conjunction with HYD-SAL results in the optimal parameter $\mathbf{p}$ which is actually the mean vector for $\mathbf{p}$ based on the F.O.S.M. method.

When the skew for the rainfall, evaporation, ungauged inflows, and tide flows are large, Rosenblueth's point estimate method should be used to incorporate the skew in the determination of the mean of random parameter $\mathbf{p}$. Rosenblueth's point estimate method is used to generate a number of "concentrated" points with its probability mass for the random vector consisting of the rainfall, evaporation, ungauged inflows, and tide flows. For each point, the Gauss–Newton minimization algorithm is used to find the optimal parameter vector. It may be noted that each point is a vector. For each iteration in the Gauss–Newton algorithm, the HYD-SAL must be solved $N + L$ times in order to compute the $N + L$ columns of the Jacobian matrix

$$\mathbf{J}_D(\mathbf{p}^k) = \begin{bmatrix} \frac{\partial s_1}{\partial p_1} & \frac{\partial s_1}{\partial p_2} & \cdots & \frac{\partial s_1}{\partial p_{N+L}} \\ \frac{\partial s_2}{\partial p_1} & \frac{\partial s_2}{\partial p_2} & \cdots & \frac{\partial s_2}{\partial p_{N+L}} \\ \vdots & \vdots & & \vdots \\ \frac{\partial s_{30}}{\partial p_1} & \frac{\partial s_{30}}{\partial p_2} & \cdots & \frac{\partial s_{30}}{\partial p_{N+L}} \end{bmatrix}_{(30 \times N+L)} \tag{14.4.6}$$

After the optimal parameter vector is found for each of those "concentrated" points, the mean of the random parameter vector **p** can be found.

If a new random vector **X** is defined as

$$\mathbf{X} = \begin{bmatrix} \text{rainfall} \\ \text{evaporation} \\ Q_1 \\ Q_2 \\ \vdots \\ Q_{K_{\text{ungauge}}} \\ T_1 \\ T_2 \\ \vdots \\ T_{K_{\text{tide}}} \end{bmatrix} \tag{14.4.7}$$

and the dimension of **X** is very large, then Rosenblueth's point estimate method may require a significantly large computation and effort. If the dimension for the random vector **X** is very large and the skew for this random vector is fairly small, Harr's point estimate method is preferred, as it requires less computation than Rosenblueth's point estimate method.

14.4.3 Discrete-Time, Stochastic, Linear, Quadratic Feedback, Optimal Control with HYD-SAL

Consider the Lavaca-Tres Palacios Estuary. Define the state vector, control vector, and random vector respectively as

$$\mathbf{s}_t = \begin{bmatrix} s_{1,t} \\ s_{2,t} \\ \vdots \\ s_{30,t} \end{bmatrix} \tag{14.4.8}$$

$$\mathbf{q}_{c,t} = \begin{bmatrix} q_{\text{Lavaca},t} \\ q_{\text{Colorado},t} \end{bmatrix} \tag{14.4.9}$$

$$\mathbf{w}_t = \begin{bmatrix} r_t \\ e_t \\ Q_{u,t} \\ T_t \end{bmatrix} \tag{14.4.10}$$

in which $s_{i,t}$ the salinity value for the ith test site in the Lavaca-Tres Palacios Estuary at the beginning of the tth month, $i = 1, 2, \ldots, 30$, $t = 1, 2, \ldots, 12$, $q_{\text{Lavaca},t}$ is the monthly inflow to the estuary from the Lavaca-Navidad River during month t, $q_{\text{Colorado},t}$ is the monthly inflow to the estuary from the Colorado River during month t, r_t is the monthly precipitation rate, e_t is the monthly evaporation rate, $Q_{u,t}$ is the total monthly discharge from the eight ungauged creeks, and T_t is the sum of the gulf tide amplitudes over two cycles, each of which is 25 hr.

The perturbation system equation for HYD-SAL is written as

$$\delta\mathbf{s}_{t+1} = \mathbf{A}_t\delta\mathbf{s}_t + \mathbf{B}_t\delta\mathbf{q}_{c,t} + \mathbf{C}_t\delta\mathbf{w}_t + \boldsymbol{\varepsilon}_t \tag{14.4.11}$$

in which

$$\delta\mathbf{s}_t = \mathbf{s}_t - \mathbf{s}_t^D \tag{14.4.12}$$

$$\delta\mathbf{q}_{c,t} = \mathbf{q}_{c,t} - \mathbf{q}_{c,t}^D \tag{14.4.13}$$

$$\delta\mathbf{w}_t = \mathbf{w}_t - \overline{\mathbf{w}}_t \tag{14.4.14}$$

$$\mathbf{A}_t = \left.\frac{\partial\mathbf{s}_{t+1}}{\partial\mathbf{s}}\right|_{\mathbf{s}=\mathbf{s}_t^D} \tag{14.4.15}$$

$$\mathbf{B}_t = \left.\frac{\partial\mathbf{s}_{t+1}}{\partial\mathbf{q}}\right|_{\mathbf{q}=\mathbf{q}_{c,t}^D} \tag{14.4.16}$$

$$\mathbf{C}_t = \left.\frac{\partial\mathbf{s}_{t+1}}{\partial\mathbf{w}}\right|_{\mathbf{w}=\overline{\mathbf{w}}_t} \tag{14.4.17}$$

$$\boldsymbol{\varepsilon}_t = \mathbf{f}(\mathbf{s}_t^D, \mathbf{q}_{c,t}^D, \overline{\mathbf{w}}_t) - \mathbf{s}_{t+1}^D \tag{14.4.18}$$

in which the superscript D denotes the desirable values for the salinity vector or monthly inflows from the Lavaca River and Colorado River given by the decision-makers, $\mathbf{s}_t$ is the salinity vector at the *beginning* of month t, $t = 1, 2, \ldots, 12$, $\mathbf{q}_{c,t}$ is the control vector *during* the time interval t, and $\mathbf{w}_t$ is the random vector *during* the time interval t with mean vector $\overline{\mathbf{w}}_t$ and covariance matrix $\text{cov}(\mathbf{w}_t)$. It should be mentioned that $\mathbf{f}(s_t^D, \mathbf{q}_{c,t}^D, \overline{\mathbf{w}}_t)$ and the matrices $\mathbf{A}_t$, $\mathbf{B}_t$, and $\mathbf{C}_t$ are computed by solving the HYD-SAL.

The quadratic performance index to be minimized is

$$J(\delta\mathbf{s}_t) = E[[\delta\mathbf{s}_{t_f}^T\mathbf{Q}_{t_f}\delta\mathbf{s}_{t_f}] + \sum_{t}^{t_f-1} [\delta\mathbf{s}_t^T\mathbf{Q}_t\delta\mathbf{s}_t + \delta\mathbf{q}_{c,t}^T\mathbf{R}_t\delta\mathbf{q}_{c,t}]] \quad (14.4.19)$$

in which E denotes the statistical expectation operator, the superscript T is the transpose of a matrix or a vector, t_f is the terminal time, the weighting matrix $\mathbf{Q}_t$ for the salinity vector is symmetric and positive semidefinite, and the weighting matrix $\mathbf{R}_t$ for the monthly inflows from the Lavaca-Navidad River and the Colorado River is symmetric and positive definite. The purpose of the stochastic optimal control is to "steer" the distributed-parameter system by controlling $\mathbf{q}_{c,t}$ such that Eq. (14.4.19) is minimized.

Therefore, the optimal feedback monthly inflows *during* month t are expressed as a linear function of the salinity vector $\mathbf{s}_t$ measured at the 30 test sites in the estuary at the *beginning* of the month t,

$$\mathbf{q}_{c,t}^* = \mathbf{M}_t[\mathbf{A}_t[\mathbf{s}_t - \mathbf{s}_t^D] + \boldsymbol{\varepsilon}_t] + \mathbf{q}_{c,t}^D \quad (14.4.20)$$

in which

$$\mathbf{M}_t = -\mathbf{L}_t^{-1}\mathbf{B}_t^T\mathbf{K2}_{t+1} \quad (14.4.21)$$

$$\mathbf{L}_t = \mathbf{R}_t + \mathbf{B}_t^T\mathbf{K2}_{t+1}\mathbf{B}_t \quad (14.4.22)$$

$$\mathbf{K2}_t = \mathbf{Q}_t + \mathbf{A}_t^T\mathbf{K2}_{t+1}\mathbf{A}_t - \mathbf{A}_t^T\mathbf{K2}_{t+1}\mathbf{B}_t\mathbf{L}_t^{-1}\mathbf{B}_t^T\mathbf{K2}_{t+1}\mathbf{A}_t \quad (14.4.23)$$

and

$$\mathbf{K2}_{t_f} = \mathbf{Q}_{t_f} \quad (14.4.24)$$

The minimized cost-to-go function can be computed by using Eqs. (14.4.20)–(14.4.24). Figure 14.5 shows the flowchart for the real-time, stochastic, feedback, optimal control with parameter estimation for a distributed-parameter estuary system.

14.4.4 Numerical Examples for Parameter Estimation with HYD-SAL

Numerical examples for parameter estimation are performed for the Lavaca-Tres Palacios Estuary in Texas (Zhao, 1994). The parameter estimation is based on the Gauss–Newton minimization algorithm in conjunction with the numerical model HYD-SAL. The uncertainty analysis methods are incorporated in the parameter estimation to find the mean vector of the parameter vector.

The parameter vector to be estimated has been defined in Eq. (14.4.3). Based on the prior information on the Manning's roughness coefficient for

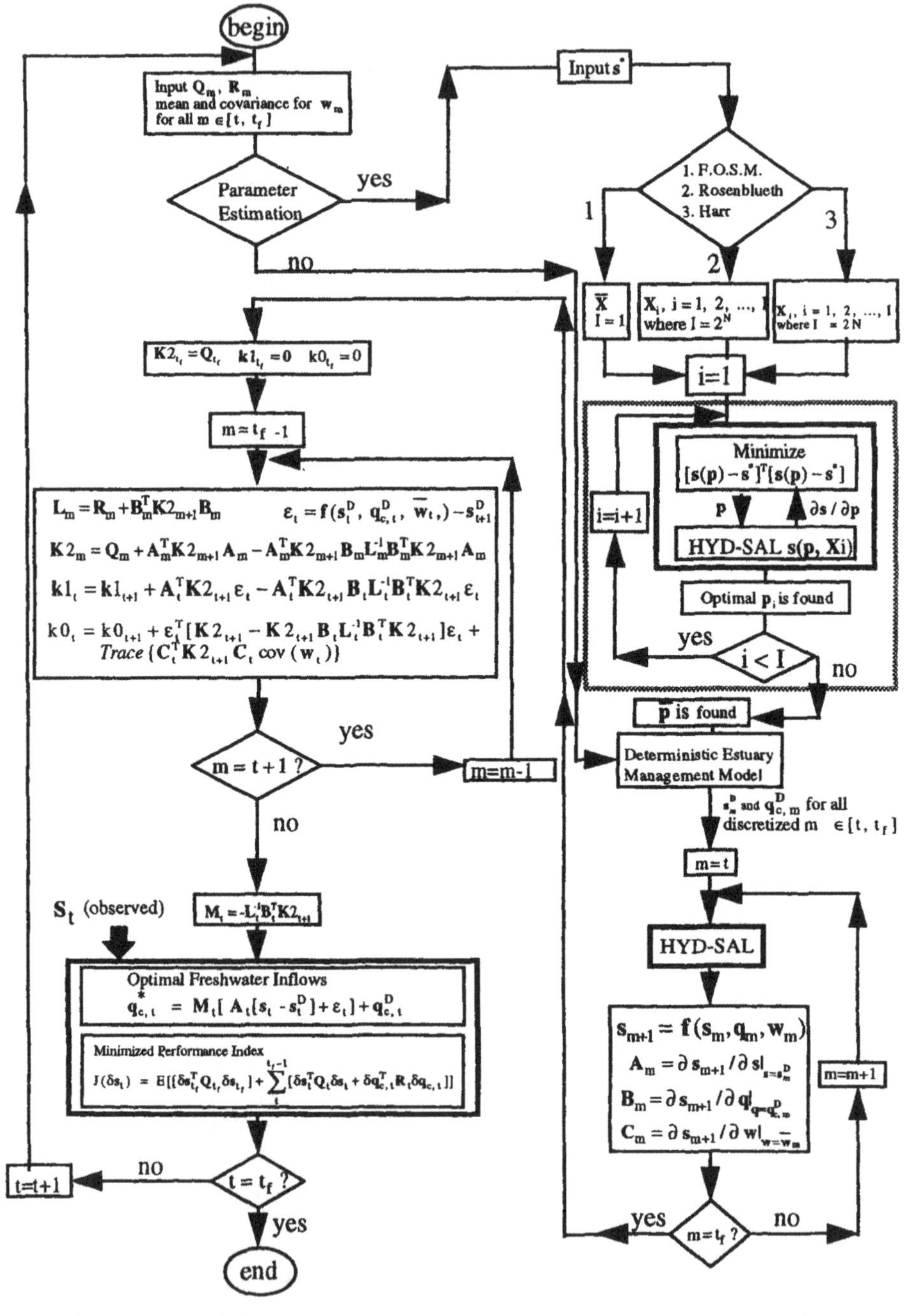

Figure 14.5 Flowchart of computation for real-time, stochastic feedback, optimal control with parameter estimation with uncertainty analysis method for a distributed-parameter estuary system (Zhao, 1994).

the Lavaca-Tres Palacios Estuary, there are 17 different values for the Manning's roughness coefficient for all the cells in the spatial domain. The values of Manning's roughness coefficient range from 0.0 to 0.054. If one uses 0.003 as the difference among different cells to regroup those 17 different values, the number of different values for Manning's roughness coefficient reduces to 6. Because the numerical model HYD-SAL uses one value for the dispersion coefficient for all cells in the spatial domain, the numerical examples herein also consider the number of different dispersion coefficients as one. Therefore, the parameter vector **p** to be estimated is

$$\mathbf{p} = \begin{bmatrix} p_1 \\ p_2 \\ \vdots \\ p_7 \end{bmatrix}_{(7\times 1)} \tag{14.4.25}$$

in which $p_1, p_2, \ldots, p_6$ are the Manning's roughness coefficients and p_7 is the dispersion coefficient.

The uncertainty analysis methods are incorporated in the Gaussian–Newton minimization algorithm and HYD-SAL to find the mean vector of the parameter vector **p**. The employed uncertainty analysis methods are the F.O.S.M., Rosenblueth's point estimate, and Harr's point estimate. Computer programs in FORTRAN have been developed for the Gauss–Newton minimization algorithm in conjunction with the numerical model HYD-SAL. The computation was performed on a 100 Mbit/sec workstation cluster which combines several workstations at Arizona State University. The statistics for the random vector **X** defined in Eq. (14.4.5) is assumed.

As discussed in Sec. 14.3, when the F.O.S.M method is used, HYD-SAL uses the mean values for the random vector **X**. The mean values of the parameter vector **p** by the F.O.S.M is found by directly using the Gauss–Newton algorithm in conjunction with HYD-SAL. The optimal value for the parameter vector **p** is also the mean vector for the parameter vector **p** based on the F.O.S.M. The starting point for the parameter vector is from a previous study. The difference between the observed salinity and the computed salinity for each test site is assumed with the starting point as the parameter vector in HYD-SAL. Five iterations are needed for convergence for the example. The CPU time required for the Gauss–Newton algorithm in this example is 12 min, 28.86 sec.

When Rosenblueth's method is used to find the mean of the parameter vector **p** in conjunction with the Gauss–Newton algorithm and the HYD-SAL, 16 (2^4) "concentrated" points and their probability masses must be generated based on Rosenblueth's algorithm. Each of the 16 points is used in the parameter estimation to yield an optimal parameter vector **p**. Then, the mean of the parameter vector **p** is computed. For each of the 16 points, the optimal parameter vector **p** is found by the Gauss–Newton minimization algorithm in conjunction with the HYD-SAL. The CPU time for computing the optimal parameter vector for each of the 16 points is between 10 min and 30 min in the workstation cluster.

When Harr's point estimate method is used to find the mean of the parameter vector, the eigenvalue–eigenvector decomposition is first performed to give four eigenvectors and four eigenvalues. Then, eight points are generated. For each of the eight points, the Gauss–Newton minimization algorithm in conjunction with the numerical model HYD-SAL is used to estimate the optimal parameter vector **p**. The CPU time for the optimal parameter for each of the eight points on a workstation cluster is between 10 min and 30 min.

14.4.5 Numerical Examples for Stochastic, Linear, Quadratic Feedback, Optimal Control with HYD-SAL

Numerical examples are performed for the Lavaca-Tres Palacios Estuary to illustrate the proposed discrete-time, stochastic, linear, quadratic optimal control estuarine management model for a distributed-parameter system governed by a set of partial differential equations (Zhao, 1994). The proposed model is essentially a feedback control which feeds back the salinity values measured at the 30 test sites in the estuary to the determination of the monthly inflow from the Lavaca-Navidad River and the Colorado River.

The matrices $\mathbf{A}_t$, $\mathbf{B}_t$, $\mathbf{C}_t$, and $\mathbf{f}(\mathbf{s}_t^D, \mathbf{q}_{c,t}^D, \overline{\mathbf{w}}_t)$ required for the perturbation model are computed by solving the numerical model HYD-SAL. The CPU time for computing these Jacobian matrices by central difference on a workstation cluster at Arizona State University is 4 hr, 45 min, 28.22 sec. The workstation cluster at Arizona State University links several workstations together with 100 Mbit/sec. After those matrices are computed, the optimal monthly inflow during each month from the Lavaca-Navidad River and the Colorado River can be directly computed after the salinity values at the 30 test sites at the beginning of each month are observed. The computations of the optimal monthly inflows are performed on a 486-PC. The CPU time for computing 11 optimal monthly inflows is 19 sec on a 486-PC-33Mhz. The associated number of floating-point operations is about 2.8×10^7. Because a real-world system is often unpredictable due to many large uncertainties, it is necessary to study the feedback control law under various kinds of observed salinity patterns.

Three general observed salinity patterns are assumed. The first pattern is that the observed salinity value at the beginning of each month is the same as the desirable salinity value. This pattern will be considered as a reference for two other patterns when the comparison of the optimal inflows are made among various kinds of situations. The second pattern of the observed salinity values is that the observed salinity values change with time (month). The third pattern is that different magnitudes of the salinity values are observed and they are independent of time. For each assumed pattern, the freshwater inflows from the Colorado and Lavaca-Tres Palacios rivers are computed by using the feedback control law. It has been found that the optimal freshwater inflows reasonably respond to the different observed salinity patterns.

REFERENCES

Bao, Y. X., Methodology for Determining the Optimal Freshwater Inflows into Bays and Estuaries. Ph.D. Dissertation, Department of Civil Engineering, University of Texas at Austin, Austin, 1992.

Bao, Y. X., and Mays, L. W., Optimization of Freshwater Inflows to the Lavaca-Tres Palacios Estuary, *Journal of Water Resources Planning and Management*, Vol. 120, No. 2, pp. 199–217, 1994a.

Bao, Y. X. and Mays, L. W., New Methodology for Optimization of Freshwater Inflows to Estuaries, *Journal of Water Resources Planning and Management*, Vol. 120, No. 2, pp. 218–236, 1994b.

Brion, L. M., Methodology for Optimal Operation of Pumping Stations in Water Distribution Systems, Ph.D. Dissertation, Department of Civil Engineering, University of Texas at Austin, Austin, 1990.

Brion, L. M. and Mays, L. W., Methodology for Optimal Operation of Pumping Stations in Water Distribution Systems, *Journal of Hydraulic Engineering*, Vol. 117, No. 11, pp. 1551–1569, 1989.

Butkovsky, A. G., Optimum Processes in Systems with Distributed Parameters, *Automation Remote Control*, Vol. 22, pp. 13–21, 1961.

Butkovsky, A. G., The Broadened Principle of the Maximum for Optimal Control Systems, *Automation Remote Control*, Vol. 24, pp. 292–304, 1963.

Butkovsky, A. G., The Method of Moments in the Theory of Optimal Control Systems with Distributed-Parameters, *Automation Remote Control*, Vol. 24, pp. 1106–1113, 1964.

Carriaga, C. C. and Mays, L. W., Optimization of Reservoir Releases to Control Sedimentation Using Differential Dynamic Programming. 21st Annual Conference, ASCE Water Resources Planning and Management Division, 1994.

Chang, L., Shoemaker, C. A., and Liu, P. L., Optimal Time-Varying Pumping Rates for Groundwater Remediation: Application of a Constrained Optimal Control Algorithm, *Water Resources Research*, Vol. 28, No. 12, pp. 3157–3173, 1992.

Culver, T. B. and Shoemaker, C. A., Dynamic Optimal Control for Ground Water Remediation with Flexible Management Periods, *Water Resources Research*, Vol. 28, No. 3, pp. 629–641, 1992.

Culver, T. B. and Shoemaker, C. A., Optimal Control for Groundwater Remediation by Differential Dynamic Programming with Quasi-Newton Approximations, *Water Resources Research*, Vol. 29, No. 4, pp. 823–831, 1993.

Dodd, C. W., An Approximate Method for Optimal Control of Distributed Parameter Systems, Ph.D. Thesis, Arizona State University, 1968.

Fattorini, H. O., Boundary Control Systems, *SIAM Journal of Control*, Vol. 6, No. 3, pp. 349–385, 1967.

Graham J. W. and D'Souza, A. F., Optimal Control of Distributed Parameter Systems Subject to Quadratic Loss, 10th Joint Automatic Control Conference, 1969, pp. 1703–1709.

Harr, M. E., *Reliability-Based Design in Civil Engineering*, McGraw-Hill, New York, 1987.

Harr, M. E., Probabilistic Estimates for Multivariate Analysis, *Applied Mathematical Modelling*, Vol. 13, No. 5, pp. 313–318, 1989.

Jacobson, D. and Mayne, D., *Differential Dynamic Programming*, Elsevier, New York, 1970.

Kim, M. and Erzberger, H., On the Design of Optimum Distributed Parameter System with Boundary Control Function, *IEEE Transactions on Automatic Control*, Vol. AC-12, pp. 22–28, 1967.

Lansey, K. E., Optimal Design of Large Scale Water Distribution Systems Under Multiple Loading Conditions, Ph.D. Dissertation, Department of Civil Engineering, University of Texas at Austin, Austin, 1987.

Lansey, K. E. and Mays, L. W., Optimization Model Water Distribution System Design, *Journal of Hydraulic Engineering*, Vol. 115, No. 10, pp. 1401–1418, 1989.

Lasdon, L. S. and Waren, A. D., *GRG2 User's Guide*, Department of General Business, University of Texas, Austin, 1989.

Li, G. and Mays, L. W., Differential Dynamic Programming for Estuarine Management, *Water Resources Planning and Management*, Vol. 121, No. 6, pp. 455–462, 1995.

Mays, L. W. (principal investigator), Bao, Y., Li, G., Zhao, B. (major contributors), LeBlanc, L., Mao, N., Shi, W., Siebert, J., Tuncok, I. K. (other contributors), Methodologies for Determining the Optimal Freshwater Inflows into Bays and Estuaries, Final Report for National Science Foundation, Grant No. BCS-9014406, 1994.

Mays, L. W. and Tung, Y. K., *Hydrosystems Engineering and Management*, McGraw-Hill, New York, 1992.

Masch, F. D. and Associates, Tidal Hydrodynamic and Salinity Models for San Antonio and Matagorda Bays, Texas, A Report to Texas Water Development Board, Austin, 1971.

Murray, D. M. and Yakowitz, S. J., Constrained Differential Dynamic Programming and its Application to Multireservoir Control, *Water Resources Research*, Vol. 15, No. 5, pp. 1017–1027, 1979.

Omatu, S. and Seinfeld, J. H., *Distributed Parameter Systems: Theory and Applications*, Oxford University Press, New York, 1989.

Ray, W. H., Some Applications of State Estimation and Control Theory to Distributed Parameter Systems, in *Control Theory of Systems Governed by Partial Differential Equations*, A. K. Aziz, J. W. Wingate, and M. J. Balas (eds.), American Press, New York, 1977.

Robinson, A. C., A Survey of Optimal Control of Distributed-Parameter Systems, *Automatica*, Vol. 7, pp. 371–388, 1971.

Rosenblueth, E., Point Estimates for Probability Moments, *Proceedings of the National Academy of Science*, Vol. 72, No. 10, pp. 3812–3814, 1975.

Rosenblueth, E., Two-Point Estimates in Probabilities, *Applied Mathematical Modelling*, Vol. 5, No. 5, pp. 329–335, 1981.

Ten Brummelhuis, P. G. J., Parameter Estimation in Stochastic Hydraulic Models, in *Control of Distributed Parameter Systems*, M. Amourous and A. Eljai (eds.), 1989.

Ten Brummelhuis, P. G. J., Parameter Identification in Tidal Models with Uncertain Boundary Conditions, *Stochastic Hydrology and Hydraulics*, Springer-Verlag, Vol. 4, 1990.

Teo, K. L. and Wu, Z. S., *Compuational Methods for Optimizing Distributed Systems*, Academic Press, New York, 1984.

Tung, Y. K. and Yen, B. C., Some Recent Progress in Uncertainty Analysis for Hydraulic Design, in *Reliability and Uncertainty Analyses in Hydraulic Design*, B. C. Yen and Y. K. Tung (eds.), ASCE, New York, 1993.

Unver, O., Simulation and Optimization of Real-Time Operations of Multireservoir Systems Under Flooding Conditions, Ph.D. Dissertation, Department of Civil Engineering, University of Texas at Austin, Austin, 1987.

Unver, O., Mays, L. W., and Lansey, K., Real-Time Flood Management Model for the Highland Lakes System, *Journal Water Resource Planning and Management*, Vol. 13, No. 5, pp. 620–638, 1987.

Wanakule, N., A Model for Determining Optimal Pumping and Recharge of Large-Scale Aquifers, Ph.D. Dissertation, Department of Civil Engineering, University of Texas at Austin, Austin, 1984.

Wanakule, N., Mays, L. W., and Lasdon, L. S., Development and Testing of a Model for Determining Optimal Pumping and Recharge of Large-Scale Aquifers, Technical Report CRWR-217, University of Texas at Austin, Austin, 1985.

Wanakule, N., Mays, L. W., and Lasdon, L. S., Optimal Management of Large-Scale Aquifers: Methoddology and Applications, *Water Resources Research*, Vol. 22, No. 4, pp. 447–465, 1986.

Wang, P. K. C., Control of Distributed Parameter System, in *Advances in Control Systems Theory and Applications*, C. T. Leondes (eds.), Academic Press, New York, 1964, Vol. 1.

Whiffen, G. J. and Shoemaker, C. A., Nonlinear Weighted Feedback Control of Groundwater Remediation Under Uncertainty, *Water Resources Research*, Vol. 29, No. 9, pp. 3277–3289, 1993.

Yakowitz, S. and Rutherford, B., Computational Aspects of Discrete-Time Optimal Control, *Applied Mathematics and Computation*, Vol. 15, pp. 29–45, 1984.

Yeh, W. W-G., Review of Parameter Identification Procedures in Groundwater Hydrology: The Inverse Problem, *Water Resources Research,* Vol. 22, No. 2, pp. 95–108, 1986.

Zhao, B., Stochastic Optimal Control and Parameter Estimation for an Estuary System, Ph.D. Dissertation, Department of Civil Engineering, Arizona State University, Tempe, 1994.

Index

Adaptive shift procedure, 214
Admissible control, 9
ARMAX, 305
Augmented Lagrangian method, 55-61
  augmented Lagrangian function, 57
  complementary slackness conditions, 61
  convergence, 61
  second-order conditions, 60
  updating formula, 61

Backward algorithm, 195
Backward recursion, 208-209
Backward sweep, 212
Basic optimal control problem, 208
Basic variables, 50
Basis changes, 51

Closed-loop control system, 301-302
Complementary slackness conditions, 61
Concavity, 40-42

Constrained optimization, 47-61
Convergence, 61
Convexity, 40-42
Convex regions, 41

Differential dynamic programming (DDP), 207-231
  adaptive shift procedure, 214
  algorithm, 211-213
  backward recursion, 208-209
  backward sweep, 212
  basic optimal control problem, 208
  convergence, 213-214
  derivation of coefficients, 225-230
  first-order necessary condition, 210
  forward sweep, 211
  for groundwater control, 218-221
  for groundwater reclamation, 221-225
  for multireservoir operation, 215-218
  objective of, 208
  quadratic convergence, 213
  successive approximation linear quadratic regulator (SALQR), 222
Discrete-time optimal control using generalized reduced gradient (GRG), 61-67
  generalized reduced gradient algorithm, 66-67
  problem statement, 61-62

[Discrete-time optimal control using generalized reduced gradient (GRG)]
reduced gradient procedure, 64-66
reduced objective function, 62
reduced objective problem, 62-66
Dual variables, 49
DWOPER, 98, 106, 110, 113, 116
Dynamic programming, 193-196
backward algorithm, 195
basic features, 195
decision variables, 194
disadvantages of, 196
forward algorithm, 195
recursive equation, 195
stages, 194
stage transformation, 194
state return, 194
state transition, 194
state variables, 194

Estuarine management models, 163-192, 233-254
differential dynamic programming approach, 233-254
hydrodynamic transport simulator, 165-169
finite difference equations, 168-169
simulator equations, 166
HYD-SAL, 182, 186-188, 349, 351, 353, 354
boundary conditions, 188
model linkage, 187
linear quadratic optimal control approach, 308-328
quadratic performance index, 308
mathematical programming approach, 163-192
augmented Lagrangian function, 178
chance-constraint formulation, 172-174
chance-constraint formulation, deterministic equivalent, 189
chance-constraint formulation, multiple linear regression model, 189
[Estuarine management models]
[mathematical programming approach]
constraints, 169-171
gradient approximation scheme, 183-185
gradient approximation scheme, flowchart of, 184
harvest equations, table of, 173
management model strategies, 171
mathematical statement of problem, 25
objective function, 169
optimizer-simulator interface, 176
problem formulation 165-174
problem identification, 163-165
problem solution, 174-185
reduced gradient, 179-181
reduced problem, 177-178
simulator equations, 34-36
solution procedure, 178-179
solution procedure, flowchart of, 180
stochastic linear quadratic optimal control (lumped system) approach, 308-328
estuary management application, 322-328
flowchart of method, 327
minimized cost-to-go performance index, 325
optimal control law, 325
feedback control law, 309-314
minimum cost-to-go function, 310
optimal performance index, 314
parameter estimation, 317-322
ordinary least squares, 319-322
recursive least squares (RLS), 318-319
quadratic performance index, 308
stochastic optimal control (distributed system) approach, 331-360
distributed-parameter systems, 336-339
feedback control law, 339
flowchart of, 341
minimized cost-to-go function, 339

[Estuarine management models]
[stochastic optimal control (distributed system) approach]
[distributed-parameter systems]
perturbation system equation, 336-339
estuary management application, 349-360
flowchart of, 357
parameter estimation, 340-349, 351-354
Gauss–Newton minimization, 340
numerical examples, 356-360
uncertainty analysis methods for parameter estimation, 342-349
first-order second moment, 342-344
first-order second moment, flowchart of, 344
Harr's method, 347-349
Harr's method, flowchart of, 350
Rosenblueth's method, 345-347
Rosenblueth's method, flowchart of, 348
successive approximation linear quadratic regulator approach, 233-252
bracket penalty function, 236
fish harvest equations, table of, 234
HYD-SAL interface, 239-243
objective function, 234-235
problem formulation, 233-236
SALQR algorithm, 236-239
flowchart of, 240
flowchart of HYD-SAL interface, 245

Feedback control methods, 193-205, 301-330
dynamic programming, 193-196
feedback rules, 197
for groundwater management, 200-201
linear quadratic feedback, 301-330
linear systems, 196-200
algorithm for, 199-200
[Feedback control methods]
nonlinear systems, 201-205
algorithm for, 202-205
nonlinear state equation, 202
quadratic objective function, 202
quadratic loss function, 197, 202
state equation, 196-197, 202
Feedback rules, 197
First-order second moment method, 342-344
flowchart of, 344
Flood control operation, 19-23, 31-33, 97-130
constraints, 106-107
DWOPER, 98, 106, 110, 113, 116
mathematical statement of problem, 19-23, 101
constraints, 20-22
objective, 19
multireservoir operation, 98
objective functions, 108-109
problem solution, 110-116
basis matrix elements, 124-125
reduced problem, 110-116
real-time operation, 97-130
HYD-SAL, 182, 186-188, 349, 351, 353,354
Saint-Venant equations, 102-105
simulator equations, 31-33, 102-106, 121-124
Forward algorithm, 195
Forward sweep, 211

Gauss–Newton minimization, 340
Generalized reduced gradient (GRG) method, 49-54
basic variables, 50
basis changes, 51
general algorithm, 51
Kuhn–Tucker multiplier vector, 52-53
nonbasic variables, 50
optimality conditions, 53-54
reduced gradient, 51-53
reduced objective, 50
reduced problem, 50
superbasic variables, 50

Gradient, 39-40
Gradient vector, 39-40
Groundwater management systems, 14-18, 30, 200-201, 218-225
  feedback control method, 200-201
    advantages of, 201
    control vector, 200
    deterministic control problem, 200
    feedback coefficients, 201
    feedback rule, 201
    state variable vector, 200
  groundwater reclamation models, 221-225
  groundwater recharge [see also Soil aquifer treatment (SAT) system operation], 277-300
  GWMAN, 84-85
  hydraulic management models, 71-72
    embedding, 71
    unit response, 71
  hydraulic management models using differential dynamic programming (DDP), 218-221
  HYDRUS, 278-279, 282
  mathematical statement of problem, 14-18, 76-77
    constraints, 16-17, 76-77
    objective, 14, 77
  problem solution, 77-84
    algorithm flowchart, 84
    augmented Lagrangian function, 82
    basis matrix elements, 88-94
    reduced gradient, 79-80
    reduced problem, 78
  reclamation models using differential dynamic programming (DDP), 221-225
  simulator equations, 30-31, 73-76
Groundwater recharge [see also Soil aquifer treatment (SAT) system operation], 277-301
Groundwater reclamation models, 221-225
GWMAN, 83-84

Harr's multivariate point estimate method, 347-349
  flowchart of, 350
HEC-6, 256, 259, 264, 265
Hessian, 40
Hessian matrix, 39-40
Hydrosystems problems, 13-37
  general optimization framework, 13-14
  general problem formulation, 30
  optimal control problems, 2-30
  process simulation equations, 13
HYDRUS, 278-279, 282
HYD-SAL, 182, 186-188, 349, 351, 353, 354
Hyperbolic penalty function, 224

Kuhn–Tucker conditions, 49
KYPIPE, 136, 147, 157

Lagrange multiplier method, 49
Lagrange multipliers, 48
Lagrangian function, 57
Linear quadratic feedback, 301-330
Linear quadratic Gaussian (LQG), 304
Linear quadratic regulator (LQR), 305-307
  deterministic system, 305, 306
    optimal control law, 306
    quadratic performance index, 306
  stochastic system, 306-307
    optimal control law, 306
    quadratic performance index, 306
Linear quadratic tracker (LQT), 307-308
  deterministic system, 307
    optimal control law, 307
    quadratic performance index, 307
  stochastic system, 307-308
    optimal control law, 307
    quadratic performance index, 308
Linear systems, 196-200, 301-330

MATLAB, 326
Matrix algebra, 39-42
  concavity, 40-42
  convexity, 40-42

[Matrix algebra]
convex regions, 41
gradient, 39-40
gradient vector, 39-40
Hessian, 40
Hessian matrix, 39-40
nonconvex regions, 41

Newton–Raphson method, 121-123
Nonbasic variables, 50
Nonconvex regions, 41
Nonlinear optimization methods, 39-69
computer software, 68-69
GAMS, 68,69
GAMS/MINOS, 69
GAMS/ZOOM, 69
GINO, 68
GRG2, 68
MINOS, 68
ZOOM, 69
constrained optimization, 47-61
augmented Lagrangian method 55-61
complementary slackness conditions, 61
convergence, 61
second-order conditions, 60
updating formula, 61
dual variables, 49
exterior penalty, 56
generalized reduced gradient method, 49-54
Kuhn–Tucker conditions, 49
Lagrange multiplier method, 49
Lagrange multipliers, 48
Lagrangian, 48
optimality conditions, 47-49
penalty function methods, 54-61
penalty terms, 57
penalty weights, 56-57
quadratic penalty function, 55
matrix algebra, 39-42
concavity, 40-42
convexity, 40-42
convex regions, 41
gradient, 39-40
[Nonlinear optimization methods]
[matrix algebra]
gradient vector, 39-40
Hessian, 40
Hessian matrix, 39-40
nonconvex regions, 41
unconstrained nonlinear programming, 42-47
basic concepts, 42-44
line search, 43-45
multivariable methods, 45-47
one-dimensional search, 43-45
search directions, table of, 46
stopping rules, 44
unimodal functions, 43
Nonlinear programming software, 68-69
GAMS, 68,69
GAMS/MINOS, 69
GAMS/ZOOM, 69
GINO, 68
GRG2, 68
MINOS, 68
ZOOM, 69

Open-loop control system, 301-302
Optimal control problems
admissible control, 9
continuous form, 9
definition of, 8-10
discrete form, 10
discrete-time optimal control using generalized reduced gradient (GRG), 61-67
generalized reduced gradient algorithm, 66-67
problem statement, 61-62
reduced gradient procedure, 64-66
reduced objective function, 62
reduced objective problem, 62-66
objective function, 9
performance index, 8-9
problem statement, 9
real-time optimal control, 301-330
state equations, 9-10
Ordinary least squares, 319-322

Parameter estimation, 317-322
ordinary least squares, 319-322
recursive least squares (RLS), 318-319
uncertainty analysis methods for, 342-349
first-order second moment, 342-344
flowchart of, 344
Harr's method, 347-349
flowchart of, 350
Rosenblueth's method, 345-347
flowchart of, 348
Performance index, 8-9
Penalty terms, 57
Penalty weights, 56-57
Penalty function methods, 54-61

Quadratic loss function, 197, 202
Quadratic penalty function, 55

Real-time optimal control, 301-328
Recursive least squares (RLS), 318-319
Reservoir system operation (water supply)
differential dynamic programming formulation, 215-218
mathematical statement of problem, 23-24
constraints, 23-24
objective, 23
simulator equations, 33-34
stochastic linear quadratic feedback control, 315-316
Reduced gradient, 51-53
Reduced objective, 50
Rosenblueth's multivariate point estimate method, 345-347
flowchart of, 348

Saint-Venant equations, 31, 102-106
SCADA, 146
Sediment control, 25-27, 255-276
differential dynamic programming algorithm, 264-265
diagram of, 265
schematic of, 264
HEC-6, 256, 259, 264, 265
mathematical statement of problem, 25-27
[Sediment control]
optimal control problem, 257-258
penalty function method, 261-263
advantages of, 262-263
hyperbolic penalty function, 261, 273-275
problem formulation, 255-263
reservoir operation constraints, 259
river hydraulic constraints, 259-260
sediment constraints, 260
problem solution, 263-265
derivative evaluation, 263
Soil aquifer treatment operation model (SATOM), 278, 285
Soil aquifer treatment (SAT) system operation, 27-30, 36-37, 277-300
bound selection, 290
convergence procedure, 284-285
derivative calculations, 295-299
HYDRUS, 278-279, 282
mathematical formulation, 279-281
mathematical statement of problem, 27-30
model application, 286-295
optimizer-simulator relation diagram, 283
problem formulation, 277-278
constraints, 280-281
control variables, 279
objective function, 279
state variables, 279
problem solution, 281-285
shift parameters, 287
simulator calls, 288-289
simulator equations, 36-37
soil aquifer treatment operation model (SATOM), 278, 285
successive approximation linear quadratic regulator (SALQR) algorithm, 281-285
successive approximation linear quadratic regulator (SALQR) interface with HYDRUS, 281-283
flowchart of 284
Stages, 194
Stage transformation, 194

State feedback, 4
State return, 194
State space analysis, 3
State transition, 194
State variable models, 2-8
  continuous-time models, 3-5
    output equation, 4
    state equation, 4
  controllability, 7
  deterministic models, 3-7
    output equation, 6
    state equation, 5
  observability, 7
  output equation, 3
  stability, 7
  state feedback, 4
  water resources applications, 7-8
State variables, 194
Stochastic linear quadratic optimal control approach, 308-328
  application to hydrologic routing, 316-317
  application to reservoir management, 315-316
Successive approximation linear quadratic regulator (SALQR), 222
  diagram of algorithm, 223
  hyperbolic penalty function, 224
System control, 301-302
  closed-loop control, 301-302
    diagram of, 302
  open-loop control, 301-302
    diagram of, 302
Systems, 1-7
  additivity, 2
  classical approach, 3
  classification of, 1-2
  concepts of, 2-7
  continuous-time system, 2
  definition of, 1
  distributed-parameter system, 2
  feedback, 1
  homogeneity, 2
  lumped-parameter systems, 2
  modern system theory, 2
  separation theorem, 304
[Systems]
  state, 3
  state space analysis, 3
  state variable model, 3-8
  static system, 1
  superposition theorem, 2
  system boundary, 1
  time-invariant system, 2
  transfer function, 1

Time-invariant system, 2
Transfer function, 1
Uncertainty analysis methods for parameter estimation, 342-349
  first-order second moment, 342-344
  first-order second moment, flowchart of, 344
  Harr's method, 347-349
  Harr's method, flowchart of, 350
  Rosenblueth's method, 345-347
  Rosenblueth's method, flowchart of, 348
Uncertainties, 303
  economic, 303
  hydraulic, 303
  hydrologic, 303
    inherent, 303
    model, 303
    parameter, 303
  structural, 303
Unconstrained nonlinear programming, 42-47
Unimodal functions, 43
Updating formula, 61

Water distribution system operation model
  augmented Lagrangian formulation, 138
  basis elements, 150
  constraints, 132-135
  Darcy–Weisbach equation, 148
  energy cost, 132
  extended period simulation, 133
  fixed grade nodes, 133
  flowchart optimization model, 141
  general formulation, 135
  Hazen–Williams equation, 148

[Water distribution system operation model]
KYPIPE, 136, 147, 157
Lagrange multipliers, 144
linear method, 149
mathematical statement of problem, 24, 132-135
objective function, 132
optimization-simulation linkage, 137
[Water distribution system operation model]
problem formulation, 132-135
problem identification, 131-132
pump characteristics, 157
pump efficiency, 157-158
reduced problem, 135-140
simulation model, 147-150
updating formula, 140

For Product Safety Concerns and Information please contact our EU representative GPSR@taylorandfrancis.com Taylor & Francis Verlag GmbH, Kaufingerstraße 24, 80331 München, Germany

Printed and bound by CPI Group (UK) Ltd, Croydon, CR0 4YY
01/05/2025
01858546-0003